PROGRESS IN CLINICAL AND BIOLOGICAL RESEARCH

RECENT TITLES

Vol 70: **Reproductive Immunology,** Norbert Gleicher, *Editor*

Vol 71: **Psychopharmacology of Clonidine,** Harbans Lal and Stuart Fielding, *Editors*

Vol 72: **Hemophilia and Hemostasis,** Doris Ménaché, D. MacN. Surgenor, and Harlan D. Anderson, *Editors*

Vol 73: **Membrane Biophysics: Structure and Function in Epithelia,** Mumtaz A. Dinno and Arthur B. Callahan, *Editors*

Vol 74: **Physiopathology of Endocrine Diseases and Mechanisms of Hormone Action,** Roberto J. Soto, Alejandro De Nicola, and Jorge Blaquier, *Editors*

Vol 75: **The Prostatic Cell: Structure and Function,** Gerald P. Murphy, Avery A. Sandberg, and James P. Karr, *Editors.* Published in 2 volumes:
Part A: **Morphologic, Secretory, and Biochemical Aspects**
Part B: **Prolactin, Carcinogenesis, and Clinical Aspects**

Vol 76: **Troubling Problems in Medical Ethics: The Third Volume in a Series on Ethics, Humanism, and Medicine,** Marc D. Basson, Rachel E. Lipson, and Doreen L. Ganos, *Editors*

Vol 77: **Nutrition in Health and Disease and International Development: Symposia From the XII International Congress of Nutrition,** Alfred E. Harper and George K. Davis, *Editors*

Vol 78: **Female Incontinence,** Norman R. Zinner and Arthur M. Sterling, *Editors*

Vol 79: **Proteins in the Nervous System: Structure and Function,** Bernard Haber, Joe Dan Coulter, and Jose Regino Perez-Polo, *Editors*

Vol 80: **Mechanism and Control of Ciliary Movement,** Charles J. Brokaw and Pedro Verdugo, *Editors*

Vol 81: **Physiology and Biology of Horseshoe Crabs: Studies on Normal and Environmentally Stressed Animals,** Joseph Bonaventura, Celia Bonaventura, and Shirley Tesh, *Editors*

Vol 82: **Clinical, Structural, and Biochemical Advances in Hereditary Eye Disorders,** Donna L. Daentl, *Editor*

Vol 83: **Issues in Cancer Screening and Communications,** Curtis Mettlin and Gerald P. Murphy, *Editors*

Vol 84: **Progress in Dermatoglyphic Research,** Christos S. Bartsocas, *Editor*

Vol 85: **Embryonic Development,** Max M. Burger and Rudolf Weber, *Editors.* Published in 2 volumes:
Part A: **Genetic Aspects**
Part B: **Cellular Aspects**

Vol 86: **The Interaction of Acoustical and Electromagnetic Fields With Biological Systems,** Shiro Takashima and Elliot Postow, *Editors*

Vol 87: **Physiopathology of Hypophysial Disturbances and Diseases of Reproduction,** Alejandro De Nicola, Jorge Blaquier, and Roberto J. Soto, *Editors*

Vol 88: **Cytapheresis and Plasma Exchange: Clinical Indications,** W.R. Vogler, *Editor*

Vol 90: **Beta-Carbolines and Tetrahydroisoquinolines,** Floyd Bloom, Jack Barchas, Merton Sandler, and Earl Usdin, *Organizers*

Vol 91: **Membranes in Growth and Development,** Joseph F. Hoffman, Gerhard Giebisch, and Linda Bolis, *Editors*

Vol 92: **The Pineal and Its Hormones,** Russel J. Reiter, *Editor*

Vol 93: **Endotoxins and Their Detection With the Limulus Amebocyte Lysate Test,** Stanley W. Watson, Jack Levin, and Thomas J. Novitsky, *Editors*

See pages following the index for previous titles in this series.

ENDOTOXINS AND THEIR DETECTION WITH THE LIMULUS AMEBOCYTE LYSATE TEST

ENDOTOXINS AND THEIR DETECTION WITH THE LIMULUS AMEBOCYTE LYSATE TEST

Proceedings of an International Conference on Endotoxin Standards and Limulus Amebocyte Lysate Use With Parenteral Drugs, Held at the Woods Hole Oceanographic Institution, Woods Hole, Massachusetts, September 1981

Editors

Stanley W. Watson, PhD
Woods Hole Oceanographic Institution
Woods Hole, Massachusetts

Jack Levin, MD
Department of Medicine
The Johns Hopkins University School of Medicine and Hospital
Baltimore, Maryland

Thomas J. Novitsky, PhD
Associates of Cape Cod, Inc.
Woods Hole, Massachusetts

ALAN R. LISS, INC., NEW YORK

Address all Inquiries to the Publisher
Alan R. Liss, Inc., 150 Fifth Avenue, New York, NY 10011

Printed in the United States of America.

Library of Congress Cataloging in Publication Data

International Conference on Endotoxin Standards and
Limulus Amebocyte Lysate Use With Parenteral
Drugs (1981 : Woods Hole, Mass.)
Endotoxins and their detection with the limulus
amebocyte lysate test.

(Progress in clinical and biological research ;
v. 93)
Includes index.
1. Injections—Microbiology—Standards—Congresses. 2. Limulus test—Congresses. 3. Endotoxin—Analysis—Congresses. 4. Parenteral therapy—Standards—Congresses. I. Watson, Stanley Willard, 1921- II. Levin, Jack, 1932- III. Novitsky, Thomas J. IV. Title.
V. Series. [DNLM: 1. Endotoxins—Analysis.
2. Limulus test. 3. Infusions, Parenteral. W1
PR668E v.93 / QW 630 I59e 1981]
RS190.I5I57 1981 615'.1901 82-8967
ISBN 0-8451-0093-9 AACR2

Contents

Contributors

Kiyoto Akama [301]
Department of General Biologics Control, The National Institute of Health, Tokyo, 141, Japan

Bruce R. Albaugh [183]
Flow Laboratories, Inc., McLean, VA 22102

J.P. Aldred [357]
Armour Pharmaceutical Company, Kankekee, IL 60901

James R. Allen [287]
Center for Infectious Diseases, Centers for Disease Control, Public Health Service, U.S. Department of Health and Human Services, Atlanta, GA 30333

Roger L. Anderson [287]
Center for Infectious Diseases, Centers for Disease Control, Public Health Service, U.S. Department of Health and Human Services, Atlanta, GA 30333

Frederik B. Bang [1]
Department of Pathobiology, The Johns Hopkins University School of Hygiene and Public Health, Baltimore, MD 21205

K.-P. Becker [385]
Abt. f. Immunologie und Serologie, Institut fur Hygiene und Med. Mikrobiologie, Klinikum Mannheim der Univ. Heidelberg, D6,5,68 Mannheim, West Germany

Leandro Bellentani [329]
Hoechst Italia Sud, Scoppito 67019, L'Aquila, Italy

Joyce Bohon [65, 247, 365]
Travenol Laboratories, Inc., Morton Grove, IL 60053

Carl W. Bruch [39]
Bureau of Medical Devices, Food and Drug Administration, Silver Spring, MD 20910

Loretta A. Carson [217]
Centers for Disease Control, Phoenix, AZ 85014

Marilyn J. Case [161]
Associates of Cape Cod, Inc., Woods Hole, MA 02543

Charles B. Chandler [183]
Associates of Cape Cod, Inc., Woods Hole, MA 02543

James F. Cooper [55]
Medical University of South Carolina, Charleston, SC 29425

A.M. Cundell [269]
Blood Derivatives Program, New York Blood Center, Inc., Melville, NY 11747

Roger Dabbah [65, 231]
Travenol Laboratories, Inc., Morton Grove, IL 60053

The bold face number in brackets following each contributor's name indicates the opening page of that author's paper.

B. Ditter [381, 385]
Abt. f. Immunologie und Serologie, Institut für Hygiene und Med. Mikrobiologie, Klinikum Mannheim der Univ. Heidelberg, D6,5,68, Mannheim, West Germany

Georgiana Donohue [365]
Travenol Laboratories, Inc., Morton Grove, IL 60053

Peter C. Fink [281, 393]
Medizinische Hochschule Hannover, Institute for Clinical Chemistry, Karl-Wiechert-Allee 9, D-3000 Hannover, West Germany

Joseph R. Firca [121]
Abbott Diagnostics Division, Abbott Laboratories, North Chicago, IL 60064

Jeffrey Fleishman [131]
Difco Laboratories, Inc., Detroit, MI 48232

Freddie Fowlkes [131]
Difco Laboratories, Inc., Detroit, MI 48232

P. Friberger [195]
Kabi AB, Peptide Research, S-43122 Mölndal, Sweden

Yoshimasa Fujita [173]
Seikagaku Kogyo Co., Ltd., 1253, Tateno-3-chome, Higashiyamato, Tokyo, Japan, 189

Dennis E. Guilfoyle [79]
U.S. Food and Drug Administration, Brooklyn, NY 11232

Elsie J. Helme [101]
Schering-Plough Corporation, Union, NJ 07083

Anita K. Highsmith [287]
Center for Infectious Diseases, Centers for Disease Control, Public Health Service, U.S. Department of Health and Human Services, Atlanta, GA 30333

H. Donald Hochstein [141]
U.S. Food and Drug Administration, Bureau of Biologics, Bethesda, MD 20205

D. Hoffman [357]
Armour Pharmaceutical Company, Kankakee, IL 60901

Reiko Homma [301]
Department of General Biologics Control, The National Institute of Health, Tokyo, 141, Japan

Laurel J. Karr [207]
Department of Pediatrics, University of Alabama, School of Primary Medical Care, Huntsville, AL 35801

R.R. Kleszynski [357]
Armour Pharmaceutical Company, Kankakee, IL 60901

M. Knös [195]
Kabi AB, Peptide Research, S-43122 Mölndal, Sweden

Kazuo Kuratsuka [301]
Department of General Biologics Control, The National Institute of Health, Tokyo, 141, Japan

Jack Levin [xvii, 7]
Formerly of Department of Medicine, The Johns Hopkins University School of Medicine and Hospital, Baltimore, MD 21205; currently of Departments of Laboratory Medicine and Medicine, University of California School of Medicine, San Francisco, CA 94143

Karen Zink McCullough [91]
Microbiological Special Services Section, Beecham Laboratories, Piscataway, NJ 08854

L. Mellstam [195]
Kabi AB, Peptide Research, S-43122 Mölndal, Sweden

Helmut Müller-Calgan [343]
Fachtierarzt für Pharmakologie und Toxikologie, Pharma-Qualitätskontrolle Biologie, E. Merke, D-6100 Darmstadt, West Germany

Terry E. Munson [25, 79]
U.S. Food and Drug Administration, Rockville, MD 20857

Chizuko Nakahara [173]
Seikagaku Kogyo Co. Ltd., 1253, Tateno-3-chome, Higashiyamato, Tokyo, Japan, 189

Thomas J. Novitsky [xiii, 161]
Associates of Cape Cod, Inc., Woods Hole, MA 02543

A. Odeneal [357]
Armour Pharmaceutical Company, Kankekee, IL 60901

Aubrey S. Outschoorn [33, 115]
Drug Standards Division, U.S. Pharmacopeial Convention, Inc., Rockville, MD 20852

Frederick C. Pearson III [65, 231, 247, 365]
Travenol Laboratories, Inc., Morton Grove, IL 60053

Norman J. Petersen [217]
Centers for Disease Control, Phoenix, AZ 85014

Suresh C. Rastogi [49]
Bureau of Biologics, Food and Drug Administration, Bethesda, MD 20205

H. Ronneberger [261]
Research Laboratories of Behringwerke AG,D 355 Marburg, West Germany

Virginia C. Ross [39]
Bureau of Medical Devices, Food and Drug Administration, Silver Spring, MD 20910

Jon A. Rudbach [121]
The Stella Duncan Memorial Research Institute, Department of Microbiology, University of Montana, Missoula, MT 59812

Ragnar Rylander [399]
Department of Environmental Hygiene, University of Gothenburg, 400 33 Gothenburg, Sweden

G. Alan Sabey [153]
The Wellcome Foundation Ltd., Dartford, Kent DA1-5AH, England

Klaus D. Schultze [393]
Medizinische Hochschule Hannover, Institute for Clinical Chemistry, Karl-Wiechert-Allee 9, D-3000 Hannover, West Germany

Edward Shanbrom [105]
Mallinckrodt, Inc., St. Louis, MO 63134

John L. Sloyer, Jr. [207]
Department of Pediatrics, University of Alabama, School of Primary Medical Care, Huntsville, AL 35801

S. Spiegelman [269]
Blood Derivatives Program, New York Blood Center, Inc., Melville, NY 11747

Thomas E. Stoneycypher [207]
Department of Pediatrics, University of Alabama, School of Primary Medical Care, Huntsville, AL 35801

M.H. Stryker [269]
Blood Derivatives Program, New York Blood Center, Inc., Melville, NY 11747

Robert S. Tutunjian [319]
Industrial Marketing, Amicon Corporation, Danvers, MA 01923

B. Urbaschek [381, 385]
Abt. f. Immunologie und Serologie, Institut für Hygiene und Med. Mikrobiologie, Klinikum Mannheim der Univ. Heidelberg, D 6,5,68 Mannheim, West Germany

R. Urbaschek [381, 385]
Abt. f. Immunologie und Serologie, Institut für Hygiene und Med. Mikrobiologie, Klinikum Mannheim der Univ. Heidelberg, D6,5,68 Mannheim, West Germany

Janet Vesterlund [399]
Department of Environmental Hygiene, University of Gothenburg, 400 33 Gothenburg, Sweden

Marlys E. Weary [65, 231, 247, 365]
Travenol Laboratories, Inc., Morton Grove, IL 60053

Rolf G. Wolkersdorfer [281]
Medizinische Hochschule Hannover, Division of Immunology and Transfusion Medicine, Karl-Wiechert-Allee 9, D-3000 Hannover, West Germany

PREFACE

An international conference was held on two timely subjects related to use of *Limulus* amebocyte lysate (LAL) by the pharmaceutical industry. The meeting covered (a) the characteristics required of an endotoxin standard for use with LAL and (b) current applications of LAL in the pharmaceutical industry with special reference to LAL use with parenteral drugs.

The International Conference held September 21-24, 1981 at the Woods Hole Oceanographic Institution, Woods Hole, Mass. was opened by remarks from Dr. Stanley W. Watson, organizer of the conference.

It was fitting that the person responsible for the "discovery" of the clotting properties of LAL, the late Dr. Frederik B. Bang, Johns Hopkins University School of Hygiene and Public Health, presented the keynote address. Dr. Jack Levin who along with Dr. Bang elaborated the biochemistry of LAL and devised a procedure for quantitating endotoxins using LAL, served as editor of this volume and also presented the conference summary.

The spirit and enthusiasm of the conference attendees was dampened with the news of the deaths of two professionals active in the area of LAL, Dr. Rose Hartley, WHO, London, England, and Mr. Philip Griffin, Mallinckrodt, Inc., St. Louis. Dr. Hartley was one of the organizers of the WHO study on endotoxin standards and LAL. Mr. Griffin was Business Manager of LAL manufacture at Mallinckrodt, Inc. and was one of the first people to encourage Dr. Watson to hold the conference. It is also with sadness that we note the death of Dr. Bang shortly after the conference ended. It is to his memory that this volume is dedicated.

Most of the success of the conference can be attributed to the participants who shared their knowledge, enthusiasm and ideas not only during presentation but also in the other lively discussions that followed. Unfortunately, this volume does not reflect the science, friendship, and new ideas exchanged outside the formal conference sessions.

Thomas J. Novitsky, Ph.D.
Co-Editor

Acknowledgments

The Conference and Proceedings were made possible through the generous support of:

Hermal-Chemie, Federal Republic of Germany
The Wellcome Foundation, England
Travenol Laboratories, Morton Grove, IL
Abbott Laboratories, North Chicago, IL
Miles Laboratories, Elkhart, IN
Dr. Guiliamo Ramadori, Federal Republic of Germany
Associates of Cape Cod, Inc., Woods Hole, MA

Special acknowledgements go to the following session chairmen for their work:

Terry Munson; H. Donald Hochstein, Dr. P.H.; Thomas J. Novitsky, Ph.D.; Frederick Pearson III, Ph.D.; Kiyoshi Tsuji, Ph.D.; and Stanley W. Watson, Ph.D.

The Woods Hole Oceanographic Institution staff, especially Mr. Donald Souza and Dr. Hartley Hoskins for their excellent technical help with the audio and video equipment, and the late Mr. George Conway and his staff for their invaluable assistance with the planning and organization of the Conference.

Mrs. Rhona Mullin and Ms. Earlene MacDowell were superior in their many and varied tasks prior to, during, and following the conference. Mrs. Mullin handled all the registrations and accommodations and Ms. MacDowell manned the registration desk during the conference. Without their assistance in handling manuscripts and revisions publication of the Proceedings would have been intolerably delayed.

The Alan R. Liss, Inc., staff for their help with the publication.

Dr. Stanley W. Watson, the organizer and co-editor wishes to extend a special thank you to all the conference participants. It is of course the participants who determine the scientific quality of a conference such as this and judging from the papers presented, the discussions that followed, and the comments of several of those who attended the sessions the conference was an unqualified success.

Frederik Barry Bang

1916-1981

Frederik Barry Bang

The proceedings of this conference are dedicated to Frederik Bang, who, had he lived, would have been a co-editor of this volume. Sadly, our colleague and friend died unexpectedly less than two weeks after this symposium was held in Woods Hole.

One year after graduation from the Johns Hopkins University School of Medicine in 1939, Dr. Bang became a National Research Council Fellow in Medicine at the Vanderbilt University School of Medicine and subsequently was on the staff of the Rockefeller Institute in Princeton, New Jersey from 1941-1947. He returned to Johns Hopkins in 1947, where he was Professor and Chairman of the Department of Pathobiology in the School of Hygiene and Public Health at the time of his death.

Dr. Bang's over 250 publications only begin to indicate the wide range of his interests and his contributions to a variety of fields including virology, parasitology, immunology, marine biology and invertebrate pathology. My own personal impression of the remarkable scope of his scientific interests can best be summarized by the years during which he and Mrs. Bang spent their summers (on a rotating basis) at the Marine Biological Laboratory in Woods Hole; at the Station Biologique in Roscoff, France; and at the Johns Hopkins University Center for Medical Research in Calcutta, India and Dacca, Bangladesh, where he was the director.

It was, of course, most appropriate that Dr. Bang gave the opening address at this international conference since his initial observations of the effects of bacterial infection on Limulus polyphemus, the horseshoe crab, provided the basis for the studies that led to the development of the Limulus test for bacterial endotoxins. It also was appropriate that the topic of his presentation was "Rules and Regulations Regarding Serendipity". Serendipity is the combination of chance and sagacity, which makes use of the collected body of previous knowledge without being blinded by precedent. Dr. Bang provided three rules: Take a fresh look at old phenomena; remain naive but carry knowledge of past experiences; and ask new questions and seek new answers. He also indicated that serendipity can be promoted by forcing people into new territories of exploration.

His presentation was a reflection of the man, and it is appropriate to recount some examples of Dr. Bang's application of the rules of serendipity in the field of marine biology which led to significant biomedical applications of his observations.

Firstly, Dr. Bang's paper in 1956 that described the effects of bacterial infection on Limulus polyphemus set the stage for our studies of the role of amebocytes in the coagulation of Limulus blood and the effects of bacterial endotoxin on the blood coagulation mechanism. Unexpectedly, these investigations led to the recognition that the blood coagulation mechanism of Limulus was exquisitely sensitive to bacterial endotoxin. Subsequently, our observations resulted in the development of the Limulus amebocyte lysate test for bacterial endotoxin. Secondly, in the 1960's Dr. Bang observed that the coelomocytes of Asterias forbesi, the sea star, aggregated after particulate matter was introduced into the coelomic fluid. He reasoned that the initially affected coelomocytes released a substance that attracted other cells and resulted in increased aggregation. He found that a lysate of sea star coelomocytes caused aggregation when injected into the coelomic fluid. Subsequently, Robert Prendergast demonstrated that this lysate also attracted vertebrate macrophages to the site of injection, in vivo. Furthermore, the lesion was essentially identical to that of delayed hypersensitivity. The lysate has been shown by Dr. Prendergast to have many of the properties of mammalian lymphokines.

During the same decade, Dr. and Mrs. Bang continued their scientific collaboration using an experimental model based on the urn cell complex of the marine worm, Sipunculus nudus, which was available at the Station Biologique at Roscoff. They demonstrated that mucus production by the urn cell complex is influenced by factors in human lacrimal fluids, stool, and blood, and excitingly that some serum fractions of children with cystic fibrosis have effects upon mucus secretion by these cells which differ from those of normal children.

In each instance, Frederik Bang applied his rules of serendipity to himself. His ability to see biologic phenomena in a fresh light was complemented by his avoidance of complex technology to make his observations. All he seemed to need was a sea table, microscope, some dissecting equipment, and of course, his enquiring and open mind. His recognition of the importance of the collective body of knowledge explains, in part, his humility and lack of any need to attract attention to his own considerable accomplishments. His capacity to repeatedly identify marine models suitable for original biomedical research documents the truth of the statement that chance favors the prepared mind. His successes were not accidental.

Perhaps Dr. Bang's most important achievement was his ability to successfully encourage people, like myself, to enter new biomedical territories and to challenge their potential creativity. His numerous original observations and his many students provide adequate evidence of his extraordinary accomplishments and his generosity to his scientific colleagues.

Jack Levin, M.D.
The Johns Hopkins Hospital
Baltimore, Maryland

Endotoxins and Their Detection With the
Limulus Amebocyte Lysate Test, pages 1–5

RULES AND REGULATIONS REGARDING SERENDIPITY*

Frederik B. Bang, M.D.

Department of Pathobiology, The Johns Hopkins University School of Hygiene and Public Health
Baltimore, Maryland U.S.A.

In choosing a subject such as "Rules and Regulations Regarding Serendipity", I wish to indicate the necessity of this method of research, and at the same time gently nudge many of you who are or are about to become administrators into realizing that this is a common methodology. Serendipity is merely that which is used by biomedical investigators whenever they are searching for correlations between observations. I think that one of the reasons that various governmental agencies really can't deal with serendipity is that there aren't sufficient rules and regulations concerning it. Therefore, I will try to review the story of serendipity and choose three examples that come from the sea so that we can obtain some idea as to what constitutes this particular phenomenon. Serendipity is really an everday phenomenon and is associated with big discoveries and small discoveries.

The first example is the famous but poorly remembered story of anaphylaxis. It was actually "first discovered" several times. The second example is the discovery of artificial pearls by the Japanese oyster industry. The third, of course, is the story of Limulus. But first we must define serendipity. It is not merely chance that we go back, I think, directly to Horace Walpole, the 18th

*This presentation was adapted and prepared for publication by Dr. Jack Levin, from a transcript of Dr. Bang's address.

century English author, who was famous for his correspondence. In one of his letters to a friend, he spoke of a process that he had recently found which, he stated, "I shall call serendipity, a very expressive word". Walpole found the process very interesting.

"I once read a silly fairy tale called the Three Princes of Serendip. As their Highnesses traveled, they were always making discoveries, by a combination of accident and sagacity, of things to which they were not accustomed." I will emphasize what Walpole himself did, and what was true of the original story from which Walpole derived his word. Serendipity is a combination of chance and sagacity. Subsequently, I will point out that sagacity is often a collective effort. In the case of the Princes, the three of them worked together.

My first example of serendipity is the discovery of anaphylaxis. At the turn of the century, the Prince of Monaco founded a Biological Station, which was actually on his ship, The Princess Alice. The ship traveled over the Mediterranean Sea. Richet and Portier were asked by the Prince of Monaco to go along on one of his expeditions, because he wanted them to study the Portuguese man-of-war. The little balloon-like body on top of the Portuguese man-of-war contains carbon monoxide. How this occurs and how the animals resist the effects of carbon monoxide I cannot tell you. Its tentacles are very toxic. The Prince apparently had been affected by this toxin and therefore, he asked these two famous physiologists to come along and attempt to discover how the toxin produced its effects. In so doing, they inoculated a series of animals, i.e., guinea pigs, fishes, and frogs, and soon found that the material that was derived by alcohol extraction of the tentacles was highly toxic. It paralyzed frogs, caused respiratory distress in pigeons, and made ducks so thirsty that they drank seawater. They wished to continue their studies after returning to shore, but were unable to obtain any more Portuguese men-of-war. Therefore, they began to use two types of sea anemones which can be found on the coast of Brittany or the coast of southern France. They are not quite as toxic as the Portuguese man-of-war. Whether this made any difference in their results, we don't know, but as they studied these animals and produced a glycerin extract of the tentacles, they found again that they could kill a variety of animals.

They also were interested in whether or not one could immunize an animal against the toxin(s). Accordingly, they immunized pigeons, dogs, etc. Subsequently, it gradually became obvious to them that they were not immunizing the animals. On the contrary, the animals seemed to become more sensitive and if they were injected a second time, they were more likely to suffer toxic symptoms than they were following the first inoculation. Richet and Portier set aside two dogs (that should be famous) called Galatea and Neptune and studied the phenomenon in their laboratory in Paris.

They started to immunize the dogs and it gradually occurred to them that they should wait for a period of time between the first inoculation and the second inoculation. After a period of approximately 20 days, Neptune and Galatea were reinoculated; they developed the dreadful, sudden reaction which is now recognized as characteristic of anaphylaxis. Surely in this is a hint of the fact that we really can not tell just where we are going until we get started.

The second example, which I mentioned at the start, is a story of the pearl industry in Japan. The formation of pearls is the basis of the first example of invertebrate pathology because the development of natural pearls in the pearl oyster is apparently due to the presence of a parasite or possibly a grain of sand within the soft mantle of the oyster. In 1878, Kokichi Mikimoto, age 20, owned a noodle shop in Mie. Intrigued by the pearl oysters brought up by the young female divers, he tried for 27 frustrating years to produce cultured pearls until "by chance and sagacity", he achieved the first round pearl. Modern cultured pearls are produced by skillful insertion of a little piece of epithelium from Mississippi River mussel shells into the soft mantle of Japanese pearl oysters to create a nucleus for formation of a pearl. It requires two years to form the mature cultured pearls, which now are so famous. No artificial nucleus will induce pearl formation.

In recounting the story of Limulus, I wish to correct any misconception that I had a remarkable inspiration. Limulus polyphemus, the horseshoe crab, is an animal that has been investigated by many people for a long period of time. However, it was Leo Loeb, at the Marine Biological

Laboratory, who became interested in the behavior of their blood cells many years ago. Leo Loeb is the person who was first responsible for the transplantation of hormonally dependent tumors. During his studies of Limulus at Woods Hole, he failed to observe that gram-negative endotoxin could produce gelation of Limulus blood, despite the fact that he recognized that staphylococci produce a coagulum when they are placed in goose blood. So chance played a role in Loeb's conclusions.

During the course of some naive experiments I was performing at Woods Hole about 30 years ago, I turned from oysters on the half shell to injecting Limulus with different bacteria. I attempted to determine whether one could observe phagocytosis of bacteria after they were injected into the Limulus. It soon became obvious that another type of bacteria was present in the animals and that this Vibrio species was capable of producing intravascular coagulation.

Now let us return to a consideration of sagacity, which is often collective sagacity. I had previously been in Korea to study hemorrhagic fever. Hemorrhagic fever is probably a viral disease which causes extensive intravascular coagulation. I had this in mind when I was working with Limulus. Secondly, a characteristic of Woods Hole is that you can sit on a beach and talk with your friends. Jules Freund, famous for the Freund's adjuvant, described to me the Shwartzman phenomenon, (i.e., massive intravascular coagulation following injection of endotoxin into a rabbit). This was particularly striking in older animals, but less so in younger rabbits. This was exactly what was happening with Limulus. Then, Jack Frost, a medical student, dug around in the library and found an old reference of Lankester, a famous British biologist, that reported that the gill leaflets of the book gills of Limulus could be examined microscopically. If one examines these in cross section one can see the circulating blood cells. When the leaflets were injected with the toxic material which I was using, the cells rapidly changed their shape and lost their granules. So it really was collective sagacity (which is one of the great joys of Woods Hole) that formulated various ideas.

To further evaluate this phenomenon, I added some

bacteria to a sample of Limulus blood, which contained blood cells, to determine whether the bacteria would be killed. Within an hour or so, a coagulum occurred which became whitish in color. This coagulum remained solid for the remainder of the summer. It was inescapable that a coagulation process had occured. Ten years later, Jack Levin applied his particular expertise in hematology and blood coagulation to an investigation of coagulation in Limulus. Subsequently, the sequence of events occurred that led to the studies of the reaction between Limulus blood and endotoxin with a light scattering method. The unique carefulness with which he studied the reaction demonstrated the great sensitivity of Limulus coagulation to bacterial endotoxins. You are all aware of the value of the Limulus test that resulted from these studies. Therefore, I wish to reemphasize the role of collectivity in the story.

In conclusion, I wish to state some rules concerning serendipity. Firstly, one needs to take a fresh look at old phenomena. The three Princes, who had been expelled from Serendip by their father, did look abroad and ask new questions. In asking those new questions, they made many discoveries. Secondly, concerning sagacity, remain naive but carry the knowledge of past experience. Thirdly, ask new questions and seek new answers. How can an administrator promote serendipity? Firstly, like the Princes' father, the King, he can force people into new territories. Encourage young persons to go into areas that have not been examined before. He can support environments for endeavor, such as Woods Hole, where the collective sagacity of a group can recognize new phenomena as they occur and consider their potential application. Finally, in this day of cost benefits, perhaps the administrator may recognize that although the cost of serendipity is moderately high and failure is not infrequent, the benefits are many times greater than those that follow the regular procedures that result from completion of forms for the NIH.

Endotoxins and Their Detection With the
Limulus Amebocyte Lysate Test, pages 7-24

THE LIMULUS TEST AND BACTERIAL ENDOTOXINS: SOME PERSPECTIVES

Jack Levin, M.D.

Department of Medicine, The Johns Hopkins University School of Medicine and Hospital
Baltimore, Maryland U.S.A.

It is gratifying that a research project that was originally designed to determine whether there were similarities between mammalian blood platelets and the blood cells of a primitive invertebrate and their respective relationships to blood coagulation has resulted in a practical and useful test for the detection of bacterial endotoxins. The Limulus test is an example of the unanticipated benefit that may occur from what would have been described as "basic research." It is difficult to summarize the many components of this meeting and I will be able to refer to only a relatively few of the excellent studies in this book.

The introduction of a new assay for the detection of a substance, which is more sensitive than the previous assay, is always difficult since one is left without an independent measure of the validity of the assay system when it purports to identify concentrations (e.g., of endotoxin) that are lower than those which could be previously measured. I had anticipated that a chemical determination such as the measurement of β-hydroxymyristic acid, which is apparently specific for bacterial endotoxin, would help in this regard but it appears that this technology is not as sensitive as the Limulus test (Maitra, et al, 1978). This provides an example that biological assays often are more sensitive than chemical determinations, because of the ability of biological systems to amplify that which one is attempting to measure. I think this is the case with bacterial endotoxins.

Considering the large number of effects of bacterial endotoxin, the different types of organs and cells which provide the basis for a wide variety of available assays, and that the Limulus test is an in vitro assay which has been repeatedly compared to an in vivo assay(i.e., the rabbit pyrogen test), I think that the correlations between the Limulus test and rabbit pyrogenicity are remarkable. Studies in our(Cooper, et al, 1971; Tomasulo, et al, 1977) and in other laboratories(Elin, et al, 1976; Tomasulo, 1979; Müller-Calgan, 1982; Pearson, et al, 1982a)have shown an excellent correlation between the Limulus test and pyrogenicity assays. Concordance between the Limulus test and other assays, such as mitogenicity and complement activation also has been documented (Young, 1975; Tomasulo, et al, 1977; Elin, et al, 1976; Tomasulo, 1979; Levin, 1979). For bacterial endotoxins, I think the correlations are as good as we could reasonably wish. However, we should not expect too much of an assay for endotoxin because of the heterogeneity of the functional effects upon which the assays are based. Although correlations of the Limulus test with rabbit pyrogen tests are important because of the regulatory implications, in analyzing their relationship, it is necessary to discriminate between the need for regulation and what might be considered a scientific analysis of the problem. The variety of bioassays for endotoxins, and the variability of endotoxins, i.e., their physical structure, chemical composition, solubility, and particulate nature, should prepare us to anticipate that different assays may produce different results.

Since we are dependent upon functional assays for bacterial endotoxins, it is important to recognize that the definition of endotoxin therefore depends on the assay that is employed. This is, of course, why the term "pyrogen" has become synonymous with endotoxin. However, it is obvious that not all endotoxins are equally pyrogenic

(Weary, et al, 1982; Pearson, et al, 1982a) and certainly not all pyrogens are endotoxins. Therefore, what we define and attempt to quantify as endotoxin is importantly related to how the measurement has been performed. This is true in many other biomedical fields; definitions are operational. Pyrogenicity is the way endotoxin has been defined in the past because of the assay that was used, but we now have to seriously consider that a new definition of endotoxin may be necessary and that the definition of endotoxin as necessarily being pyrogenic may not always be appropriate.

Two problems have been appropriately considered during this meeting. The first is the specificity of the Limulus test. The second is its sensitivity and the effect of our new ability to detect lower concentrations of endotoxin on future acceptable levels of endotoxin or endotoxin-like material in drugs or medical devices. It is obvious that it is impossible to claim that a biological test is specific, and therefore it is possible that substances other than endotoxin react with the Limulus test. However, one must be cautious about reports that some substances produce false positive Limulus tests. For example, if different laboratories disagree about the ability of a reagent to react with amebocyte lysate, and one report reliably demonstrates that the substance fails to produce a positive test, it is reasonable to conclude that the material does not produce false positive tests but that some preparations are either contaminated with endotoxin or produce technical artifacts. These possibilities apparently have occurred when dithiols were evaluated with the Limulus test (Platica, et al, 1978; Scully, et al, 1981).

I would like to put in proper perspective the literature about false positives. For many years it has been known that concanavalin A can produce a variety of biological effects which resemble some of those produced by bacterial endotoxin. Evaluation of ten commercially available concanavalin A preparations demonstrated that nine of ten were Limulus positive and also were able to produce fever in rabbits (Rickles, et al, 1979). The single batch of concanavalin A that was Limulus negative also was non-pyrogenic, although it retained many other biological activities attributed to concanavalin A. Therefore, it is reasonable to conclude that concanavalin A does not produce false positive Limulus tests but rather that essentially all batches of concanavalin A have been contaminated with bacterial endotoxin. It is much easier to contaminate material with endotoxin than to produce an endotoxin-free product. An often quoted paper reports that peptidoglycans, components of the bacterial wall of gram-positive bacteria, cause gelation of amebocyte lysate (Wildfeuer, et al, 1974). Usually overlooked is that the concentrations of peptidoglycan required were 1,000 to 400,000 times higher than the threshold concentration of endotoxin required to produce a positive Limulus test. Although the authors attempted to rule out contamination by endotoxin, as little as 0.1%-0.00025% contamination could have caused their results. These important quantitative relationships are almost invariably ignored.

Currently, one can not unequivocally conclude whether endotoxin is present if the Limulus test is positive and the rabbit test is negative, because we have no independent, adequately sensitive, specific measurement. Hopefully in the future we will. At the moment, the best we can do is to perform multiple assays for endotoxin, analyze the correlations,establish reasonable conclusions, and then evaluate the nature of the problem. Experience with various medical devices will establish whether materials with Limulus positive but rabbit negative assays for endotoxin are associated with pyrogenic or other troublesome effects. However, after exposure of endotoxin to certain conditions as occurs during preparation of dialysis membranes (Carson, 1982; Pearson, et al, 1982c), we cannot expect endotoxin to demonstrate parallel activity in different biological assays. Gradual "detoxification" of endotoxin has been shown to variably and unpredictably alter the activity of endotoxin in a variety of bioassays; potency can decrease in one assay while remaining unchanged or even increased in others(Cundy and Nowotny, 1968).

A characteristic that is used to evaluate whether a positive test for endotoxin is due to endotoxin is the stability of the activity. Therefore, it has been concluded that if the Limulus positivity of a sample is destroyed by heating to 100°C, the positive reaction was not due to endotoxin but rather to the reagent under study. However, when endotoxin is in complex solutions, particularly in the presence of protein, heating may destroy its activity. This may be due to binding to proteins or endotoxin may precipitate with denatured protein. The observation that pasteurization of plasma proteins reduces detectable endotoxin concentrations is most interesting (Cundell, et al, 1982), because heating to 60°C is not believed to destroy endotoxin. Therefore, the disappearance of Limulus positivity after heating a complex solution does not rule out the presence of endotoxin. In this regard, are there any substances in blood or other body fluids that produce positive Limulus tests? I do not think that anyone has convincingly demonstrated that any soluble component of body fluids produces a positive Limulus test. The observation that white cells may cause a positive Limulus test requires additional study (Elin, et al, 1978; Fink, et al, 1981; Fink and Wolkersdorfer, 1982). The data that positive Limulus tests of joint fluid correlate with the presence of leukocytes have to be interpreted cautiously (Elin, et al, 1978), because the study also indicated a correlation between Limulus positivity and pyrogenicity. Hollingsworth and Atkins (1965) have shown that leukocytosis in joint fluid is a highly sensitive indicator of bacterial endotoxins.

The Limulus test is based on a reaction between bacterial endotoxin and an enzyme or series of enzymes in the coagulation system of the horseshoe crab and therefore reflects a relatively direct result of the reaction between endotoxin and the coagulation mechanism. In contrast, the production of fever after administration of endotoxin to a rabbit is indirect since the rabbit must first process the endotoxin in some manner. Therefore, the solubility and particulate nature of the endotoxin is important; endotoxin presumably must reach the reticuloendothelial system or some circulating cell where it results in the production of endogenous pyrogen. Fever finally occurs only after endogenous pyrogen (which does not produce a positive Limulus test) reaches the hypothalamus. Pyrogenicity has been successful as an indicator of endotoxin because it is a measurable effect of bacterial endotoxin and because temperature elevation has been historically used and appropriately recognized as a sign of infection or inflammation. However, this is not necessarily the most important effect of endotoxin. Shanbrom(1982)has emphasized that temperature, per se, usually is not a dangerous effect of bacterial endotoxin. It is possible that the Limulus test measures a more important function of bacterial endotoxin, i.e., its ability to activate various enzymatic systems(other examples are the mammalian coagulation, complement, and kallikrein systems), inappropriate activation of which can produce pathophysiological effects and cause death. A definition of endotoxin based on a positive Limulus test may be of greater importance than whether it is pyrogenic. One is often bound by historical precedent; diseases and materials previously have been inappropriately defined by the first available test.

The Limulus test is approximately 5 to 100 times as sensitive as the rabbit pyrogen test (Tomasulo, et al, 1977; Müller-Calgan, 1982; Pearson, et al, 1982a; Weary, et al, 1982; McCullough, 1982). Importantly, the relationship between minimum detectable concentrations in the Limulus and pyrogen test differs among endotoxins (Weary, et al, 1982). Therefore, there are many instances in which the Limulus test is positive and the rabbit test is negative. There will be materials and drugs which are Limulus positive but the amounts of endotoxin will not be clinically significant. I hope that we do not turn the advantage provided by the greater sensitivity of the Limulus test into a problem. It is easy to determine the sensitivity of a batch of amebocyte lysate and then to use this information to ascertain if the endotoxin concentration of the tested substance is above or below acceptable levels for that drug or device. If endotoxin is present within acceptable limits, the quantitative nature of the Limulus test can be utilized as an indicator of a production problem. This unfortunately is not a benefit that can be derived from the rabbit pyrogen test. The understandable tendency of regulatory agencies to lower acceptable levels of endotoxin to provide a wider margin of safety should be balanced by our knowledge of levels of endotoxin that can be safely tolerated by humans. The sensitivity of the Limulus test should not become the basis for arbitrarily lower acceptable levels. Additional information is required about the relationship between partially purified endotoxins used as standards and native endotoxins (i.e., pieces of the cell wall of gram-negative bacteria)in regard to pyrogenicity and perhaps other more important, but previously ignored, pathophysiologic effects of endotoxin before the acceptable concentration(s) of endotoxin, as defined by the Limulus test, is soundly established.

The concept of the endotoxin unit now seems secure (Sabey, 1982; Outschoorn, 1982; Cooper, 1982). We have stated repeatedly during the past that one does not know the actual concentration of endotoxin in blood or any other fluid based on currently available bioassays,since functional assays have been arbitrary in their choice of standards. It is obvious that different standards may indicate greatly different concentrations of endotoxin in the material being tested. Endotoxins differ widely in their biological activity when assayed for pyrogenicity or with the Limulus test(Pearson, et al, 1982a, 1982b; Weary, et al, 1982; Fleischman and Fowlkes, 1982). Therefore, I hope we will take advantage of what I think is a propitious time to establish an international reference standard for endotoxin. This step can prevent many future problems, not the least of which is the inability of investigators to interpret each others' results. We have used endotoxin obtained from Difco for many years, because it is commercially available and stable. My data could be confirmed by other investigators using the same endotoxin.

There are many applications of the Limulus test (Table 1). Furthermore, with proper controls and consideration of whether a tested material is inhibitory, applications of the Limulus assay are quite unlimited. The availability of a simple, inexpensive, sensitive, and rapid assay has made it feasible for investigators to study the role of bacterial endotoxin both in clinical settings and in experimental laboratories, which previously was difficult and impractical when a bioassay required intact animals. I expect that an increasing number of studies of the role of bacterial endotoxins will utilize the Limulus test, especially as quantification of endotoxin concentration is made easier by the use of synthetic substrates or turbidimetric measurement. In the absence of inhibitors in the test system, false-negative assays should be rare.

The use of the Limulus test to detect endotoxin in blood remains difficult. This is a problem for the Limulus and other assays because there are inhibitors of endotoxin in blood, which were recognized before the Limulus test was available. We described this problem during the past 10 years and suggested the use of either chloroform or dilution, while recognizing that neither of these is an ideal solution(Levin, et al, 1970a, 1970b). Chloroform extraction must be done vigorously and requires hours to complete. Dilution is sometimes adequate, but the process of dilution may decrease the endotoxin concentration below the detectable level. Dilution combined with heating also has been successfully used(Cooperstock, 1975; Friberger, et al, 1982). The presence of inhibitors and the necessity to manipulate a sample of blood before testing for endotoxin makes interpretation of positivity in terms of endotoxin concentration (or units)difficult, if not impossible. The choice

Table 1

Applications of the Limulus Test

- Pharmaceutical preparations
  - large volume parenterals
  - small volume parenterals
  - drugs
  - vaccines
  - biological products
- Radioisotopes
- Water supplies
- Plasma protein fractions
- Medical devices
- In process monitoring
- Body fluids
  - blood
  - cerebrospinal fluid
  - urine

of an endotoxin standard is critical and it must be evaluated in both saline and blood. Another important variable generally not considered is that levels of inhibitors of endotoxin probably vary not only from person to person but also in a patient during various stages of illness associated with gram-negative infection.

Despite these limitations, there are many papers which demonstrate a good correlation between endotoxemia and gram-negative infection (Levin, 1979). However, it is unrealistic to expect the Limulus test or any other bioassay for endotoxin to correlate perfectly with positive blood cultures because it should not do so. The presence of endotoxin in blood is not usually a direct consequence of the bacteria which are in the circulation at the time the sample is obtained, but more commonly is due to bacteria outside of the blood stream that have generated endotoxin that subsequently gains access to the circulation. When it occurs, bacteremia in humans is a momentary phenomenon that lasts seconds to minutes. Therefore, the claim that a positive test for endotoxin in the blood is a false positive if a concomitant blood culture is negative is erroneous. It is more appropriate to determine if there is evidence of a significant gram-negative infection anywhere in the body and whether other laboratory observations suggest that there is a gram-negative infection. It is not the bacteria, per se, in the blood that cause death but the pathophysiological effects of their components.

There is a precedent for the concept that bacterial products can be present in blood (or other body fluids)without concomitant viable organisms. Sensitive immunoelectrophoretic methods have demonstrated capsular antigens in body fluids after those fluids have become sterile and the patients have recovered from acute infection. There is no operational difference between antigen or endotoxin in the blood. Both are components of the cell wall of bacteria. Since bacterial antigens can persist in the blood for days or weeks after clinical recovery has been achieved, it is not surprising that one can detect endotoxin in the blood of appropriate patients with a sufficiently sensitive test. Our current obligation when using the Limulus test for bacterial endotoxins clinically is to accumulate as much information as possible and then determine what correlations exist between the Limulus test and other abnormalities which we can reasonably attribute to the effects of bacterial endotoxin.

The technical performance of the Limulus test clearly affects the results. We originally described the reaction between endotoxin and amebocyte lysate as the formation of a gel, following an initial increase in turbidity and viscosity (Levin and Bang, 1964). Beginning in 1968, our series of papers demonstrated that the rate of reaction between bacterial endotoxin and amebocyte lysate, whether measured by gelation time or optical density is proportional to the concentration of endotoxin(Levin, et al, 1970a; Young, et al, 1972; Levin and Bang, 1968). These findings have been confirmed by many laboratories. The presence or absence of gelation(as a positive or negative result)has been the most commonly used endpoint for the Limulus test. Unfortunately, this does not take into account the time of gelation, which provides a highly reproducible means to semiquantify the concentration of bacterial endotoxin. In my opinion, at least in certain circumstances the introduction of gelation time as a factor in the assay would avoid what I believe are the time consuming, repetitive steps required for the preparation of serial dilutions, with their associated loss of precision.

Therefore, I would like to suggest that one use the gelation endpoint as a positive/negative test to screen large numbers of samples. The gelation time greater than which clinically significant concentrations of endotoxin are not present can be easily established for each batch of lysate, using an endotoxin standard, the biological activity of which had been compared to the national or international endotoxin reference standard(Cooper et al, 1971). The tests are read at the determined time. Samples that produce a significantly positive test(i.e.,gelation times that are shorter than acceptable)can then be re-evaluated by determining the exact gelation time. A simple standardization curve permits quantification of the amount of endotoxin in the sample. This approach also allows serial measurements to be useful in determining when endotoxin contamination is increasing or decreasing.

There are other ways to perform the Limulus test, including turbidimetric measurement and assays based on the use of synthetic substrates (Friberger, et al, 1982; Fujita and Nakahara,1982; Case and Novitsky, 1982; Albaugh and Chandler, 1982; Sloyer et al, 1982). They offer certain advantages. Scientifically, they are particularly important because they measure the initial stage of the reaction between endotoxin and amebocyte lysate. However, the reactions of different endotoxins may have different characteristics(Pearson, et al, 1982b; Homma, et al, 1982; Firca and Rudbach, 1982), although one report demonstrated great similarity between different endotoxins when tested at constant concentrations (Case and Novitsky, 1982). The characteristics of an unknown endotoxin may not resemble the standard. Since we are dealing with an infinite number of endotoxins, it is important to be careful that as procedures for performance of the Limulus test become more precise and as the measured aspect of the reaction becomes limited, we not make the test too difficult for general use and at the same time perhaps generate quantitative data(based on either micrograms of endotoxin or endotoxin units)that are clinically meaningless. It is necessary to discriminate between research, commercial, and general clinical applications. Furthermore, apparent differences in results using similar assay techniques may reflect the type of commercial lysate that is used. Similarly, different results may be observed when methods used for preparation of lysate also change.

Table 2

Problems for Future Consideration

1. The maximum acceptable endotoxin concentration or burden in drugs and medical devices.

2. The definition(s) of bacterial endotoxin.

3. The component of the reaction between endotoxin and amebocyte lysate that will be used to detect and quantify endotoxin.

4. Inhibitors of endotoxin in blood(and possibly other body fluids).

5. Better understanding of the biochemical basis of the reaction between endotoxin and amebocyte lysate.

6. Instability of reconstituted commercial preparations of amebocyte lysate.

7. Different techniques for preparation of lysate and their effects on the sensitivity of lysate in different assay systems.

Table 2 indicates some major problems. The future definition of endotoxin is important and this will depend in part on the assay used to detect endotoxin. Will performance of the Limulus test involve the initial rate of the reaction or will we utilize a more obvious manifestation of the reaction, such as gelation or increased viscosity? Will measurement of the duration of the latent period or of the maximum rate of reaction provide the most useful information? Certain preparations of lysate may not be suitable for some assay techniques. Different endotoxins may produce different kinetic patterns of reaction but identical endpoints. The relationships of the functional characteristics of unidentified endotoxins to their concentrations will be unknown as will be the relative potencies of an unknown endotoxin and the reference standard.

A great deal more must be learned about inhibitors in blood if this test is to become as valuable as I hope it can be for the detection of endotoxin in blood. A full understanding of the reactions(s) that is the basis of the test has been difficult to achieve, because of the ease with which the components of amebocyte lysate are contaminated by endotoxin as they are purified. Perhaps when we finally fully comprehend the biochemistry of the reaction, we will understand better why preparations of lysate are variable and be able to correct this important problem. Different techniques of preparation clearly have an effect on stability because the lysate prepared in our laboratory is stable for at least 7 years at 4°C, without any loss of sensitivity. An important benefit of stable lysate is that it provides us with an ability to measure the stability of our endotoxin standards. One cannot accurately determine the stability of an endotoxin standard unless a stable lysate is also available. (Stability and uniformity are not characteristics of rabbit colonies(Bellentani, 1982; Pearson, et al 1982b)). Nevertheless, considering the progress of the past decade, I am most optimistic about the future, because of the large number of talented and motivated people studying these problems throughout the world.

ACKNOWLEDGMENT

Supported in part by Research Grant HL 01601 from the National Heart, Lung, and Blood Institute, National Institutes of Health, Bethesda, Maryland.

Request for Reprints

Jack Levin, M.D., Hematology Division, The Johns Hopkins Hospital, Baltimore, Maryland 21205 USA.

REFERENCES*

Albaugh BR, Chandler CB (1982). Automated methodology for the Limulus amebocyte lysate (LAL) assay using the multiskan microplate reader. In Watson SW, Levin J, Novitsky TJ (eds): "Endotoxins and Their Detection with the Limulus Amebocyte Lysate Test," New York: Alan R. Liss. Subsequent references that appear in this book will be cited as (this volume).

Bellentani L (1982). Cyclic and chronobiological considerations when employing the rabbit fever test. (this volume).

Carson LA, Petersen NJ (1982). LAL-Reactive material associated with hemodialysis membranes. (this volume).

Case MJ, Novitsky TJ (1982). Comparison of several endotoxin standards with a kinetic LAL test. (this volume).

Cooper JF, Levin J, Wagner HN Jr (1971). Quantitative comparison of in vitro and in vivo methods for the detection of endotoxin. J Lab Clin Med 78:138.

Cooper JF (1982). Formulae for the determination of valid LAL test specimens. (this volume).

Cooperstock MS, Tucker RP, Baublis JV (1975). Possible pathogenic role of endotoxin in Reye's syndrome. Lancet I:1272.

Cundell AM, Spiegelman S, Stryker MH (1982). Use of the Limulus amebocyte lysate endotoxin assay in plasma fractionation. (this volume).

Cundy KR, Nowotny A (1968). Comparisons of five toxicity parameters of Serratia marcescens endotoxins. Proc Soc Exp Biol Med 127:999.

Elin RJ, Sandberg AL, Rosenstreich DL (1976). Comparison of the pyrogenicity, Limulus activity, mitogenicity and complement reactivity of several bacterial endotoxins and related compounds. J Immunol 117:1238.

*References that appear in this book (Watson SW, Levin J, Novitsky TJ (eds): "Endotoxins and Their Detection with the Limulus Amebocyte Lysate Test," New York: Alan R. Liss (1982) will be cited as (this volume).

Elin RJ, Knowles R, Barth WF, Wolff SM (1978). Lack of specificity of the Limulus lysate test in the diagnosis of pyogenic arthritis. J Infect Dis 137:507.
Fink PC, Lehr L, Urbaschek RM, Kozak J (1981). Limulus amebocyte lysate test for endotoxemia: investigations with a Femtogram sensitive spectrophotometric assay. Klin Wochenschr 59:213.
Fink PC, Wolkersdorfer RG (1982). Leucocyte activation of the Limulus amebocyte lysate. (this volume).
Firca JR, Rudbach JA (1982). Reference endotoxin: a practical rationale. (this volume).
Fleishman J, Fowlkes F (1982). A comparison of pyrogenicity of bacterial endotoxins from a variety of gram negative bacteria as determined by the LAL test. (this volume).
Friberger P, Knos M, Mellstam L (1982). A quantitative endotoxin assay utilizing LAL and a chromogenic substrate. (this volume).
Fujita Y, Nakahara C (1982). Preparation and application of a new endotoxin determination kit, Pyrodick^{R}, using a chromogenic substrate. (this volume).
Hollingsworth JW, Atkins E (1965). Synovial inflammatory response to bacterial endotoxin. Yale J Biol Med 38:241.
Homma R, Kuratsuka K, Akama K (1982). Application of the LAL test and the chromogenic substrate test to the detection of endotoxin in human blood products. (this volume).
Levin J (1979). The Limulus test: a status report. In Cohen E, Bang FB, Levin J, et al (eds): "Biomedical Applications of the Horseshoe Crab (Limulidae)," New York: Alan R. Liss, p. 235.
Levin J, Bang FB (1964). The role of endotoxin in the extracellular coagulation of Limulus blood. Bull Johns Hopkins Hosp 115:265.
Levin J, Bang FB (1968). Clottable protein in Limulus: its localization and kinetics of its coagulation by endotoxin. Thromb Diath Haemorrh 19:186.

Levin J, Tomasulo PA, Oser RS (1970a). Detection of endotoxin in human blood and demonstration of an inhibitor. J Lab Clin Med 75:903.

Levin J, Poore TE, Zauber NP, Oser RS (1970b). Detection of endotoxin in the blood of patients with sepsis due to gram-negative bacteria. N Engl J Med 283:1313.

Maitra SK, Schotz MC, Yoshikawa TT, Guze LB (1978). Determination of lipid A and endotoxin in serum by mass spectroscopy. Proc Natl Acad Sci USA 75:3993.

McCullough KZ (1982). The use of LAL as an alternative to the CFR rabbit pyrogen test for disodium ticarcillin. (this volume).

Muller-Calgan H (1982). Experiences with comparative examinations for pyrogens by rabbit pyrogen test versus the LAL test. (this volume).

Outschoorn AS (1982). Characterization of the USP reference standard endotoxin. (this volume).

Pearson FC III, Weary ME, Bohon J, Dabbah R (1982a). Relative potency of environmental endotoxin as measured by the Limulus amebocyte lysate test and the USP rabbit pyrogen test. (this volume).

Pearson FC III, Weary ME, Dabbah R (1982b). A corporate approach to in-process and end-product testing with the Limulus amebocyte lysate assay for endotoxin. (this volume).

Pearson FC III, Weary ME, Bohon J, Nachum R (1982). Detection of Limulus amebocyte lysate reactive material in capillary flow hemodialyzers. (this volume).

Platica M, Harding W, Hollander VP (1978). Dithiols simulate endotoxin in the Limulus reaction. Experientia 34:1154.

Rickles FR, Levin J, Rosenthal DI, Atkins E (1979). Functional interaction of Concanavalin A and bacterial endotoxin (lipopolysaccharide): effects on the measurement of endogenous pyrogen release, human mononuclear cell tissue factor activation, lymphocyte DNA synthesis, and gelation of Limulus amebocyte lysate. J Lab Clin Med 93:128.

Sabey GA (1982). International acceptance of reference endotoxins. (this volume).

Scully MF, Newman YM, Kakkar VV (1981). Inability of dithiols to cause activation of Limulus endotoxin sensitive procoagulase. Experientia 37:685.

Shanbrom E (1982). Pyrogenicity versus endotoxicity: which is more significant? (this volume).

Sloyer JL Jr, Karr LJ, Stonecypher TE Jr (1982). Quantitative techniques for the LAL test. (this volume).

Tomasulo PA (1979). Correlation of Limulus amebocyte lysate assay with accepted endotoxin assays. In Cohen E, Bang FB, Levin J, et al (eds): "Biomedical Applications of the Horseshoe Crab (Limulidae)," New York: Alan R. Liss, p. 293.

Tomasulo PA, Levin J, Murphy PA, Winkelstein JA (1977). Biological activities of tritiated endotoxins: correlation of the Limulus lysate assay with rabbit pyrogen and complement activation assays for endotoxin. J Lab Clin Med 89:308.

Weary M, Pearson FC III, Bohon J, Donohue G (1982). The activity of various endotoxins in the USP rabbit test and in three different LAL tests. (this volume).

Wildfeuer A, Heymer B, Schleifer KH, Haferkamp O (1974). Investigations on the specificity of the Limulus test for the detection of endotoxin. Appl Microbiol 28:867.

Young LS (1975). Opsonizing antibodies, host factors, and the Limulus assay for endotoxin. Infect Immun 12:88.

Young NS, Levin J, Prendergast RA (1972). An invertebrate coagulation system activated by endotoxin: evidence for enzymatic mediation. J Clin Invest 51:1790.

Endotoxins and Their Detection With the Limulus Amebocyte Lysate Test, pages 25–32

FDA GUIDELINE FOR VALIDATION OF THE LAL TEST AS AN END-PRODUCT ENDOTOXIN TEST FOR HUMAN AND BIOLOGICAL DRUGS

Terry E. Munson
Food and Drug Administration
5600 Fishers Lane
Rockville, MD 20857

The guideline was developed to inform manufacturers of human drugs, biologicals, veterinary drugs and medical devices of procedures FDA considers necessary to validate the use of Limulus Amebocyte Lysate (LAL) as an end-product endotoxin test. A manufacturer who follows the guideline may be assured that the procedures or standards will be acceptable to FDA. This article will only deal with the validation of the LAL test for human drugs and biologicals.

A. VALIDATION OF THE LAL TEST

Validation of the LAL test as an endotoxin test for the release of human drugs and biologicals requires: 1. Determination of LAL Sensitivity and 2. Inhibition and Enhancement Tests.

1. Lysate Sensitivity

Before any official tests are performed, the variability of the testing laboratory should be assessed. A minimum of 4 and a maximum of 8 replicates should be performed by each technician using a single lot of LAL and a single lot of endotoxin. A two-fold dilution series should be used with a range suitable for yielding an endpoint with the lysate being tested. The endpoints should be expressed as ng/ml and then converted to the log and the standard deviation (S.D.) calculated.

If the S. D. is less than or equal to the value for the 99 percent limit on the S.D. for 4 replicates (See Table 1), the test is in control. If the S.D. is greater than the tabulated value, the test may be repeated using a maximum of 8 replicates and a new S.D. calculated using data from all tests. If after doing 8 replicates the S.D. exceeds the 99 percent upper limit value corresponding to 8 replicates, the test is invalid due to excessive variability. The source of the variability should be investigated and corrected before retesting.

TABLE 1

99% Upper Limit on S.D.

Size	LOG_2	LOG_{10}
4	1.214	0.365
8	1.015	0.306

After the laboratory has passed the above test for variability, the sensitivity of the LAL reagent should be checked using a series of concentrations of the Reference Standard Endotoxin (RSE) or Control Standard Endotoxin (CSE) bracketing the labeled sensitivity. The results of replicate tests are considered satisfactory if the sensitivity of the lysate is shown to differ no more than one two-fold dilution from the label value.

2. Inhibition and Enhancement Testing

The lack of product inhibition or enhancement of the LAL procedure should be validated for each drug formulation by the Preparatory test in the USP Bacterial Endotoxin Test Monograph.

In inhibition/enhancement testing, a drug concentration containing varying concentrations of a standard endotoxin which bracket the sensitivity of the lysate is compared to a series of the same endotoxin concentrations in water alone. Results of endotoxin determination in the standard series and the drug product are compared and should not differ by more than plus or minus a two-fold dilution.

In those instances when the drug is manufactured in various concentrations of active ingredient while the other components of the formulation remain constant, only the highest and lowest concentrations need be tested. If either concentration shows inhibition or enhancement then each remaining concentration should be tested.

All validation tests should be performed on undiluted drug product or an appropriate dilution. Dilutions should not exceed the Maximum Valid Dilution (MVD). The MVD is calculated as follows:

Step 1. Minimum Valid Concentration

$$MVC = \frac{\lambda M}{K}$$

Where:

λ = Sensitivity of the LAL reagent in Endotoxin Units (EU) per ml.

M = Rabbit Dose or Maximum Human Dose/Kg that would be administered in a single one hour period, whichever is more stringent.

K = 5.0 EU/Kg

Step 2. Maximum Valid Dilution

$$MVD - \frac{\text{Potency of Product}}{MVC}$$

For drugs administered on a weight-per-kilogram, the potency is expressed as mg or units/ml and for drugs administered on a volume-per-kilogram, the potency is equal to 1.0 ml/ml.

If the drug product shows inhibition or enhancement and is amenable to rabbit testing, then the rabbit test will still be the appropriate pyrogen test for that drug. At least three (3) production batches of finished product should be tested for inhibition and enhancement.

The inhibition/enhancement tests must be repeated on at least three units of a product if the manufacturing process, product formulation or the source of a particular ingredient of the drug formulation is changed. At least one unit of each product should be retested when the lysate manufacturer or lysate lot is changed.

B. TESTING OF DRUGS BY THE LAL TEST

End-product testing is to be based on the validation data. Although not specified in the USP, the LAL test may be repeated no more than twice, and the geometric mean of the accumulated values should be used for evaluating the test results. The following should be considered the tolerance limit that all parenteral human drugs and biologicals must meet if the LAL test is to be used as an end-product endotoxin test:

K/M: For any parenteral drug except those administered intrathecally, the tolerance limit is defined as K/M, which equals the amount of endotoxin allowed per mg or ml of product.

NONE DETECTABLE: For parenteral drugs that have an intrathecal route of administration. A negative response using a lysate with at least a sensitivity of 0.2 EU/ml.

A batch which fails a validated LAL release test should not be retested in the rabbit and released if it passes, unless the rabbit test is the official test.

Drugs exempted from the tolerance limits are:

1. Compendial drugs for which endotoxin limits have been established.

2. Drugs covered by NDAs, or Antibiotic Form 5 and Form 6 or Biological Product License where different limits have been approved by the Agency.

3. Investigational drugs or biologicals for which an exception has been filed and approved.

4. Drugs or biologicals which cannot be tested by the LAL method.

C. REGULATORY REQUIREMENTS

The regulatory requirements that must be met by a firm before using the LAL test as an end-product test are as follows:

1. Drugs Subject to NDA's or ANDA's

This class of drugs requires that a supplement be submitted to provide for LAL testing. However, under the provisions of 21 CFR 314.8(d)(3) changes in testing of a drug that gives increased assurance that the drug will have the characteristics of purity it purports or is represented to possess may be placed into effect at the earliest time. Therfore, if a firm validates the LAL test for a particular drug product covered by a new drug application by procedures set forth in this guideline, the change may be made concurrently with the submission of the supplement providing for such change. If a validation procedure differs from that set forth in this guideline a supplement must be submitted and approved before using the LAL as an end-product test.

2. Drugs Not Subject to Premarket Approval

A manufacturer of an injectable drug for human use that is not subject to premarket approval would be able to use the LAL test as an end-product test for endotoxins without submitting any information to the agency. Current Good Manufacturing Practice Regulations (GMPs) require the manufacturer to have data on file to validate the LAL test for each product for which it is being used.

3. Antibiotics

A manufacturer making an antibiotic drug must submit under the provisions of 21 CFR 431.16 an amendment and obtain approval before changing to the use of LAL as an end-product test irrespective of the validation procedure used.

4. Biological Products for Human Use

Under the provisions of 21 CFR 601.12 significant changes in the manufacturing methods of biological products are required to be reported to the agency and may not become effective until approved by the Director, Bureau of Biologics.

MUNSON SPEAKER
TITLE: INTRODUCTION TO THE FDA GUIDELINES

SIDOTE: Can you go over when the rabbit test would be the official test?
MUNSON: If the rabbit test is the pyrogen test in your approved New Drug Application or in the compendial monograph, it is the official test for determining the pyrogenicity of the product.
MORRISON: In all your formulations for MVC and MVD you didn't define an endotoxin unit. Will you define that for us?
MUNSON: An endotoxin unit is the activity equivalent to 0.2 ng of EC-2 which was the original lot of the U.S. Reference Standard Endotoxin.
WEARY: Since the USP allows the use of alternate tests that are the equivalent or better than the official test, can we release a monograph product using lysate and use the rabbit test as the referee test in case of dispute?
MUNSON: Yes.
PEARSON: I have a question concerning inhibition retesting when the lysate manufacturer or lot is changed. Travenol is in a unique position in that we manufacture lysate for our own use. In all of our testing, we have no data to suggest that this testing would really be worthwhile. I am curious as to what kind of data you would be looking for to justify not doing that kind of extra retesting.
MUNSON: In the case of lysate manufacturer changes, since you manufacture your own lysate, this will not be a problem for you. But in our testing, we have seen cases where changing the lysate manufacturer made a substantial difference in the inhibition endpoint. For lot changes, we want to see data showing that when you changed lots, the inhibition endpoint did not change. The data should represent at least four lot changes and must be on a product-by-product basis. The two lambda positive product controls can be used to detect any inhibition changes when lysate changes are made. Thus, extra testing is not required unless the positive product control is negative in which case the product must be revalidated.

PEARSON: I think there are things that become even more unwieldy if you have a manufacturing change. I can cite one example. We have one product that has undergone maybe 20 to 25 changes in a very short period. So, in essence, we would have to test three times every change we made. I think that there are some apparent things that must be discussed in more detail. For example, what is a manufacturer change? Some are minor and some are major. There is a lot of testing which these changes would involve.

MUNSON: I do not think the amount of testing is prohibitive. If there are manufacturing changes, i.e. process, formulation or source of ingredient, you only have to retest once using three units product. Of course, if the retest shows that a change has taken place, then a full validation must be done. I also think that we have to use discretion concerning whether it is a major or minor change or a change which might theoretically cause a problem.

PEARSON: How do you feel about taking a look at the raw material rather than the finished product in this case.

MUNSON: I would prefer to see it in the finished product. The raw material by itself might be non-inhibitory, but once it is added to the formulation and interacts with the other components of the products, the finished product might be inhibitory.

Endotoxins and Their Detection With the Limulus Amebocyte Lysate Test, pages 33–38

THE USP BACTERIAL ENDOTOXINS TEST

Aubrey S. Outschoorn, L.M.S., Ph.D.
Senior Scientist, Drug Standards Division
U. S. Pharmacopeial Convention, Inc.
12601 Twinbrook Parkway
Rockville, MD 20852

During the 1975-1980 U. S. Pharmacopeial revision cycle, considerable interest developed in the Limulus lysate test for bacterial endotoxins which appeared to be an alternative to keeping expensive rabbit colonies with attendant costs of carrying out numbers of rabbit tests per week. The scientific aspects in relation to particular parenteral articles, however, had hardly been explored. A notice had been published in the Federal Register by the Bureau of Biologics (January 12, 1973) stating that such lysates which were used for detecting or measuring endotoxins were biological products, under Section 351 of the Public Health Service Act. The notice stated that the value of the product was recognized, and standards for lysates were in course of development. It was pointed out however, that the test was not then accepted as an alternative to the pyrogen test using rabbits. However, it could be used as an in-process test in manufacture. Proposed additional standards for Limulus lysates were then published (September 18, 1973). A further publication was made (November 13, 1974) to the effect that as a General Safety Test for Inactivated Influenza Vaccine, the lysate test could be used, by comparing the product with a Reference Influenza Virus Vaccine. The Bureau of Biologics subsequently organized many collaborative studies, and an interesting resumé of experience with the test using lysates was recently presented (Hochstein 1981). The restriction to the use of the lysate test as an alternative to the rabbit pyrogen test was removed by a further Federal Register Notice (November 4, 1977), in respect to biologics and medical devices. The final rule Additional Standards for

Limulus Amebocyte Lysate have now been published in the Federal Register (May 16, 1980).

It is difficult to appreciate the haste for such permissive Federal Register notices since current regulations already permit the use of alternative methods - by alternatives is meant alternative to existing procedures in regulations, existing NDAs, etc. No alternatives are still acceptable, unless adequately validated. The compendia had not as yet introduced the test, or even made a proposal for one. However, it was the case that many interested parties approached USP for a statement that the lysate test may be used instead of the pyrogen test using rabbits. It is not the practice of the compendia - as observed over the last 160 years - to publish alternative methods, and it is the responsibility of users of alternate methods to validate such alternatives. Further, no data had been submitted relating to particular compendial articles which might warrant the specification of a bacterial endotoxins test rather than a rabbit pyrogen test.

The Subcommittee concerned of the Committee of Revision undertook a detailed review of available information, and set up principles for introducing the test into the compendia. The summary of developments which follows is my own description of what I believe to have taken place. It is not necessarily a stated position of the Committee of Revision. The materials for making lysate tests included lysate itself, endotoxins, and other reagents. Since lysate products had been declared biologics, under the law, any standards for such would be the responsibility of the FDA Bureau of Biologics. Endotoxin preparations, not then nor now licensed products, would serve as references in assays. The task for the compendia resolved to specifying a test procedure, which may require controls in the form of Reference Standards, and specifications for individual articles. Any Reference Lysate would be that established by the Bureau of Biologics. But there was a clear case for a USP Reference Standard of Endotoxin, and the establishment of one was to be undertaken along with a review of compendial articles for which a Bacterial Endotoxins Test would be applicable. Such articles included existing ones for which the Pyrogen Test (of the USP) was already specified, as well as other possible articles for which such a test might be useful, but for which a rabbit test could not be used owing to the nature of the articles to be tested. The first step taken was to formulate a test procedure.

A particularly useful collaborative study (Rastogi, Hochstein and Seligmann, 1977) was to determine laboratory-to-laboratory variability of lysate sensitivity using a reference endotoxin. Among other measures this indicated the tests which should be carried out in each laboratory to assess that the quality of testing was within sufficient control to yield true values. This required a number of replicate tests with each lot of lysate, using a reference endotoxin. A proposal to introduce this into the compendial description was objected to on the grounds that it constituted laboratory qualification, a topic not within the scope of the compendia. However, the official test does require confirmation of the label potency of the lot of lysate used, with conformity to the upper limit of the standard deviation specified in the federal regulations for licensing the LAL reagent. This is obligatory in view of the fact that the test in the USP is a referee test, and some preparatory testing is specified, not to determine laboratory competence, but to assure that when the test material is evaluated, it is by means of a valid test. This is current practice in the USP for biological assays, where the result obtained as well as the confidence limits must be within the specifications.

Several drafts of the test procedure were made, and rejected. This was because of the great variation in procedural details between lysates, if one were to follow the manufacturer's recommendations. These were largely stabilized later, and it was possible to develop a reasonable description (USP XX, 1980) which was adopted after prior publication in Pharmacopeial Forum. Among the preparatory procedures, it is required to demonstrate that the article under test does not itself interfere with the test - whether by inhibiting or by enhancing the reaction. Appropriate controls, both positive and negative, are indispensable. The description mentions not only the gel formation type of test, but it would be necessary for any other principle which is used, e.g., turbidity of the reactant mixture to be validated. Originally it was required that not less than 4 sets of dilution tubes of the material under test be set up. By Supplement, however (USP XX, Supplement 2), this was reduced to 2 sets of dilutions, in the light of experience and custom. The reference endotoxin, of course, should be set up in 4 sets of dilutions, but such could serve for a number of test materials, run concomitantly. The test is to serve either for determining a compliance limit, i.e., a pass/fail result, or for an actual assay of the concentration of endotoxin in the material under test. In either case, the limit or the limiting concentration must be expressed in terms of a Reference Standard Endotoxin. The RSE selected was that already established by the FDA Bureau of Biologics as the U. S. Reference Standard Endotoxin (Lot EC-2) with a defined Unit of potency.

Since the USP Bacterial Endotoxins Test requires that all endotoxin determinations be in units of potency equivalent to those defined by the U. S. Reference Endotoxin, the compendia introduced the Control Standard Endotoxin (CSE) concept. That is a preparation which has been calibrated in terms of the U. S. Reference Standard Endotoxin. This covers several parameters and requires primarily that its potency be evaluated with lysate. A description is given of the scope of such a comparison. In addition a CSE, to be adequately characterized, if it is to serve as such, must also be studied for such parameters as stability, uniformity (i.e. between aliquots, whether measured each time a test is done or the preparation is put up in vials) and most important, by its reaction in rabbits. The U. S. Reference Endotoxin has been widely studied in those ways, and it is perhaps the only material serving as a reference standard for which there is information on its behavior in humans (Elin, Wolff, Chedid and Bernard, 1981). It therefore has valid properties to serve as an RSE, and any RSE established by the USP in its Reference Standards Program should have been similarly studied. A later presentation will elaborate on this.

A parallel activity, undertaken by the FDA, was to formulate guidelines for the use of the LAL test as an end-product test for release of pharmaceutical articles. This is not a referee role, as for compendial tests. Proposed guidelines have been released, their availability having been notified in the Federal Register of January 18, 1980. Revisions thereto have been under development for some time. However, what was done in the proposed guidelines was the adoption of the USP test procedure, quoted by reference, as the guideline procedure.

With the publication of USP XX, the Bacterial Endotoxins Test became official on July 1, 1980. However, that does not mean that it can be used indiscriminately instead of the Pyrogen Test using rabbits, wherever the latter is specified. A decision will have to be made, for each individual monograph, and a particular limit of endotoxin determined, for that article. This must be by evaluation of data. Such data should include at least results of endotoxin evaluations of a series of batches, preferably from several sources, and data to show the relationships of the amounts found to results in rabbits. Until this can be done, it is not envisaged that a particular limit applicable in general, e.g., for all parenterals, would be specified. In specifying the limit of endotoxin for each article it is necessary to take into consideration the dose, i.e., the amount of the drug which the patient will receive and the time over which the maximum human dose may be administered. The limit would generally relate to the maximum the patient would have to cope with. Some of the candidate articles for consideration include radiopharmaceuticals, particularly owing to difficulties in maintaining safety considerations in a testing facility, and vehicles for parenteral products. A prime candidate of these is Water for Injection. This article, however, poses certain problems. Since Water for Injection can be used, with further processing, to make either small volume parenterals (e.g., 50 ml or less) or large volume injections (e.g., 1 litre) how can the endotoxin limit to be specified be determined? Would there be different limits when the article is used for different final products? Those questions have not yet been settled. Other articles include toxic drugs, e.g., anti-tumor agents, for which a rabbit Pryogen Test cannot now be required.

Elin RJ, Wolff SM, Chedid L, Bernard C (1981) J Infect Dis (in press).

38 FR 1404 (January 12, 1973).

38 FR 26130 (September 8, 1973).

39 FR 40016 (November 13, 1974).

42 FR 57749 (November 4, 1977).

45 FR 3668 (January 18, 1980).

45 FR 32290 (May 16, 1980).

Hochstein HD (1981). The LAL test versus the rabbit-pyrogen test for endotoxin detection. Pharm Tech 5 (8):36.

Rastogi SC, Hochstein HD, Seligmann EB Jr (1977). II. Quality control procedures for testing limulus amebocyte lysate sensitivity using reference endotoxin 1b. Bull Parent Drug Assoc 3(3):120.

The United States pharmacopeia 20th rev. (1980). Easton: Mack, p. 888.

The United States pharmacopeia 20th rev., Suppl. 2 (1981). Easton: Mack, p. 160.

Endotoxins and Their Detection With the Limulus Amebocyte Lysate Test, pages 39–48

ENDOTOXIN TESTING OF MEDICAL DEVICES WITH LAL: FDA REQUIREMENTS

Virginia C. Ross, M.S., and Carl W. Bruch, Ph.D.

Bureau of Medical Devices, FDA
Silver Spring, Maryland 20910

In a notice published in the FEDERAL REGISTER of November 4, 1977 (42 FR 57749), the Food and Drug Administration (FDA) stated conditions under which the LAL test could be used for end-product testing of medical devices for endotoxin as follows:

> With respect to the use of LAL to test for pyrogens in medical devices, the Bureau of Medical Devices will consider a device to be misbranded and/or adulterated as to pyrogenicity unless the manufacturer has submitted to the Bureau of Medical Devices data establishing that the LAL test the manufacturer proposes to use is at least equivalent to the rabbit test, and the manufacturer has obtained written approval to use the LAL test from the Director of the Bureau.

Based on BMD's experience in approving petitions to use the LAL test in place of the rabbit pyrogen test, the Bureau of Medical Devices (BMD) issued a draft guideline on March 26, 1979 advising device firms what procedures to follow for validation of the LAL test as an end-product endotoxin test. This original draft guideline was made available to all interested parties and underwent several changes as a result of comments received from the Health Industry Manufacturers Association (HIMA), FDA staff from BMD, other FDA Bureaus, and other organizations. The USP Rabbit Pyrogen Test, the USP Transfusion and Infusion Assemblies Test, and a HIMA

Collaborative Study were used as the basis for selection of a suitable pass/fail point for LAL testing of medical devices (HIMA Document, 1979).

HIMA conducted a collaborative study to evaluate the pyrogenicity of Difco *E. coli* 055:B5 endotoxin in the USP Rabbit Test. The Biological Testing Branch, Bureau of Biologics and ten industrial laboratories participated in the study. This study established with 95% confidence that a laboratory will obtain 50% pass/fail results at some concentration above 0.098 ng/ml of Difco *E. coli* 055:B5 endotoxin administered at a dose of 10 ml/kg. The study recommends that any laboratory using the LAL test in medical devices should be able to detect levels of 0.1 ng/ml of Difco *E. coli* 055:B5 endotoxin with the LAL test. When an LAL failure occurs, manufacturers have the choice of retest in rabbits or quantitating with the LAL test. After reviewing this study and the literature, the Bureau of Medical Devices accepted this sensitivity level (0.1 ng/ml of *E. coli* 055:B5 endotoxin) as adequate for LAL testing of devices (Greisman, Hornick 1969; Rudbach, *et al*. 1976; Wachtel, Tsuji 1977). To be consistent, the Bureau now recommends that any rabbit colony used for retest of devices should show a pyrogenic response at 0.1 ng/ml of the above endotoxin.

In the LAL test, the concentration of endotoxin in a specific device rinse volume will determine whether or not a product is pyrogenic. The Bureau of Medical Devices uses the USP rinsing requirements for testing of transfusion and infusion assemblies as the basis for rinsing or eluting of devices for endotoxin testing. This procedure specifies rinsing each of 10 devices per lot with 40 ml and pooling the effluent for a total of 400 ml. If this volume must be increased appreciably because of the size or configuration of a device, then consideration should be given to using a more sensitive LAL endpoint in order to assure detection of a pyrogenic product.

Validation of the LAL test for end-product release of medical devices involves three phases: 1) the sensitivities of the tests to be used for endotoxin must be established; 2) product inhibition or activation must be determined; and 3) representative lots of medical devices must be tested using the LAL test.

SENSITIVITY TESTING

If a firm intends to test devices solely by the LAL test they need only demonstrate a positive LAL reaction of at least 0.1 ng/ml. Any method for LAL determination that is reproducible can be used, although the gel clot method is the most commonly employed. Laboratory personnel should refer to the USP Bacterial Endotoxin Test.

The sensitivity of the LAL test should be determined using a series of endotoxin concentrations bracketing the endpoint of the lysate. Two-fold dilutions of endotoxin made with LAL-negative water or saline should bracket the lysate endpoint sensitivity. Aliquots of these endotoxin dilutions are added to tubes containing a specified amount of lysate preparation. Tubes are incubated and the endpoint is determined according to the technique selected. Results must be reproducible and differences between replicate tests must not be statistically significant. Routine performance of the LAL test should include a standard series (run in duplicate) and a negative control. Running an endotoxin series is essential for checking lysate sensitivity, the competence of the technician, and for identifying other problems such as the contamination of glassware.

Since the LAL reagent is usually sensitive to endotoxin levels less than 0.1 ng/ml, a manufacturer may determine the concentration of endotoxin in the device by serial dilution and release the lot if less than 0.1 ng/ml of endotoxin is present.

As an alternative, if an LAL positive has been obtained using an LAL test more sensitive than 0.1 ng/ml, then the Bureau will accept a provision for retest of device lots in the USP Rabbit Test provided that a manufacturer demonstrates that its rabbit colony is sensitive to 0.1 ng/ml endotoxin. The procedure for performing the USP Rabbit Pyrogen Test is explained in the USP XX. Sufficiently small dilutions (five-fold or less) of endotoxin bracketing the endpoint should be used to determine the sensitivity of the rabbit colony.

INHIBITION TESTING

Before using the LAL test for pyrogenicity of devices, lack of product inhibition or activation of the LAL test must be shown for each type of device. Consideration should be given to possible inhibition from different chemical components of similar devices. Manufacturers can logically divide their device products into groups of products according to common chemical formulation; it is then only necessary to qualify a representative product from each such group. Ideally, the product chosen from each group would be the one with the largest surface area contacting the body or fluids for administration to a patient.

At least three production lots of each product type must be tested for inhibition. In general, use of the sampling technique selected should result in a random sampling of a finished production lot. The number of devices in any finished production lot may vary and individual devices may differ in size, configuration, composition, expense, and their criticality in terms of pyrogenicity. It is recommended that a sample of at least 10 test units per production lot be used. For small lots the Bureau recommends testing two devices per lot for lot sizes under 30, three devices per lot for lot sizes 30-100, and 3% of lots above size 100 up to a maximum of 10 devices per lot.

The process of preparing an eluate/extract for pyrogen or inhibition testing may vary for each device. Some medical devices can be flushed, some may have to be immersed in the LAL-negative rinse solution (water or saline), while others may be tested by disassembling or by cutting the device up into pieces prior to extraction by immersion. In general, for devices being flushed, the LAL-negative rinse solution should be held in the fluid pathway for 1 hour at room temperature (above 18°C); effluents should be combined. If a device is to undergo extraction, a minimum extraction time should be 15 minutes at 37°C, 1 hour at room temperature (above 18°C). Other techniques demonstrated to be equivalent to the above can be employed.

Guidelines for rinse volumes include the following:

1. Each of the 10 test units should be rinsed with 20-40 ml of LAL-negative saline or water depending on the size of the device.

2. For unusually small or large devices, the surface area of the device which comes in contact with the patient may be used as an adjustment factor in selecting the rinsing or extracting volume.

The rinsing scheme must not result in a greater dilution of endotoxin than used previously in USP rabbit testing of transfusion and infusion assemblies. For inhibition testing, both the rinsing/extraction solution and the device extract must be spiked with the same endotoxin levels used in a standard series bracketing the sensitivity of the lysate; negative controls for both solutions must be included. Results of endotoxin determination in the standard series and device extracts must not differ significantly.

TESTING OF DEVICES BY THE LAL TEST

After obtaining a suitable eluate/extract pool from a finished production lot, this pooled extract is kept under conditions appropriate for endotoxin stability until it is tested. Testing must be done on three finished product lots for each group of devices using rinsing/eluting and sampling techniques similar to those described for inhibition testing. As in inhibition testing, sampling can be adjusted for special situations.

For each product group, five copies of the protocol and test results should be compiled and submitted to the Document Control Center, HFK-20, Bureau of Medical Devices, 8757 Georgia Avenue, Silver Spring, Maryland 20910. If review by the Bureau determines the data to be adequate, a letter from the Bureau giving the firm permission to use the LAL test for end-product testing of the devices included in the submission will be issued.

IN PROCESS TESTING WITH LAL

Based on the Bureau's experience reviewing data from LAL testing of devices for most medical devices, contamination with levels of endotoxin detectable by the gel clot technique is rare. The high temperatures used in processing of plastic materials used in devices are probably responsible for this general lack of contamination. For these types of devices, manufacturers may apply for approval to use in process control of endotoxin contamination. Recently, the Bureau approved two submissions for reduced end-product testing when the history of pyrogen or LAL testing demonstrated a lack of endotoxin contamination.

Submission of a scheme for process control of endotoxin contamination is required. The manufacturer must demonstrate adequate control of the production process with routine checks for endotoxin at key stages of production except where it has been shown that no possibility of contamination exists. The total endotoxin contamination possible when components of a device are being tested separately should not be more than the amount necessary to give a device failure in conventional end-product testing. If contamination is detected at any point in process, full end-product testing of the lot is required.

SPECIAL CONSIDERATIONS

When tested with LAL, devices used in contact with cerebrospinal fluid should have less than 0.04 ng/ml of endotoxin in the rinse fluid since this anatomical area is more sensitive to endotoxin contamination.

Devices prepared from processed biological tissues may be naturally contaminated with endotoxin. These products are reviewed case by case to assess the risks and benefits involved in their use.

PRODUCT CLEARANCE STATUS

As of May 1981, the Bureau had approved 168 products from thirty firms for LAL end-product testing for endotoxin as follows:

PRODUCTS APPROVED FOR LAL END-PRODUCT TESTING	
Gastro/Urology and General Use	73
Cardiovascular	61
Neurological	23
Ob/Gyn and Radiology	6
Surgical and Rehabilitation	4
Opthalmic, Ear, Nose, Throat and Dental	1
	168

These products represent many generic classes of devices and include transfusion and infusion assemblies as well as other products used in the blood stream or as implants.

Currently, the Bureau of Medical Devices requires products used in the blood stream to be nonpyrogenic. Other products which are labeled nonpyrogenic must also be tested routinely by the LAL or USP Rabbit Pyrogen Test. If the Bureau obtains information concerning a continuing problem with adverse effects of endotoxin contamination on implants, then a policy requiring implants to be nonpyrogenic may be considered.

Greisman SE, Hornick RB (1969). Comparative Pyrogenic Reactivity of Rabbit and Man to Bacterial Endotoxin. Proc Soc Exp Biol Med 131:1154.

HIMA Collaborative Study for the Pyrogenicity Evaluation of a Reference Endotoxin by the USP Rabbit Test (1979). HIMA Document No. 7, Vol 1.

Rudbach JA, Akiya FI, Elin RJ, Hochstein HD, Luoma MK, Milner ECB, Milner KC, Thomas KR (1976). Preparation and Properties of a National Reference Endotoxin. J Clin Microbiol 3:21.

Wachtel RE, Tsuji K (1977). Comparison of Limulus Amebocyte Lysates and Correlation with the United States Pharmacopeial Pyrogen Test. Appl and Env Microbiol 33:1265.

ROSS SPEAKER
TITLE: INTRODUCTION TO THE FDA REQUIREMENTS FOR TESTING MEDICAL DEVICES.

PEARSON: I would like to ask questions concerning the rabbit retest and the value that you established as 0.1 ng in previous guidelines. I understood you to say that if the manufacturer can indeed qualify the rabbit colony you suggest that the manufacturer should do so. Is this mandatory within the scope of the guidelines or is it suggested within the scope of the guidelines?
ROSS: We suggest that a rabbit colony used for retest after a positive initial LAL test should be sensitive to 0.1 ng/ml of endotoxin (10 ml/kg). Different endpoints in LAL and rabbit retests would be inconsistent. Of course, a guideline does not have the legal power of a regulation.
PEARSON: Secondly, if one were to have a product which was not amenable to LAL testing, would the rabbit colony have to also qualify at 0.1 ng.
ROSS: We have not addressed that problem so far.
PEARSON: Have you any data concerning the number of rabbit colonies that can be qualified at the 0.1 ng level?.
ROSS: I would have to examine all the submissions before I could answer that question, however we know that some colonies can detect this level.
URBASCHEK: We have some experience with I.V. injections of endotoxins into volunteers. We used E.coli endotoxin 0111 (Boivin extraction) from our own laboratory. This endotoxin was very toxic. It would be of interest to compare the acute toxicity of this endotoxin with the FDA-standard, especially in regards to the differences in toxicity of different batches. We injected the volunteers with 100 ng of endotoxin per 70 kg of body weight. A person of 70 kg has about 5,500 ml of blood. This represents an amount of 20 pg of endotoxin/ml of blood. Fever, headache, chills, nausea, and restless legs are observed. The characteristic changes in platelet and leukocyte counts were observed in these volunteers.

ROSS: I'd like to see your data at some point. I would like to point out for medical devices the situation is slightly different than for drugs. The guideline calls for 40 ml rinse per device for 10 devices for a total of 400 ml. In the worst case if the contamination is limited to only 1 of the 10 devices a person would be receiving 40 ng. Less than 4 ng per device would be possible if the contamination is spread out among the 10 devices. Of course, you have to realize that in the manufacture of devices you have more of a chance for non-uniform contamination than in LVP drug products. We believe that we are in a safe area with a limit of 0.1 ng/ml as described in the medical device guideline.

URBASCHEK: If you maximally inject 40 ng, then I believe you are in a safe area. However, it is questionable, whether the quanitity measured with the LAL test provokes the same toxic activities that are induced with corresponding concentrations of an injected endotoxin preparation.

ROSS: We are considering converting to the FDA standard at some point and there will be another collaborative study going on to compare the 055:B5 endotoxin with the FDA, US standard. If you relate your endotoxin standard to the US standard then we could compare our results.

SABEY: I think there is a little confusion brought into this by referring to the concentration of 0.1 ng/ml for the LAL test and 0.1 ng/kg for the rabbit test.

ROSS: No, it is 0.1 ng/ml given 10 ml/kg for rabbits.

ENZINGER: You mentioned that the rabbit colony should show a 0.1 ng/ml sensitivity. How do you define this sensitivity, do you define it as the average pyrogenic dose or the PD_{50} ?

ROSS: We define it as conducting the USP rabbit test and finding the sensitivity according to the definition of that test.

MORRISON: In the event that both the LAL test and the rabbit pyrogen test would be employed and one test is positive and one test negative at a higher concentration, which test would predominate? Let's say the LAL test is positive at 100-fold dilution of test sample and the pyrogen test is negative when used undiluted.

ROSS: I haven't seen any situation like that and I have to do some thinking about that.
MORRISON: Assuming you were confronted by such a situation, which test would the FDA favor?
ROSS: I'm not going to go into that at the moment. I think that selection of the appropriate test would depend on the product involved.
STOTT: Could you provide us with more information on the requirements from biological preparations?
ROSS: If a manufacturer is interested in using the LAL test on a device derived from biological material he would use the same type of data collection that was described for other devices. These products usually contain endotoxins, thus you would also submit all of your data showing us how much endotoxin is contained in the product as well as the product dose. You should provide data on many lots, detailing the endotoxin content and what procedures you have employed to remove the endotoxin and show that these methods did not alter your product. We would then consider setting limits for this product based on its intended use in the body and the data submitted.
STOTT: Did you say there were no endotoxin limits for implants?
ROSS: Officially the bureau does not have any requirements for non-pyrogenicity of those products but we do require that if the manufacturer labels the products as non-pyrogenic, each lot should be tested either by the rabbit test or the LAL test. Depending on the intended use of a new implant product, FDA reviewers may indicate that it should be non-pyrogenic. Permission from the Bureau of Medical Devices, FDA is required before using the LAL test as an end product test for endotoxin.

Endotoxins and Their Detection With the Limulus Amebocyte Lysate Test, pages 49–54

STATISTICAL PROCEDURE FOR STANDARDIZING CONTROL STANDARD ENDOTOXIN AGAINST U.S. REFERENCE STANDARD ENDOTOXIN USING LIMULUS AMEBOCYTE LYSATE TEST

Suresh C. Rastogi, Ph.D.

Bureau of Biologics
Food and Drug Administration
8800 Rockville Pike
Bethesda, Maryland 20205

The U.S. Reference Standard Endotoxin (RSE) is required as a reference for the determination of bacterial endotoxins. It is issued by the Bureau of Biologics of the Food and Drug Administration and has a defined potency in Endotoxin Units (EU) per unit weight or per unit volume. The RSE is supplied to the licensed manufacturers of Limulus Amebocyte Lysate (LAL) without charge by the Bureau. A control standard endotoxin (CSE) is an endotoxin preparation that has been standardized against the RSE. The laboratories or the manufacturers use their CSE for either intermediate testing or for in-house quality control. The users of the LAL test are required to confirm the labeled sensitivity of their LAL reagent. Their CSE is used for this purpose. In this paper, we present a statistical procedure for standardizing a CSE against the U.S. RSE using the LAL test. We also give a procedure of checking the quality control of the testing laboratory.

The LAL gel test is used here in developing the procedure. But the procedure can be easily modified for other types of LAL test. Let us assume that the RSE is in homogeneous vials with a defined potency of E EU/vial. A CSE lot may be in homogeneous vials containing a certain weight of CSE per vial or it may be a bulk lot from which a unit of weight is taken for testing. It is assumed in the rest of the paper that CSE is in vials, each vial containing 1 ug of endotoxin content. A licensed lysate with a labeled sensitivity against RSE is used.

Figure 1 respresents a typical form for recording the results of LAL testing. Initial dilution factors A for CSE and B for RSE are to be obtained by doing some preliminary LAL testing using the given lysate. Only tenfold dilutions of CSE and RSE are made for this purpose. Once the end-points using tenfold dilutions are established, five twofold dilutions are made using A for CSE and B for RSE as starting points. The labeled sensitivity of lysate may be used as a guide for getting end-points in the middle of twofold dilutions.

CSE versus RSE Testing Form

Date: ____________ Technician Name: ____________________

U.S. RSE Lot: EC-X Units: E EU/vial

CSE Lot: 100X Testing Unit: Vial containing 1 ug

Lysate: Lot No. Sensitivity: ____________________

CSE vial No	←CSE Dil. Factors→ A	2A	4A	8A	16A	←RSE Dil. Factors→ B	2B	4B	8B	16B	Ratio R
1	+	+	+	–	–	+	+	–	–	–	R1
2	+	+	+	–	–	+	+	+	–	–	R2
⋮											⋮
⋮											⋮
⋮											⋮

Indicate reaction for every dilution tested as either a + or –. R = CSE end-point dil factor/RSE end-point dil factor.

For establishing statistical parameters in a laboratory a minimum of 16 vials of CSE are tested in parallel with RSE. A parallel test is an assay where CSE and RSE are tested side by side under identical conditions. Generally contents of one vial of RSE will be enough for this parallel testing. The ratio of CSE end-point dilution factor with RSE end-point dilution factor is obtained for each parallel test. Vials of CSE may be parallel tested more than once, say 4 times each, to have more confidence in their ratios. If the CSE vials are tested more than once, the geometric mean (GM) of ratios for each vial is calculated. Therefore, each vial of CSE will have only one ratio value associated with it.

A parallel test is considered invalid if, for either of the two endotoxins, results are of the following type:

(i)	+	+	+	+	+
(ii)	-	-	-	-	-
(iii)	+	-	+	-	-

In situation (i), starting dilution factor is moved to the right, while in (ii) it is moved to the left. In situation (iii), an improvement is needed in the technical precision.

Now suppose that N (>= 16) vials of CSE have been parallel tested with RSE and N ratios, R1, R2,...., RN have been calculated. The ratios are converted to the logarithm to the base ten. Mean and standard deviation (SD) are calculated. The antilog of the mean is the GM of the ratios. The product of this GM ratio and the labeled potency of RSE is the potency of CSE per unit weight (1 ug) of testing. That is,

Potency of CSE = GM ratio x E EU/ug

The variability of parallel testing CSE with RSE is going to be estimated very precisely by a large study on the proposed Reference Endotoxin. This study was undertaken by the Bureau of Biologics, FDA and USP in September 1981. To fix ideas, let us assume that the overall SD of log ratios obtained from the BoB/USP Study is 0.2. Tables 1 and 2 are generated using the chi-square and normal tables (Snedecor and Chochran 1980) for different number of parallel tests from 4 to 28 in increments of 4.

Table 1

99% Fiducial Upper Limit for a SD of N Log Ratios

N	99% Upper Limit
4	0.39
8	0.32
12	0.30
16	0.29
20	0.28
24	0.27
28	0.26

Table 2

2-SD (95%) Limits for a GM of N Ratios Assuming Theoretical Ratio of One Between CSE and RSE Potencies

N	Lower Limit	Upper Limit
4	0.63	1.58
8	0.72	1.38
12	0.77	1.30
16	0.79	1.26
20	0.81	1.23
24	0.83	1.21
28	0.84	1.19

If the Laboratory tested only 16 vials of its CSE in parallel with the RSE and found that:

GM of ratios = 0.5; SD of log ratios = 0.25

and if

Potency of RSE = 6000 EU/vial

then

Potency of CSE = 0.5 x 6000
= 3000 EU/ug
= 3 EU/ng

Since a SD of 0.25 is less than the upper limit shown in Table 1 for N=16, the laboratory is within acceptable variability and is in control. The laboratory can assign to its CSE a potency of 3000 EU/ug of 3 EG/ng. If the CSE lot is in homogeneous vials with 1 ug per vial, than each vial is labelled 3000 EU/vial.

The laboratory should regularly check whether its CSE is maintaining its potency. For this purpose Table 3 is generated by multiplying the values of Table 2 by the potency of CSE, 3 EU/ng in this case. Four vials of CSE are tested in parallel with RSE using the LAL test. The CM of ratios and SD of log ratios are calculated. The potency of CSE in EU/ng is calculated using this CM ratio. To verify the acceptable variability, it is checked whether this SD is less than the limit shown in Table 1 for N=4. If the calculated potency value is within the 2-SD limits given in Table 3 for N=4, the CSE is maintaining the labeled potency. If it is less than then lower limit then the SCE has lost potency significantly. If necessary, more tests may be performed in increments of 4 parallel tests. Results of all valid tests are included in all calculations.

Table 3

95% Limit for the Potency of CSE in EU/ng

N	Lower Limit	Upper Limit
4	1.89	4.74
8	2.16	4.14
12	2.31	3.90
16	2.37	3.78
20	2.43	3.69

References

Rastogi SC, Hochstein HD, Seligmann EB Jr., Dawson JH, Farag LG, Marquina RE (1979). Statistical procedure for evaluating the sensitivity of Limulus amebocyte lysate by using a reference lysate. J Applied and Envir Microbiol 6:911-915.

Snedecor GW, Cochran WG (1980). Statistical methods 7th ed Iowa State University Press Ames.

Endotoxins and Their Detection With the Limulus Amebocyte Lysate Test, pages 55–64

FORMULAE FOR MAXIMUM VALID DILUTION.

James F. Cooper

Medical University of South Carolina
Charleston, South Carolina 29425

The long-term acceptance and success of the rabbit bioassay as a screen for pyrogenic contamination in sterile products dictated that the Limulus test had to have significant advantages over the former before it could be seriously considered as an alternative test. The greater sensitivity of the Limulus Test was recognized early, but if it was to be widely applied, the new test could not be so sensitive as to fail products which contain clinically insignificant amounts of endotoxic pyrogen. At the encouragement of academia and industry, the concept of an endotoxin tolerance limit was developed within the **U. S. Pharmacopeia (USP)** and Food and Drug Administration (FDA) to accommadate the unique sensitivity of the Bacterial Endotoxins Test.

To understand the concept of maximum valid dilution (MVD) is to recognize that the formula attempts to equalize the sensitivity of two tests that measure endotoxin presence in a different manner; the in vitro test is concentration dependent whereas the in vivo test is dose-dependent. When a drug is diluted to the MVD, a negative **LAL** Test, in the absence of interference, implies that there is a low probability that the test specimen contains sufficient endotoxin to cause rabbit or human pyrogenicity.

A forerunner of the MVD formulae was termed the Limits Dilution Factor, LDF (Cooper, Neely 1980). By this approach, the test product would be diluted, as determined by the LDF calculation, and tested in a pass/fail manner as a qualitative test. The LDF is calculated as follows:

$$LDF = C_e \times 1/\lambda$$

where C_e is the maximum permissible endotoxin concentration (EU/ml) as established by the manufacturer or **USP** monograph, and λ is the sensitivity of the **LAL** reagent (EU/ml). The test specimen for assay could then be diluted by the following ratio:

test specimen = undiluted drug:LDF

where one part of specimen is diluted to LDF parts of **LAL** reagent diluent. When a drug product yields a negative **LAL** assay after being diluted to the calculated extent, the product would meet the requirements of the Bacterial Endotoxins Test. The limitation of this approach was the lack of a uniform means for selecting the tolerance limit. At the time the LDF formula was proposed, we suggested that the limit would be that determined by manufacture interaction with the FDA or that specified by the **USP**.

In response to the first draft FDA Guideline for **LAL** Testing, which would have established limits based solely on endotoxin concentration, a PDA Task Force addressed the issue of endotoxin limits and suggested a dose-related formula (Parenteral Drug Association 1980). The PDA proposed that the **LAL** Test for a given product should be designed to at least detect a threshold pyrogenic dose (TPD) or quantity of endotoxin. The TPD was defined as the lower 95% confidence limit of the average pyrogenic dose. This value approaches 1 ng/kg (approximately 5 EU/ml) for two purified endotoxins of **E. coli** as described by several convincing studies (Greisman, Hornick 1969; Dabbah, et al 1980; Tsuji, et al 1980). Therefore, the PDA proposal suggested that the limits test be designed to detect at least 1 ng of endotoxin per kg of the administered dose. For example, if the dose of a drug product was 10 mg per kg, such as gentamicin, the limits test should be designed to detect at least 1 ng or 5 EU of endotoxin (EC-2 or equivalent) per 10 mg of gentamicin.

$$\frac{\text{TPD (amt. of endotoxin)}}{\text{10 mg of gentamicin}} = \frac{\text{1 ng or 5 EU EC-2 equivalent}}{\text{10 mg of gentamicin}}$$

The rabbit test dose per kg would be used except in those cases where the usual human dose exceeded the rabbit dose on a weight basis. Hydrocortisone sodium succinate and lincomycin hydrochloride are examples of the latter.

The current draft **LAL**-Test Guidelines described earlier in this session proposes calculation of the maximum valid dilution (MVD) as the basis for a limits test. The calculation is a two-step process as follows:

1) Minimum Valid Concentration, MVC = $\lambda M/K$

WHERE: λ = LAL Reagent Sensitivity (EU/ml)

M = Rabbit or Maximum Human Dose per Kg, whichever is higher. Assume 70 Kg man

K = Tolerance Limit, 5 EU/Kg (1 ng/Kg)

2) Maximum Valid Dilution, MVD = $\frac{\text{Potency of Product}}{\text{MVC}}$

Finally, the MVD, a unit-less number, is applied by diluting one part of the drug product to the MVD value (parts) with **LAL**-Reagent Water or Saline.

In our laboratory, the calculation was simplified to a single expression:

$$\text{MVD} = \frac{\text{C (Drug conc.) x 5 EU/Kg (Limit)}}{\lambda \text{ (Reagent sen.) x M (Max. dose/Kg)}}$$

where C represents the concentration or potency of the drug product; C would be unity (one) for drugs where dosage is determined by volume. Definition of the other units remain the same. The relationship of CK/M is a fixed value for a specific drug formulation. This expression represents the tolerance limit per unit volume of the drug product. Only λ , the **LAL** reagent sensitivity, is a variable in the MVD determination for each product. Therefore, a parenterals manufacturer would have to specify an acceptable range of **LAL**-reagent sensitivity in test procedures to accommodate small lot-to-lot variations in **LAL**-reagent sensitivity. As an example, let us calculate the MVD for phenytoin sodium, an anticonvulsant. The human dose is 15 mg/kg as stated in the **USP** Dispensing Information (U.S. Pharmacopeial Convention 1981). This SVP is commercially available only at a concentration of 50 mg/ml. The MVD for this product is calculated as follows:

GIVEN: Phenytoin Sodium, 50 mg/ml

$$M = 15 \text{ mg/Kg (USP-DI)}$$

$$\lambda = 0.2 \text{ EU/ml}$$

$$MVD = \frac{C \times K}{\lambda \times M} = \frac{50 \text{ mg/ml} \times 5 \text{ EU/Kg}}{0.2 \text{ EU/ml} \times 15 \text{ mg/Kg}} = 83$$

Therefore, dilute 1 part of phenytoin to 83 parts with WFI.

NOTE: MVD = 50 when λ = 0.3 EU/ml

The MVD value is 83 for a reagent sensitivity of 0.2 EU/ml, or 50 for 0.3 EU/ml. Therefore, it would be prudent for a manufacturer to specify a MVD value in its **LAL**-Test specifications that is considerably less than the maximum, provided low-level contamination is not a problem. This would permit the use of a range of **LAL**-reagent sensitivities.

In the case of phenytoin sodium, a MVD value of 50 would be a logical choice for the manufacturer. This value would permit a reagent sensitivity up to 0.3 EU/ml. Also, this value would allow for modification of the undiluted product to overcome interference. Dilution is important because phenytoin sodium is a basic drug which requires pH adjustment and dilution to at least 1:25 before interference is overcome. The MVD calculations for other drugs are presented in **Table 1** to provide further examples. The maximum dose, M, for NSA 5% and heparin were taken from CFR requirements for rabbit testing, whereas the dose selected for the radiopharmaceutical was based on actual use in medical practice.

TABLE 1

Drug Product	M	K/M	MVD*
Thallium Tl-201 1 mCi/ml	0.1 ml/kg	50 EU/ml	250
NSA 5%	10 ml/kg	0.5 EU/ml	2.5
Heparin Na 5000 Units/ml	2000 U/kg	0.0025 EU/U	62.5

*λ = 0.2 EU/ml

The major limitation to the MVD calculation is the lack of uniformity in labeling **LAL**-Test reagent and Control Standard Endotoxin (CSE). To exemplify this problem, the sensitivity of **LAL** reagents from three suppliers was determined in a sensitivity-verification experiment. The **LAL**-reagent sensitivity was labeled as 0.125, 0.06, and 0.023 ng/ml with respect to a U.S. standard endotoxin. Although the labeling suggested that the three reagents varied greatly in sensitivity, the geometric mean sensitivities determined for the three, under the same conditions and with the identical endotoxin, fell in a range of 0.02 to 0.03 ng/ml with reference to **E. coli**, 055:B5 (Difco). Thus, this experiment indicated that there was no significant difference in the reagent sensitivities. Inadequate labeling and the lack of reproducibility between EC-2 and EC-4 endotoxins resulted in this misleading labeling.

To add to the confusion, the CSE endotoxins were inadequately labeled except for one supplier. Although the total quantity of the endotoxin in its container was listed, only one supplier labeled the expected reactivity between the supplied reagent and endotoxin. This data is not presented as a criticism of **LAL**-reagent manufacturers, but as a stimulus for industry and government to expedite a solution to the reagent and endotoxin reference labeling problems.

The FDA proposal states a tolerance limit in Endotoxin Units (EU) even though this value has not been assigned for the EC-5, the future U.S. Reference Standard Endotoxin to be maintained by the USP. A worthy objective of this conference would be the development of a plan, involving industry, the FDA's Bureau of Biologics (BOB), and the USP, which would expedite the labeling of all **LAL**-Test materials in Endotoxin Units (EU). The mechanism for implementing label changes could be developed while awaiting the results of the EC-2/EC-5 comparison study now in progress.

A second problem with the MVD calculation is the assumption that the endotoxin tolerance limit provides the equivalent margin of safety for pyrogen which occurs in its native or environmental state as it does for purified endotoxins. The FDA tolerance limit was based on comparison studies using purified endotoxins of **E. coli**, endotoxins noted for their potency for inducing mammalian pyrogenic response. There is an increasing body of evidence that suggests a lack of equivalence between so-called purified and environmental endotoxins in that the latter may be much less potent in rabbits. However, **LAL** titers are apparently more equivalent. The study of the pyrogenicity of environmental endotoxins presented in this session by Pearson and co-workers indicated that these pyrogens were seldom detected in rabbits unless present at a dose of 8 ng/kg or greater.

Similar findings are apparent in a recent report which described the FDA's compliance program for biological products (Hochstein 1981). For purpose of discussion, the rabbit and **LAL** test data for 25% serum albumin (NSA) were rearranged and an additional column presented which represented the implied dose range of endotoxin per kg of rabbit, as classified by an **LAL** titer (Table 2). Of the 949 batches tested, the majority had an **LAL** titer of less than 4 EU/ml, representing less than 12 EU/kg per rabbit; all of these batches passed the rabbit test. About 19% of the batches had a titer of 4-20 EU/ml, which implied an equivalent dose to the rabbit of 12-60 EU/kg; only 6% of these batches failed the rabbit test. The 3% of batches classified by **LAL** titer as a range of 60-240 EU/kg of rabbit produced only 33% failure. Finally, only one-half of the batches failed the rabbit test at the highest level, 240 EU/kg or greater. A similar pattern was found for 5% NSA wherein batches at levels of 40 EU/kg or greater seldom failed the rabbit pyrogen test.

TABLE 2

COMPARISON OF LAL AND RABBIT TESTS OF 25% NSA BUREAU OF BIOLOGICS

LAL TITER (EU/ml)	NSA 25% (# of lots)	RABBIT DOSE* (EU/Kg)	FAILED (%)
<4	735	<12	NR†
4–20	178	12–60	6.7
21–80	26	61–240	30.8
>80	10	>240	50

*test dose of 3 ml/Kg; †not reported or retested by BOB

The discrepancy between the **LAL** and rabbit test data was surprising and led the author to conclude that the **LAL** test was not as reliable as the rabbit bioassay for pyrogen testing of biologics (Hochstein 1981). Further, the concern for false positives by the **LAL** test was seen as a deterrent to expanded **LAL** testing of biologics. The inability of the in vitro test to more accurately predict rabbit response to low levels of pyrogenic contamination in biologics is not necessarily an **LAL**-Test limitation. An alternative explanation for the discrepancy is that the rabbit pyrogen test is a less reliable and less sensitive measure of the true endotoxin content in biologics. We know that serum proteins have the capacity to strictly modify biological properties of endotoxin. This occurs by binding phenomena and other mechanisms which are poorly understood. The parenteral drug industry has identified many agents which interfere with the **LAL** test. Technology now exists to overcome most of these interferences; the advent of a commercial dispersing agent represents a significant advancement. However, little is known about the effect of these interfering agents on mammalian pyrogenic response. Industry should face the challenge of learning more about such interference so that a more firm basis can be established for comparing **LAL** and rabbit measurements of environmental endotoxins.

SUMMARY

The preparation of test specimens using the MVD calculation represents a reasonable approach to equating the sensitivity of the in vitro and in vivo tests for endotoxin. Labeling of **LAL** reagents and control standard endotoxins are a point of confusion for the parenterals industry. Government and reagent suppliers must work together expeditiously to provide test materials that are explicitly labeled as to reactivity in EU/ml and ng/ml. The relatively greater reactivity of **LAL** reagent to environmental endotoxin than the rabbit bioassay provides a greater margin of safety for the in vitro approach than previously anticipated, and suggests that rabbit retest of **LAL** failures may be appropriate for specific products.

REFERENCES

Cooper JF, Neely ME (1980). Validation of the LAL test for end-product evaluation. Pharm Tech (No. 6) 4:72.

Dabbah R, Ferry E, Gunther DA, Hahn R, Mazur P, Neely M, Nicholas P, Pierce JS, Slade J, Watson S, Weary M (1980). Pyrogenicity of E. coli 055:B5 endotoxin by the USP rabbit test - a HIMA collaborative study. J Parenteral Drug Assoc 34:212.

Greisman SE, Hornick RB (1969). Comparative pyrogenic reaction of rabbit and man to bacterial endotoxin. Proc Soc Exp Biol Med 131:1154.

Hochstein HD (1981). The LAL test versus the rabbit pyrogen test for endotoxin detection. Pharm Tech (No. 8) 5:36.

Information Bulletin No. 3 (1980). "Response to Food and Drug Administration draft guideline for use of Limulus Amebocyte lysate," New York: Parenteral Drug Assoc., p.7.

Tsuji K, Steindler KA, Harrison SJ (1980). Limulus amebocyte lysate assay for detection and quantitation of endotoxin in a small-volume parenteral product. Appl Environ Microbiol 40:533.

U. S. Pharmacopeia Dispensing Information (1981). "Phenytoin" Rockville, MD: U. S. Pharmacopeial Convention, Inc., p.335.

COOPER SPEAKER
TITLE: FORMULAE FOR THE DETERMINATION OF VALID TEST DOSES.

TSUJI: We know that a number of drugs mask the physiological activity of endotoxins, thus isn't it dangerous to compare the values obtained with the LAL test to those found with the rabbit fever test.
COOPER: We certainly need more information regarding the effect of certain drug formulations on the biological activity of endotoxins. I think the proposed FDA guideline is sufficiently flexible to permit a manufacturer to write standard operating procedure to include variables and risk factors.
URBASCHEK: You mentioned purified endotoxin, may I ask you what purification method you are using?
COOPER: I've only addressed the two E.coli-derived endotoxin preparations which were purified by classical methods. The biological activity of endotoxins may be proportional to the amount of LPS per unit weight of endotoxin. We know that the potency of EC-2 and the 055:B5 are similar, even though their methods of purification vary.
WATSON: Most of the endotoxins used are not pure but crude chemical extracts of bacteria. Possibly Novo-Pyrexal is the purest of all endotoxin standards being used today, next in purity would be the FDA standard which was prepared by Dr. Rudbach.
COOPER: Nevertheless, purification may not increase pyrogenicity. The problem we need to address is the differences in pyrogenic response between purified and environmental endotoxins.
WATSON: In our laboratory Mrs. Valois has examined several autotrophic bacteria, such as nitrifying bacteria, cyanobacteria and others. The activities of endotoxins may vary by five orders of magnitude. For example the activity of endotoxin (as judged by the LAL test) from E. coli can be five orders of magnitude greater than the activity of endotoxins from the autotrophic bacteria. It is important to determine if the biological activity of these various endotoxins varies to the same degree in rabbits.

RYLANDER: We have some experience in studying different endotoxins from natural environments. Most of our studies which I will be talking about Thursday are related to industries where organic materials are being processed but we also have some data from other factories. We isolated several strains of bacteria which had widely different endotoxin activity, particularly a strain of agrobacteria which was not active at all but commonly present in these environments. The endotoxin from this organism was not pyrogenic in guinea pigs. This difference in activities probably relates back to the differences in the Lipid A moiety.

MORRISON: There is a point that has come up several times in regards to crude endotoxin preparation versus purified endotoxin preparation and I think it is important to keep in mind there may be a distinction between the two depending upon how one defines purity. Endotoxins are composed of a protein, Lipid A, and polysaccharide complex. The active component we all recognize is the lipopoloysaccharide component, in particular the Lipid A region. Nevertheless, from a physical chemical point of view, one can have highly purified endotoxins which consist of protein and lipopolysaccharides and that is not the same as purified lipopolysaccharides even though both are, by definition, purified endotoxins. On a percent by weight basis, one could have moderate to very significant differences, two-to-eightfold, simply on the basis of weight of the active component. It may be a minor distinction but it becomes important when we talk about relative levels of either LAL sensitivity or pyrogen sensitivities.

COOPER: That is a good point. I recall that the collaborative pyrogen study in 1939, which generated the criteria for the USP test, used nitrogen content to monitor endotoxin concentration. Certainly LPS used in current comparative studies is more potent than that used to develop the rabbit pyrogen test.

Endotoxins and Their Detection With the Limulus Amebocyte Lysate Test, pages 65–77

RELATIVE POTENCY OF "ENVIRONMENTAL" ENDOTOXIN AS MEASURED BY THE LIMULUS AMEBOCYTE LYSATE TEST AND THE USP RABBIT PYROGEN TEST

Frederick C. Pearson III, Marlys E. Weary, Joyce Bohon, and Roger Dabbah
TRAVENOL LABORATORIES, INC.
6301 Lincoln Ave.
Morton Grove, IL 60053

When the USP rabbit pyrogen test was introduced in 1942, no attempt was made to define what levels of endotoxin were pyrogenic in rabbits or man. In 1956, Westphal indicated that the minimal pyrogenic dose of purified *Salmonella abortus equi* endotoxin on a per kg basis was comparable in man and rabbits. Keene et al. (1961) demonstrated that purified endotoxin of *Serratia marcescens* administered at 30 ng/kg gave comparable threshold fever responses in four human subjects and in rabbits. This was confirmed and extended by Greisman and Hornick (1969), who determined the relative sensitivities of rabbit and man to various endotoxin injections. Their results indicate clearly that the induction of minimal but unequivocal pyrogenic reactions to purified endotoxins requires approximately the same dosage per kg of body weight in rabbits and man. Furthermore, if total endotoxin dosage rather than per kg dosage is considered, the rabbit actually responds to lower quantities than does man.

On Jan. 18, 1980, the Bureau of Drugs drafted a "Guideline for Validation of the *Limulus* Amebocyte Lysate Test as an End-Product Test for *Human* and Veterinary Injectable Drugs and Medical Devices," and established 0.5 ng of endotoxin per kg as a final product release limit for most pharmaceutical products other than intrathecals and large volume parenterals (LVP's). There are, however, no definitive studies to support these limits. The overwhelming scientific evidence indicates that 100 pg/ml (1.0 ng/kg) is an acceptable release limit for LVP's, providing a significant safety factor over the rabbit test. Data suggest that "environmental" endotoxins are relatively impure, and the 1.0-ng/kg limit provides approximately a tenfold safety factor. An additional safety factor is inherent in the use of the 95% confidence level for pyrogenicity.

While the rabbit assay has served well as a predictor of human pyrogenicity, the *Limulus* amebocyte lysate (LAL) assay has, in large part, replaced the more cumbersome rabbit test for in-process testing and final release of medical devices and many LVP's. The LAL test is rapid, simple, inexpensive, accurate, sensitive, and can be used with substances that are toxic to animals. Parallel tests clearly demonstrate that the LAL assay is more sensitive than the rabbit test.

For about ten years, LAL studies were designed to convince the FDA of the superior sensitivity of the test. Most of these studies used purified endotoxin or lipopolysaccharide (LPS), thereby establishing a falsely low limit of LPS relative to environmental endotoxin. Although both are quite sensitive to LPS, the LAL test is approximately an order of magnitude more sensitive than the rabbit test to environmental endotoxin.

The following studies show the disparity between LAL end points and the rabbit test when environmental endotoxins are evaluated against a purified, re-formed LPS. In addition, the threshold pyrogenic dose (TPD) for rabbits is established both for purified endotoxin, other than *E. coli*, and for naturally occurring endotoxin.

MATERIALS AND METHODS

Limulus Amebocyte Lysate. Travenol LAL was used. Each vial of lyophilized material was reconstituted with 3 ml of 0.5% $MgCl_2$ $6H_2O$.

Lipopolysaccharides. *E. coli* 055:B5, *Shigella flexneri*, *S. abortus equi*, and *S. marcescens* were purchased (Difco Laboratories, Detroit, MI). *Pseudomonas* endotoxin (Piromen®, Travenol) is a crude digest. *Klebsiella* and EC-2 LPS were acquired through the Bureau of Biologics (BOB), FDA, Rockville, MD. All LAL results are referenced to *E. coli* 055:B5.

LAL Assay - Spectrophotometric. The Travenol procedure has been reported by Nandan and Brown (1977).

LAL Gel Clot Assay. Each 0.1-ml sample was mixed with an equal volume of LAL reconstituted with 100 ml of $MgCl_2$ $6H_2O$ and 50 ml of Tris/Tris HCl in water (3 ml per vial) in 10 x 75-mm, nonpyrogenic glass test tubes. The mixtures were incubated in a water bath at 37°C for 1 h.

Positive controls were spikes of endotoxin (12, 25, 50, and 100 pg/ml). Negative controls were nonpyrogenic water for injection. A positive end point was recorded when a solid gel clot remained stable after being inverted 180°. Endotoxin concentrations of samples were determined by multiplying the sensitivity of the LAL reagent by the reciprocal of the dilution factor required.

USP Rabbit Pyrogen Tests. A complete description of this test can be found in the US Pharmacopeia (1980). The Travenol test has been reported by Nandan and Brown (1977).

Evaluation of Environmental Endotoxin. Over 500,000 LAL tests conducted on raw materials, various waters, and finished products were evaluated by comparing the LAL and rabbit results. Because 50 pg of endotoxin per ml is the alert limit used by Travenol for rabbit retesting, and 200 pg/ml is the highest concentration on the LPS standard curve, rabbit results were compared with two groups of LAL results: those from 0.050-0.2 ng/ml, and those with more than 0.2 ng/ml.

Tap water was collected throughout the Travenol Corporate R&D facility and evaluated using both the gel clot end point and the rabbit test. Water samples contained 0.2-80.0 ng of endotoxin per ml.

RESULTS

Threshold Pyrogenic Dose of LPS

Studies were undertaken to define the pyrogenicity of various LPS's other than those from *E. coli*. The rabbit assay was used to measure endotoxin potency. All endotoxin preparations tested had TPD's of 1.0 ng/kg or higher. Endotoxin preparations from *E. coli* 055:B5 and *S. abortus equi* exhibited TPD's of slightly more than 1.0 ng per kg of body weight. Endotoxin preparations from *K. pneumoniae* and *S. flexneri* were less potent, exhibiting TPD's of above 2.0 ng per kg of body weight. The least potent endotoxins were from *S. marcescens* and *P. aeruginosa*, exhibiting TPD's of approximately 15 and 370 ng per kg of body weight, respectively. These results are consistent with other findings with *E. coli* (Dabbah et al., 1980; Enzinger et al., 1980; Tsuji et al., 1980), *P. aeruginosa* and *S. abortus equi* (Greisman and Hornick, 1969), and *K. pneumoniae* (Cooper et al., 1971; Castor et al., 1971; Van Noordwijk and de Jong, 1976).

Environmental Endotoxin

Review of 530,371 in-process and final release LAL tests conducted at various domestic facilities from 1977 to 1979 revealed that endotoxin content was between 0.05 and 0.2 ng/ml in 489 and over 0.2 ng/ml in 70. All samples containing more than 0.05 ng were evaluated using the 8-rabbit test. Only 13 (3%) containing 0.05-0.2 ng/ml and 8 (11%) containing more than 0.2 ng/ml failed (Table 1). The probability of obtaining a rabbit failure at 0.1 ng/ml is much less than it would be at around 0.2 ng/ml, because pyrogenicity above the threshold level is a dose-response phenomenon. Even at 0.2 ng/ml or above, most "natural" endotoxins are significantly less pyrogenic than are standard endotoxin preparations, and will pass rabbit testing at the 0.1-ng/ml level.

Table 1. Analysis of Parallel Pyrogen Tests Conducted by Travenol Domestic Plants from 1977 to 1979

LAL Tests Conducted 530,371			
50 to 200 pg/ml		More than 200 pg/ml	
489 Samples		70 Samples	
476 (97%)	13 (3%)	62 (89%)	8 (11%)
Passed Rabbit	Failed Rabbit	Passed Rabbit	Failed Rabbit

A similar analysis was conducted using parallel LAL and rabbit test data from 17 international plants during 1980. 67 LAL failures were recorded, with endotoxin concentrations ranging between 50 and 200 pg/ml. Of these, 98.5% (66) passed the rabbit test, while only 1.5% (1) failed. A total of 18 samples contained over 200 pg/ml; 94% (17) of these samples passed the rabbit assay, while only 6% (1) failed (Table 2). Most of the LAL failures involved process waters and raw materials, not finished product.

Table 2. Analysis of Parallel Pyrogen Tests Conducted by Travenol International Plants in 1980

LAL Tests Conducted			
50 to 200 pg/ml		More than 200 pg/ml	
67 Samples		18 Samples	
66 (98.5%)	1 (1.52%)	17 (94.1%)	1 (5.9%)
Passed Rabbit	Failed Rabbit	Passed Rabbit	Failed Rabbit

These data strongly suggest that a 100-pg/ml release limit for LVP's has a significant safety margin, and that there is little likelihood that a naturally occurring, environmental endotoxin will be found that is as pyrogenic as any of the standard preparations on a weight-for-weight basis. The data on environmental endotoxin are consistent with current knowledge of the chemical composition and biological activity of endotoxin. Lipid A is responsible for pyrogenicity, as well as other biological activities, and on a weight basis, purified standards contain significantly more Lipid A than the "real-world" endotoxin that occurs in the production of LVP products.

Environmental Endotoxin in Tap Water

Studies were undertaken to compare the results for LAL and rabbit tests conducted on environmental endotoxin in water, a significant source encountered in the manufacture of parenteral drugs. To ensure samples would contain a significant load of environmental endotoxin, samples were taken from common city supplies--drinking fountains, washroom faucets, etc. LAL tests were conducted using clot end points, and rabbit test failures were scored as an average temperature rise per rabbit of 0.46°C, a marginal USP rabbit pyrogen test failure.

When environmental endotoxin was administered to rabbits at doses between 2.0 and 8.0 ng/kg (ascertained using LAL), rabbit temperature responses were generally well below that for the marginal USP rabbit pyrogen test failure. These doses represent up to 800 pg/ml when the LAL results are referenced to the *E. coli* 055:B5 endotoxin standard, a dose well above the 100-pg/ml suggested limit, especially when the 95% confidence level for rabbit pyrogenicity in a cross section of rabbit test colonies is used. While rabbit test failures were frequently encountered above a dose of 10 ng/kg, passes were demonstrated at doses of 16 and 1,000 ng/kg. This is clearly a tenfold safety factor over the 100-pg/mg release limit (Table 3).

DISCUSSION

TPD's per kg are virtually identical in man and rabbit; however, the dose-response relationship for man is steeper than for the rabbit (Greisman and Hornick, 1969). At doses considerably higher than the TPD, man responds to LPS more vigorously than do rabbits. This does not change the fact that the rabbit test is a valid predictor of pyrogenic risk for man, particularly at threshold levels. The TPD's in man and rabbit were demonstrated to be 0.1-0.14 ng/kg for *S. typhosa*, 1.0 ng/kg for *E. coli*, and 50-70 ng/kg for *P. aeruginosa* (Greisman and Hornick, 1969). The low pyrogenic dose reported for *S. typhosa* was based on a highly purified preparation and bears little, if any, resemblance to the lessened pyrogenic response with the naturally occurring endotoxins encountered in pharmaceutical manufacturing.

The comparability of pyrogenic risk depends on which endotoxin is being studied, as well as the dose level (e.g., in this study, the pseudomonal preparation was relatively unpurified). Wolff (1973) concluded that man was much more sensitive than the rabbit to Lipexal, a purified endotoxin preparation from *S. abortus equi*. This finding

Table 3. Equivalency of the Rabbit Test and LAL Test Results on Environmental Endotoxin in Water

LAL Results* (pg/ml)	USP Rabbit Dose (ng/kg)	USP Rabbit Test Results
200	2.0	Pass
200	2.0	Pass
200	2.0	Pass
220	2.2	Pass
400	4.0	Pass
800	8.0	Pass
800	8.0	Pass
800	8.0	Pass
800	8.0	Pass+
800	8.0	Fail
1,600	16.0	Pass+
1,600	16.0	Fail
1,600	16.0	Fail
2,500	25.0	Fail
5,000	50.0	Fail
6,200	62.0	Fail
10,000	100.0	Pass
10,000	100.0	Fail
10,000	100.0	Fail
20,000	200.0	Fail
40,000	400.0	Fail
80,000	800.0	Fail

* LAL results are referenced to E. coli 055:B5 endotoxin.
\+ Based on original 3-rabbit test results, an additional five rabbits were employed.

is consistent with the results of Greisman and Hornick (1969) only when doses higher than the TPD are considered. Wolff's general claim is clearly in conflict with the earlier studies of Westphal (1956), who also used Lipexal and found the TPD for man and rabbits to be comparable. When 2 ng of Lipexal per kg was administered to humans, a maximum temperature rise of 1.9°C and mean 7-hour fever index of 50 cm^2 were observed. There is no question that man is more sensitive to Lipexal at these higher doses. However, these doses are clearly not TPD's at which man and rabbits respond similarly. A temperature rise of 1.9°C is just over three times that expected for a decided USP 3-rabbit failure, and 1.3°C is approximately twice that expected for a failure. Thus, doses of 2-5 ng per kg of Lipexal are above the TPD of rabbits and man. If these experiments had been conducted at lower levels, the fever responses of man and rabbit may well have been comparable.

During 1978, the Health Industry Manufacturers Association (HIMA) initiated a collaborative study to establish the pyrogenicity of E.coli 055:B5 endotoxin using the rabbit test. Testing laboratories included representatives of regulatory agencies, industry, and reference pyrogen evaluation facilities. E. coli 055:B5, a widely used commercial preparation, was used instead of the national reference standard EC-2.

The HIMA study established with 95% confidence that a statistically average laboratory will attain 50% pass/fail results at some concentration of Difco E. coli 055:B5 endotoxin above 0.098 ng/ml (0.98 ng/kg, USP dose of 10 ml/kg). From a practical standpoint, the study concluded, it was "reasonable to set 0.1 ng/ml, which is slightly above the lower confidence limit, as the standard of reference pyrogenic activity against which the Limulus Amebocyte Lysate test (LAL) is to be compared." The suggested end point is equivalent to 1.0 ng/kg when test material is delivered at a USP dose of 10 ml/kg. The study concluded that, if a laboratory could demonstrate an LAL failure rate significantly greater than 50% at 0.1 ng/ml of Difco's E. coli endotoxin, the test would be considered qualified for use under conditions of evaluation as equivalent to the rabbit test.

A position paper developed by the Parenteral Drug Association (PDA) Limulus Amebocyte Lysate Task Force provided valuable data confirming and extending the studies of Greisman and Hornick, as well as the HIMA study. Based on the latter, the task force concluded that, under the best conditions, the sensitivity of the rabbit test approaches 1 ng of commercial E. coli endotoxin per kg, and that this value could serve as a needed reference point for endotoxin limits in small volume parenteral (SVP's). Unfortunately, this task force did not deal with "large volume parenterals (LVP'S)".

The Bureau of Medical Devices (BMD) has apparently accepted 100 pg/ml (1 ng/kg referenced to E. coli 055:B5) as the pass/fail criterion for medical devices. Following receipt of the HIMA report, the BMD issued a draft, "Bureau of Medical Devices Guidelines for Adoption of the Limulus Amebocyte Lysate Test for Pyrogenicity of Devices." According to these guidelines, if a device manufacturer intends to pass or fail product lots on the basis of LAL results, the test laboratory must demonstrate a test failure rate of 90% or greater with 100 pg of E. coli 055:B5 endotoxin per ml, or 1.0 ng/kg. If data substantiate this level of sensitivity, rabbit equivalency data are not required (Weary and Pearson, 1980).

Investigations conducted by Tsuji et al. (1980) using E. coli 0127:B8 show that the average pyrogenic dose (APD) and the TPD could be calculated at 1.43 and 1.01 ng of endotoxin per kg of body weight, respectively. Similarly, the APD and the TPD were calculated for EC-2 to be 1.37 and 1.05 ng of endotoxin per kg of body weight, respectively. The study concluded that, statistically, E. coli 0127:B5 and 055:B5 endotoxins are equivalent to EC-2, and that the TPD can be defined as 1.0 ng of E. coli with reference to EC-2 per kg of body weight. It was concluded that 1.0 ng per kg of these endotoxins is a logical limit for the quantity of bacterial endotoxin for SVP's based on the highest recommended product dose, or the rabbit test, whichever is more stringent.

Most recently, Garratt et al. (1981) reported on a collaborative study undertaken by U.K. laboratories to assess the reproducibility of the LAL assay and its relative sensitivity to the rabbit test. Results were assessed according to the criteria outlined in the European Pharmacopoeia. If the average increase in temperature for 3 rabbits did not exceed 0.38, the sample was judged nonpyrogenic. These studies are consistent with the findings of other workers. The APD of the World Health Organization endotoxin standard (Shigella dysenteriae) was shown to be approximately 2.0 ng/kg. The average pyrogenic response and range of responses to various endotoxins described above strongly support 1.0 ng/kg as a valid endotoxin LAL release limit.

Although the BOB has not published an official document governing the explicit use of LAL for the final release of biological products, several BOB studies have shown that surprisingly high concentrations of endotoxin are required to elicit a pyrogenic reaction in the rabbit test for biological products. Seligmann et al. (1979), reported that the BOB has used the LAL test in recent years to evaluate all lots of 5% and 25% Normal Serum Albumin (NSA) and Purified Protein Fraction (PPF). The BOB showed that when such preparations contained 4 ng or more of endotoxin per ml (referenced to EC-2), they generally failed the rabbit test. When a 3-ml/kg dose was used to test 25% NSA and a 10-ml/kg dose was used for 5% NSA and PPF, 12-40 ng/kg were required to consistently fail the rabbit test. Steere (1978) reported that albumin preparations containing between 4 and 64 ng of endotoxin per ml did not cause pyrogenic reactions in humans. Such findings were confirmed and extended by Pearson et al. (1981), who reported that of 957 parallel tests conducted with LAL and rabbit assays, a final release limit of 1.2 ng endotoxin per ml (referenced to _E. coli_ 055:B5) was appropriate for 25% NSA.

These data suggest that environmental endotoxins are significantly less pyrogenic on a weight-for-weight basis than extracted endotoxin preparations made from _Enterobacteriaceae_, e.g., _E. coli_ and _Salmonella_ species. Thus, the standards that have been selected and evaluated in human and rabbit pyrogenicity studies are considerably more pyrogenic than endotoxins found in the "real world" of LVP manufacturing. Highly purified endotoxin preparations should be used as LAL standards, but should not serve as an absolute standard of pyrogenicity, because they do not parallel the pyrogenicity of environmental endotoxin as measured by the LAL and rabbit tests.

REFERENCES

Bureau of Drugs (1980). Guideline for validation of the _Limulus_ amebocyte lysate test as an end-product test for human and veterinary injectable drugs and medical devices. Federal Register, January 18, 1980:3668.

Bureau of Medical Devices (1980). Guidelines for adoption of the Limulus amebocyte lysate test for pyrogenicity of devices. Draft Unpublished.

Castor GB, Kantor N, Knoll E (1971). Characteristics of a highly purified pyrogenic lipopolysaccharide. J Pharm Sci 60:1578.

Cooper JF (1979). Acceptance of the Limulus test as an alternate pyrogen test for radiopharmaceuticals and intrathecal drugs. In Cohen E (ed): "Biomedical Applications of the Horseshoe Crab (Limulidae)," New York: Alan R. Liss, p 345.

Cooper JF, Levin J, Wagner H (1971). Quantitative comparison of in vitro and in vivo methods for the detection of endotoxin. J Lab Clin Med 78:138.

Cooper JF, Hochstein HD, and Seligmann EB, Jr (1972). The Limulus test for endotoxin (pyrogen) in radiopharmaceuticals and biologicals. Bull Parent Drug Assoc 26:153.

Dabbah R, Ferry E, Gunther DA (1980). Pyrogenicity of E. coli 055:B5 endotoxin by the USP rabbit test - a HIMA collaborative study. J Parent Drug Assoc 34:212.

Enzinger RM (1980). Limulus amebocyte lysate test for small volume parenterals - a response to Federal Register 45, January 18, 1980, p. 3668. PDA Informational Bulletin No. 3.

Garratt DC, Hartley RE, Mussett MV (1981). Reproducibility of the Limulus amebocyte lysate test in UK laboratories. Pharm J 31:112.

Greisman SE, Hornick B (1969). Comparative pyrogenic reactivity of rabbit and man to bacterial endotoxin. Proc Soc Exp Biol Med 131:1154.

Guzman RJ, Kuo HS (1979). Pyrogen testing of intravenous fat emulsion. In Cohen E (ed): "Biomedical Application of the Horseshoe Crab (Limulidae)," New York: Alan R. Liss, p 333.

Harrison SJ, Tsuji K, Enzinger RM (1979). Application of LAL for detection of endotoxin in antibiotic preparations. In Cohen E(ed): "Biomedical Applications of the Horseshoe Crab (Limulidae)," New York: Alan R. Liss, p 353.

Hart E, Penfold WJ (1912). A critical study of experimental fever. Proc R Soc London Ser B 85:174.

Keene WR, Silberman HR, Landry MJ (1961). Observations on the pyrogenic response and its application to the bioassay of endotoxin. J Clin Invest 40:295.

Lowry OH, Rosebrough NJ, Farr AL, Randall RJ (1951). Protein measurement with the folin phenol reagent. J Biol Chem 193:265.

Nandan R, Brown DR (1977). An improved in vitro pyrogen test: To detect picograms of endotoxin contamination in intravenous fluids using Limulus amebocyte lysate. J Lab Clin Med 89:910.

Oyama VI, Eagle H (1956). Measurement of cell growth in tissue culture with a phenol reagent (Folin-Ciocalteau). Proc Soc Expt Biol Med 91:305.

Pearson FC, Bacich J, Srigley W, Maylalang E, Graff E, Abroson F, Garcia D (1981). Applications of the Limulus amebocyte lysate assay to testing of human serum albumin. Society of Industrial Microbiology (in press).

Seibert FB (1923). Fever-producing substances found in some distilled waters. Am J Physiol 67:90.

Seligmann EB, Hochstein HD, Farag LG (1979). Problems with validation of the Limulus amebocyte lysate. Parenteral Drug Regional Meeting, Chicago.

Steere AC, Rifat MK, Seligmann EB, Hochstein HD, Friedland G, Dasse P, Ulustrock KO, Axnick KJ, Backer LF (1978). Pyrogen reactions associated with the infusion of normal serum albumin. Transf 18:102.

Tsuji K, Steindler A, Harrison S (1980). Limulus amebocyte lysate assay for detection and quantitation of endotoxin in a small-volume parenteral product. Appl Environ Micro 40:533.

US Pharmacopeia XX (1980). Rockville, MD: US Pharmacopeial Convention, Inc.

Van Noordwijk J, de Jong Y (1976). Comparison of the Limulus test for endotoxin with the rabbit test for pyrogens of the European Pharmacopeia. J Biol Stand 4:131.

Varney RF (1962). The intruders - pyrogens. Bull Parenter Drug Assoc 16:6.

Weary M, Baker B (1977). Utilization of the Limulus amebocyte lysate test for pyrogen testing large volume parenterals, administration sets and medical devices. Bull Parent Drug Assoc 31:128.

Weary M, Pearson FC (1980). Pyrogen testing with Limulus amebocyte lysate. Med Device Diagn Indus 2:34.

Westphal O (1956). In Springer GF (ed): "Polysaccharides in Biology," Trans 2nd Conf, New York.

Whittlet TD (1975). Chairman's opening address. In "International Symposium on Pyrogens," St. Louis: Mallinckrodt, p 1.

Wolff S (1973). Biological effects of bacterial endotoxins. In "Bacterial Lipopolysaccharide," Chicago: University of Chicago Press, p 251.

PEARSON SPEAKER
TITLE: THE RELATIVE PYROGENICITY OF ENVIRONMENTAL ENDOTOXIN AS MEASURED BY THE LIMULUS AMEBOCYTE LYSATE ASSAY AND THE USP RABBIT PYROGEN TEST.

ENZINGER: You presented data showing that a product containing 1,000 ng/kg endotoxin passed the USP test in one case but failed in another case. Did you by chance isolate those organisms to find out what species they were?
PEARSON: The preliminary work we did on bacterial isolation and identification was limited.
ENZINGER:- If you look at environmental isolates in manufacturing plants you would rarely find *E. coli*, *Salmonella* or *Shigella*; instead, you would find *Pseudomonas* and *Chromobacter*, or similar non-fermenters. These species probably wouldn't elicit a greater pyrogenic response in the rabbit than *E. coli*. I think this knowledge must be factored in when we are talking about safety factors in the amount of allowable endotoxin in parenteral solutions.
PEARSON: Our data on environmental endotoxins is very preliminary. I see nothing wrong with using purified endotoxins if we recognize we have created a significant safety factor when we use them.
MORRISON: Are environmental endotoxins less potent because they were derived from different species than the purified endotoxins employed? If these environmental endotoxins could be purified, would their potency increase and be equal to the present purified endotoxin standards?
PEARSON: Our data indicate *Pseudomonas* endotoxin, prior to purification, is potent in LAL but relatively impotent in rabbits. However after purification this endotoxin becomes very potent in rabbits.
MORRISON: Let me repeat the question. Are environmental endotoxins less toxic or are they on a per weight basis equally toxic but are diluted out by the fact that you have non-toxic contaminants present?
PEARSON: The pyrogenicity of *Pseudomonas sp*. is far less than you would expect based on its potency with LAL. We know if you extract the endotoxin from this species, the pyrogenicity increases greatly. This suggests the more purity the greater toxicity.

Endotoxins and Their Detection With the Limulus Amebocyte Lysate Test, pages 79–90

PROCEDURES FOR IMPROVING DETECTION OF ENDOTOXIN IN PRODUCTS FOUND INCOMPATIBLE FOR DIRECT ANALYSIS WITH LIMULUS AMEBOCYTE LYSATE.

Dennis E. Guilfoyle and Terry Munson

U.S. Food & Drug Administration
850 Third Avenue,
Brooklyn, NY 11232

U.S. Food & Drug Administration
5600 Fishers Lane,
Rockville, MD 20857

The 1980 edition of the United States Pharmacopeia/ National Formulary (USP XX/NF XV) has included the Limulus amebocyte lysate or LAL test as a method for detecting bacterial endotoxins in pharmaceutical products. Standardization of the LAL test for bacterial endotoxins allowed for Food and Drug Administration (FDA) to initiate a surveillance program using this technique for testing a wide variety of aqueous or lyophilyzed pharmaceutical products. This paper deals with our experience using this method at the FDA New York Regional Laboratory in Brooklyn. A brief description of the procedure employed and the results of the surveillance study are presented. Also described are results on the use of the surfactant called Pyrosperse, which was found to reduce product interference with the LAL reaction in certain products which could not otherwise be analyzed.

Some products, for example, antibiotics, solutions with high ionic strength, and viscous products may inhibit gel clot formation (Cooper 1977; Daoust 1976; Sullivan 1975; Work 1966). The inhibitory concentration of these products must be identified prior to running an actual sample analysis. The first step in testing for the level of product interference with the LAL reaction is to determine the sensitivity of the lysate preparation using an appropriate endotoxin standard in pyrogen-free distilled water. Next, serial dilutions of this endotoxin standard are prepared using the product as a diluent. Compatibility is indicated when gelation occurs within a twofold dilution of the determined lysate sensitivity. If undiluted product is found to be incompatible, the test is repeated to see what dilution of the product is compatible with the LAL reaction.

In 1980, the first year of the surveillance program, we noted that of the 298 samples examined at the New York Regional Laboratory, 181 or 61% were found to be incompatible when the undiluted product was tested with the LAL reagents. Before testing dilutions of these products to see if one can be found that is no longer inhibitory, it is necessary to calculate the lowest dilution that can be validly tested. This dilution is called the Maximum Valid Dilution (MVD) (Munson, 1982). The MVD is calculated as follows:

Step 1. Minimum Valid Concentration

$$MVC = \frac{\lambda \quad M}{K}$$

Where:

λ = Sensitivity of the LAL reagent in Endotoxin Units (EU) per mL.

M = Rabbit Dose (mg or units or Maximum Human Dose Kg) that would be administered in a single one hour period, whichever is more stringent.

K = 5.0 EU/Kg

Step 2. Maximum Valid Dilution

$$MVD = \frac{\text{Potency of Product}}{MVC}$$

For drugs administered on a weight-per-kilogram basis, the potency is expressed as mg or units/ml and for drugs administered on a volume-per-kilogram, the potency is equal to 1.0 mL/mL.

Product dilution procedure

Phase I

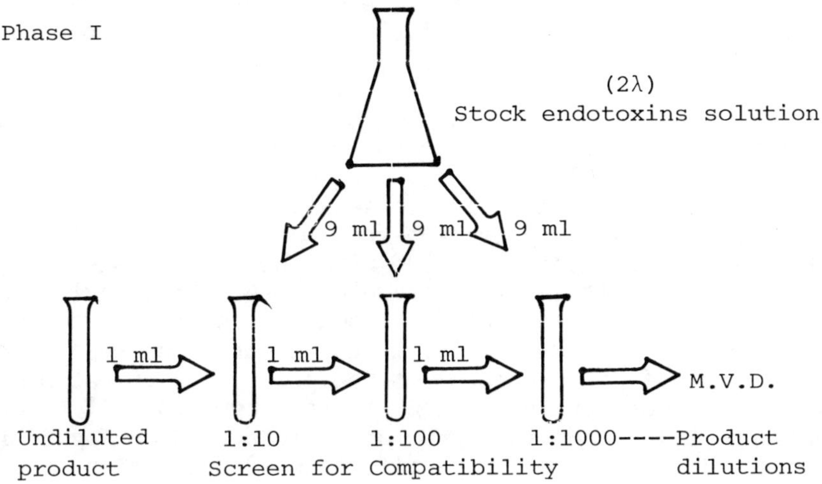

Phase II (Assume compatibility indicated at 1:100)

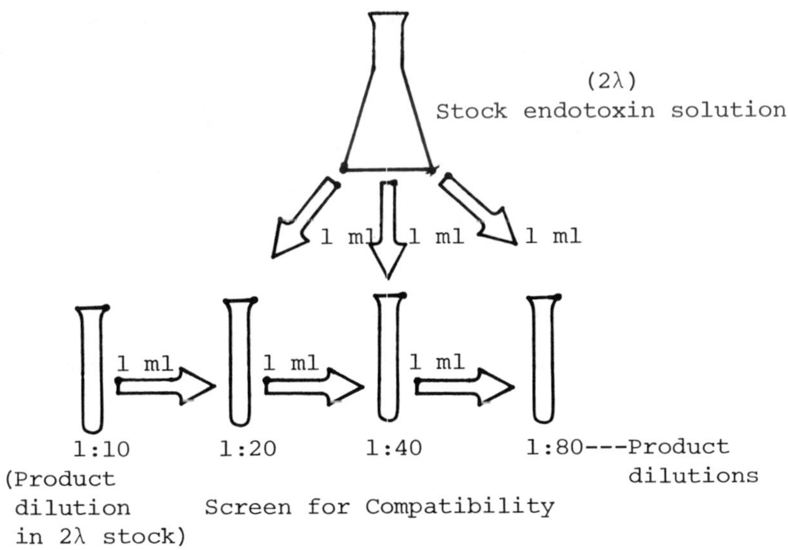

Figure 1. Product dilution procedure

It should be emphasized that the MVD really sets a limit as to how far an incompatible product can be diluted in attempting to find a dilution compatible with the LAL reagent.

An outline of the procedure for determining the lowest product dilution that is compatible with the LAL test is shown in Figure 1. A stock solution of endotoxin at the 2 λ (where λ = labeled lysate sensitivity) concentration is prepared by using pyrogen-free water as a diluent. The stock solution is tested with the LAL reagent to confirm that it will produce a positive gelation reaction. If undiluted product is found to be incompatible in the initial test, it is necessary to determine if a dilution of the product is compatible. This is done by preparing serial tenfold dilutions between 1:10 and the MVD as illustrated in Figure 1 (Phase I). Each of these dilutions of product is then tested in duplicate for compatibility with the LAL test. If a compatible dilution is found, then serial twofold dilutions are prepared in the range between that dilution and the preceding lower dilution, using the stock endotoxin solution as diluent (bottom half Figure 1). Again each of these twofold dilutions are tested in duplicate for compatibility. The lowest twofold dilution exhibiting a gelation in both tubes is then run at various endotoxin concentrations according to the preparatory testing section of the Bacterial Endotoxin test as outlined in USP and described previously. If compatibility is confirmed by this test, then this dilution is used in assaying the product for the presence of endotoxin.

The New York Regional Laboratory completed the second year of the surveillance program in September 1981. A breakdown of the samples analyzed in terms of compatibility with the LAL test is shown in Table 1. A total of 298 small volume parenteral drugs were analyzed in New York in 1980 and 289 in 1981. Of these, 32% and 11% respectively, could be tested directly and 61% and 84% of the products required dilution to be compatible. However, 7% of the products in 1980 and 5% in 1981 were not compatible even after dilution to the MVD. In addition to not testing the exact same product types in both testing years, lysate from a different supplier was used during 1981. FDA is now investigating the effect different lysates have on the inhibition endpoint.

Table 1. Number of samples of small volume parenterals found compatible or incompatible for analysis by LAL procedure.

Categories of product compatibility	Product Samples			
	(1980)		(1981)	
	No.	(%)	No.	(%)
1. Compatible undiluted	96	(32)	31	(11)
2. Compatible with dilution	181	(61)	244	(84)
3. Not compatible	21	(7)	14	(5)
Total	298	(100)	289	(100)

The FDA Bureau of Drugs has compiled data from the 5 laboratories (including NY) participating in the surveillance program. Approximately 1,400 samples representing 257 different drugs were analyzed. Of these drugs, 20% could be analyzed directly, and 75% required dilution. About 5% of the product types were inhibitory at their MVD. Table 2 lists the types of small volume parenteral drugs that were found to be incompatible with the lysate reagent, even after diluting these products to their respective MVD's. It also lists what we feel are the inhibitory agents in the product.

A second study, performed at the New York Regional Laboratory, was to determine if addition of Pyrosperse (a metallo-modified polyanionic dispersant, Mallinckrodt Inc., St. Louis, MO) to the LAL test reaction mixtures improved compatibility (decreased interference). Twenty-three of the 181 products tested in 1980 that were incompatible when undiluted (Table 1) were selected for this study. These products were analyzed with and without 2% v/v Pyrosperse. The same lot of lysate and endotoxin were used and the comparisions were run at the same time.

Eleven of the 23 incompatible products became compatible upon addition of 2% Pyrosperse to undiluted samples (Table 3). The first column of figures represents the lysate sensitivity in distilled pyrogen-free water. This value was determined each time a particular product was tested. Variations are due to differences in the lot of lysate employed. The second column shows minimum concentration of endotoxin detectable in product without addition of Pyrosperse. Addition of 2% Pyrosperse to these 11 products improved the minimum concentration of endotoxin detectable from twofold to tenfold or possibly higher. In 8 of the products, the titer improved, but it was not as high as for water. In the case of one of the lots of 0.9% sodium chloride the minimum concentration of endotoxin detectable, 19.5 pg/ml, was actually twofold less than that determined in pyrogen-free water (39 pg/ml), although one must remember that the variation in running the LAL test in $\pm$ a twofold dilution.

The remaining 12 of the 23 products showed no improvement with the addition of Pyrosperse to undiluted product only. They were then analyzed for compatibility using dilutions of product. The results comparing the concentration of endotoxin found with and without the addition of 2% Pyrosperse to the dilutions as shown in Table 4. The first column of figures, labeled water, again represents the determined lysate sensitivity in pyrogen-free distilled water. The next column labelled "Diluted product alone" are values shown for lysate sensitivity that were determined at the product dilution shown in the parentheses. These dilution factors were the highest twofold dilution which could be easily employed without exceeding the MVD for a given product.

Without the addition of Pyrosperse, none of the 12 products were found to be compatible for analysis. However, with all 12 products, the detectable level of endotoxin were markedly improved by the addition of Pyrosperse to the diluent as seen in the third column. Improvement ranged from about a twofold increase in detection for potassium chloride to a 175-fold increase for prochlorperazine.

We concluded from this work that Pyrosperse is useful for the analysis of problem products. At this time, it is not suggested that Pyrosperse be added routinely to all samples, including those that are known to be compatible without its addition. With one sample of atropine sulfate, 0.4 ng/mL (not in Tables), addition of 2% Pyrosperse to the product and diluent actually made the product less compatible for analysis and when included with one sample of paline (Table 3) detected more endotoxin than was actually present.

TABLE 2
Drug Products Incompatible with Lysate[a]

Drug Product[b]	Potency	MVD	Inhibitors
Carbozochrome Salicylate	5 mg & 10 mg/mL	119 & 238	Active Ingredient
Cyclizine Lactate	50 mg/mL	417	Active Ingredient
Diatrizoate Meglumine & Diatrizoate Sodium	285 & 291 mg/mL	5	EDTA
Doxycycline Hyclate	10 mg/mL	22	Ascorbic Acid
Edetate Disodium	150 mg/mL	10	Active Ingredient
Fluorescein Sodium	250 mg/mL	38	Active Ingredient
Heparin	20,000 units/mL	167	Active Ingredient
I^{131} Sodium Rose Bengal	- - -	ND	Active Ingredient
Ketamine	10 mg/mL	37	Active Ingredient
Meperidine HCl & Promethazine HCl	50 & 50 mg/mL	389	Promethazine HCl, EDTA, Sodium Bisulfite, Phenol
Oxacillin Sodium	1000 mg/vial	417	Active Ingredient
Peptonized Iron	5 mg/mL	4	Phenol
Sulfisoxazole Diolamine	400 mg/mL	166	Active Ingredient
Sulfobromophthalein	50 mg/mL	83	Active Ingredient

a. List compiled from data accumulated by FDA Bureau of Drugs.
b. One manufacturer for each product.

TABLE 3
Undiluted Products Plus 2% Pyrosperse

Product	LAL Detection Level (pg/mL)[a]		
	Water	Product alone	Product w/pyrosperse
Bacteriostatic Sodium chloride, 0.9%[b]	62.5	353.5	125
Bacteriostatic Sodium chloride, 0.9%[b]	62.5	250	62.5
Sodium Chloride, 0.9%[b]	19.5	78	27.5
Sodium Chloride, 0.9%[b]	39	125	19.5
Procaine hydrochloride, 1%	19.5	312	39
Travasol-M, 3.5%	39	312	78
Sterile buffered diluent	39	312	78
Dextrose injection (lot A), 50%	62.5	500	125
Dextrose injection (lot B), 50%	62.5	500	125
Hydrocortisone acetate suspension, 25 mg/mL	62.5	500	125
Naloxone hydrochloride, 0.4 mg/mL	62.5	250	62.5

a. Minimum amount of endotoxin (*E. coli* 055:B5) detectable under test conditions.
b. Similar products were from different manufacturers.

TABLE 4
Diluted Products Plus 2% Pyrosperse

Product	LAL Detection Level (pg/mL)[a]		
	Water	Diluted product alone	Diluted product w/pyrosperse
Potassium chloride injection, 3 mEq/mL[b]	62.5	176 (1:10)	62.5 (1:10)
Potassium chloride injection, 3 mEq/mL[b]	62.5	250 (1:10)	62.5 (1:10)
Procaine HCl, 2%	62.5	500 (1:10)	62.5 (1:10)
Chlorpromazine HCl, 25 mg/mL[b]	62.5	500 (1:50)	125 (1:50)
Chlorpromazine HCl, 25 mg/mL[b]	7.8	15.6(1:800)	15.6 (1:20)
Thiamine HCl, 100 mg/mL	62.5	500 (1:200)	62.5 (1:200)
Amitriptyline HCl, 10 mg/mL	62.5	125 (1:1000)	125 (1:100)
Prochlorperazine, 5 mg/mL	7.8	15.6(1:700)	15.6 (1:4)
Thiothiene HCl, 2 mg/mL	62.5	500 (1:150)	62.5 (1:10)
Cimetidine HCl, 150 mg/mL	62.5	125 (1:100)	125 (1:30)
Nafcillin sodium, 250 mg/mL	62.5	125 (1:100)	62.5 (1:80)
Minocycline HCl, 20 mg/mL	39.0	312 (1:100)	78 (1:80)

a. Minimum amount of endotoxin (*E. coli* 055:B5) detectable under test conditions.
b. Similar products were from different manufacturers.

Acknowledgment

We thank James D. Macmillan, Department of Biochemistry and Microbiology, Rutgers University, New Brunswick, NJ for his invaluable assistance in the preparation of this manuscript.

Cooper JF, Pearson SM (1977). Detection of endotoxin in biological products by the Limulus test. Develop Biol. Standard 34:7.

Daoust DR, Orlowski SJ, McMahon G, Weber CK, Shaw WC, Fisher TL, Bennett CR, Gray A (1976). Limulus amebocyte lysate test as a method for detection of endotoxins and endotoxin-like materials. Bull Parenteral Drug Assoc 30:13.

Munson TE (1982). Introduction to the FDA Guideline, "Endotoxins and Their Detection with the Limulus Amebocyte Lysate Test" Alan R. Liss, Inc., pp.

Rudbach JA, Milner KC (1968). Reaction of endotoxin and surfactants. Ill Effect of sodium lauryl sulfate on the structure and pyrogenicity of endotoxin. Canadian J of Clinical Microbiol 2:151.

Sullivan JD, Watson SW (1975). Inhibitory effects of heparin on the Limulus test for endotoxin. J of Clinical Microbiol 2:151.

The United States Pharmacopeia XX/National Formulary XIV (1980). Mack Publishing Co., Easton PA, p 888.

Work E, Know KW, Vesk M (1966). The chemistry and electron microscopy of an extracellular lipopolysaccharide from *Escherichia coli*. Annals NY Acad Sci 133:438.

GUILFOYLE SPEAKER
TITLE: PROCEDURE FOR IMPROVED DETECTION OF ENDOTOXIN IN PRODUCTS FOUND INCOMPATIBLE FOR DIRECT ANALYSIS.

LEVIN: What happens when you add endotoxin to 2% Pyrosperse?
GUILFOYLE: Pyrosperse increases the sensitivity about twofold when tested against one endotoxin standard diluted in pyrogen-free distilled water.
MORRISON: Both Pyrosperse and endotoxins are polyanionic. Presumably they can bind up residual cations. Would this suggest that cations are inhibitory to the LAL assay?
GUILFOYLE: We did not study the biochemical actions which are taking place. Literature references indicate that similar reagents break micelle formations in the product making active sites on endotoxin more available to react with the LAL.
MORRISON: Are you suggesting that Pyrosperse decreases the average molecular weight of the endotoxin?
GUILFOYLE: I don't think so. It probably disperses the endotoxins and makes it more available to react with the LAL enzyme.
URBASCHEK: Do you see an increase in the endotoxin content when you use Pyrosperse with body fluids such as urine or serum? Also, I would like to know if Pyrosperse can be used with other types of LAL?
GUILFOYLE: We haven't used Pyrosperse with any clinical specimens. We have used Pyrosperse with two different brands of LAL with similar results.
WATSON: I question your data on heparin, we have published data showing that with our LAL we find no inhibition up to 2,000 units per ml. I think that you should explain this apparent discrepancy between your work and published data which I am sure you are aware of.
GUILFOYLE: Our data on heparin was obtained in the first year of the study and we only used LAL from one manufacturer. We did not test heparin with your LAL.
WATSON: Pyrosperse certainly enhances LAL sensitivity with some products. However, all brands of LAL are not the same, with one brand of LAL you may need Pyrosperse to overcome inhibition, however using a different brand of LAL the same product may not be inhibitory. We have often found this to be the case.

ENZINGER: Also keep in mind that all heparin products are not the same. The type of preservative used may be responsible for the observed interference.
WATSON: We have found, using our LAL, that the main inhibitor in heparin can be the benzyl alcohol.
REICHLER: I wonder why you talk about the material actively changing the sensitivity of the lysate. Isn't it really the availability of the endoxtoxin you change with the addition of Pyrosperse? In reality you are not changing the sensitivity of LAL.
GUILFOYLE: That is correct. Presumably Pyrosperse only changes the level of detectability of endotoxin and not the LAL sensitivity.
STOTT: Were all the lysates shown in your last table from the same lot and supplier and were they all of the same potency?
GUILFOYLE: No. LAL from two different manufacturers was employed, however for a given product the same LAL was used with and without Pyrosperse.
TSUJI: Have you found that some products inhibit one brand of LAL but not another brand?
MUNSON: We have tested some products with lysate from two different manufacturers. With the exception of heparin we obtained the same results, i.e., within one twofold dilution, with the two lysates we have used. However, we do not have enough data to know whether all products will react identically with lysates from all manufacturers. This is why the guidelines will suggest that you retest your products when you switch lysate manufacturers.
ENZINGER: If you test a product with known concentrations of endotoxins, do you obtain the same results using LAL from two different manufacturers?
GUILFOYLE: We haven't tested enough products with LAL from different manufacturers. We hope that industry will share their findings with us so we will be in a better position to know if real differences exist between brands of LAL.
ENZINGER: Will one be able to use Pyrosperse according to the FDA ?
MUNSON: Yes, the addition of surfactants, salts or other substances to overcome the effect of inhibitors will be permitted. Removal of inhibitors by filtration or other appropriate methods will also be allowed.

Endotoxins and Their Detection With the Limulus Amebocyte Lysate Test, pages 91–100

THE USE OF LAL AS AN ALTERNATIVE TO THE CFR RABBIT PYROGEN TEST FOR DISODIUM TICARCILLIN

Karen Zink McCullough, M.S.

Microbiological Special Services Section
Beecham Laboratories
Piscataway, New Jersey 08854

INTRODUCTION

The use of Limulus Amebocyte Lysate (LAL) as an alternative test method to the compendium Rabbit Pyrogen Test has been well received by both the pharmaceutical industry and the FDA, especially for the routine testing of Water for Injection. The use of LAL for in-house control or final release criterion seems to have gained wider acceptance among the Large Volume Parenteral (LVP) manufacturers than the Small Volume Parenteral (SVP) or the antibiotic manufacturers. This may be due to the type and limited diversity of products manufactured by LVP firms as compared to the varied nature of SVP products. For example, it has been found that most antibiotics interfere with the formation of a firm clot in the gel-clot LAL methodology (McCullough and Scolnick, 1976; Newsome, 1977; Harrison, Tsuji, and Enzinger, 1979; Bondar, et.al., 1979; T. Novitsky, Personal Communication).

We, at Beecham Laboratories, a manufacturer of semisynthetic penicillins, reported that the penicillins will inhibit the LAL reaction, but that the inhibition may be overcome through dilution (McCullough and Scolnick, 1976; Newsome, 1977). Continued research on the dilution approach to resolve the inhibition problem has been successful: we have developed and validated a test method for detecting lipopolysaccharide in disodium ticarcillin, a semisynthetic penicillin. Data are presented on rabbit-LAL correlation studies and the adaptation of semiautomated test methods which have been developed for routine pyrogen analysis.

RABBIT-LAL CORRELATION STUDY

Before consideration could be given to using LAL for the in-house pyrogen testing of bulk disodium ticarcillin, it was necessary to assure that the rabbit colony was equally as sensitive to injections of endotoxin-antibiotic solutions as it was to endotoxin-water solutions. To demonstrate this equivalency, we chose three endotoxin levels for the study: 1.0 ng/kg, 0.75 ng/kg, and 0.5 ng/kg. The concentration of the test material, disodium ticarcillin, was 100 mg/ml, as prescribed in 21 CFR section 440.90a, and section 436.32. The rabbit testing was performed on the same colony that is used for routine pyrogen testing. All solutions tested were prepared at the pyrogen testing laboratory and were injected within one-half hour of preparation. Table 1 summarizes the results of these tests.

Table 1

Individual Temperature Rises:
Endotoxin-Water vs. Endotoxin-Ticarcillin

Endotoxin Level	Mean Temperature Rise °C *		Variance		Range, °C	
	H_2O	Ticar	H_2O	Ticar	H_2O	Ticar
1.0 ng/kg	0.80	0.83	0.05	0.03	0.4-1.1	0.6-1.1
0.75 ng/kg	0.53	0.80	0.07	0.06	0.0-0.9	0.5-1.1
0.5 ng/kg	0.47	0.41	0.04	0.05	0.1-0.7	0.0-0.8

* N = 9 for each group (98% confidence level)
E. coli endotoxin, lot #6 Associates of Cape Cod

A total of nine rabbits for each test solution (endotoxin-water and endotoxin-ticarcillin) and at each of the three endotoxin concentration levels were tested. Statistical analyses to compare the mean temperature increases for both test solutions at each endotoxin concentration were performed. In every case, analyses of the data showed correlation of response between the disodium ticarcillin and water solutions at each specific endotoxin concentration.

The mean pyrogenic dose (MPD), or the amount of endotoxin needed to elicit a temperature rise of 0.6°C, can be extrapolated from regression analyses of the rabbit data. The MPD for endotoxin in water is .72 ng/kg, and for endotoxin in TICAR is .63 ng/kg. As evidenced in Table 1 data and the MPD values, we have demonstrated that the rabbit colony was equally as sensitive to the endotoxin in disodium ticarcillin as to the endotoxin in water. Additionally, aliquots from each of the solutions tested on the rabbits were tested using LAL. All of the solutions which were injected into the rabbits were positive by LAL, thereby demonstrating that the LAL test is more sensitive than the CFR Rabbit Test.

INHIBITION STUDIES-ALTERNATIVE METHODS OF ANALYSIS

Disodium ticarcillin, at a concentration of 100 mg/ml, inhibits the formation of a firm clot using the gel-endpoint method of analysis. At two facilities of Beecham Inc., Beecham Research Laboratories in Surrey, England, and Beecham Laboratories in Piscataway, New Jersey, independent research by Dr. Peter Newsome and myself showed that a tenfold dilution to 10 mg/ml will overcome the inhibition caused by the presence of the disodium ticarcillin. At Piscataway, we confirmed 10 mg/ml as the optimum antibiotic concentration using two different turbidimetric automated methods of analysis described below.

Photometric Method Using the Titertek Multiskan, Flow Laboratories

The Titertek Multiskan from Flow Laboratories is an eight channel photometer. Interchangeable filters permit reading of test samples at wave lengths from 380 to 620 nm. In order to perform the LAL test using the Multiskan, equal quantities (0.1 ml) of the standard or the unknown solution and reconstituted LAL reagent are carefully mixed in each of the wells of a microtitration plate.

A standard set of two-fold endotoxin dilutions ranging from 0.5 to 0.004 ng/ml, and a series of negative control tests, are included on each plate. The microtitration plates are covered and incubated undisturbed at 37°C for one hour. Upon completion of the incubation period, the cover is removed and the optical density at 405 nm or 380 nm is simultaneously measured in all 96 wells of the plate by the Multiskan instrument. Standard curves are then plotted using the optical density measurements of the endotoxin dilutions. Our data showed the standard curve to be linear for endotoxin concentrations ranging from 0.016 ng/ml to 0.125 ng/ml. Figure 1 is a graphic representation of the linear portion of an inhibition study of disodium ticarcillin using antibiotic concentrations ranging from 5 mg/ml to 100 mg/ml.

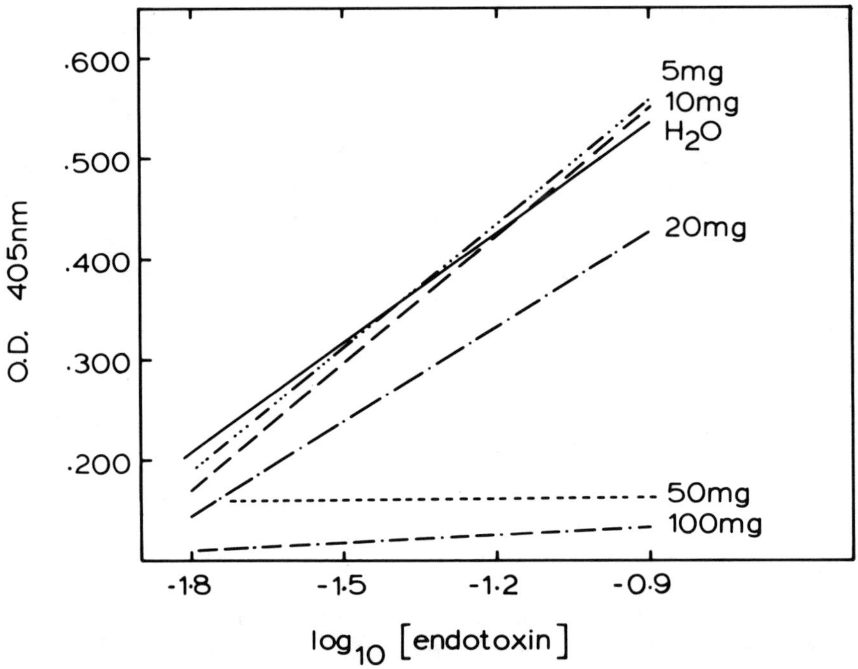

Fig. 1. Inhibition study using the Multiskan and testing disodium ticarcillin in concentrations from 5 mg/ml to 100 mg/ml. Each antibiotic solution was spiked with a control standard E. coli endotoxin ranging in concentration from 0.016 ng/ml to 0.125 ng/ml.

In order to determine if inhibition is independent of

endotoxin concentration, each solution of ticarcillin was tested at endotoxin spike levels ranging from 0.008 ng/ml to 1.0 ng/ml. The data confirm our earlier finding that a concentration of 10 mg/ml is optimum for disodium ticarcillin and further indicate that at this level, the inhibition is overcome for a wide range of endotoxin spike levels. The results obtained at 100 mg/ml and 50 mg/ml also demonstrate that at high ticarcillin concentrations, inhibition is independent of endotoxin level.

MS-2, Abbott Laboratories

The MS-2 from Abbott Laboratories is an instrument that was originally designed for use in clinical microbiology. Like the Multiskan, the MS-2 is a spectrophotometer; however, the test methodology and data interpretation for the LAL assay differ sharply between the two instruments.

In the MS-2 methodology, 0.8 ml of sample and 0.2 ml of reconstituted LAL reagent are mixed together in a disposable glass or plastic pyrogen free cuvette. The cuvettes are placed in a holder which is inserted directly into the analysis module of the instrument. Up to eleven holders are incubated undisturbed at 35.5°C in the module. The MS-2 can be programmed to measure the optical density of each sample at preset time intervals (we chose five-minute intervals). At the end of the run, the instrument prints individual optical density readings, as well as a graphic representation of the data. Graphs of the resulting data clearly illustrate the kinetics of the limulus reaction (Figure 2).

Using the MS-2, we tested disodium ticarcillin concentrations ranging from 5 mg/ml to 100 mg/ml at an endotoxin spike level of 0.1 ng/ml. Figure 2 is an enlargement of the resulting MS-2 graph. To simplify interpretation of the kinetic data, we extrapolated the incubation time required for the turbidity of each spiked sample to change 0.1 optical density unit: these times are listed in Table 2.

The data presented in Table 2 show that 10 mg/ml disodium ticarcillin concentration is optimum: this confirms previous data from both the gel-clot and Multiskan methods. The optical densities of the 100 mg/ml, 50 mg/ml, and 25 mg/ml did not change appreciably during the two hour

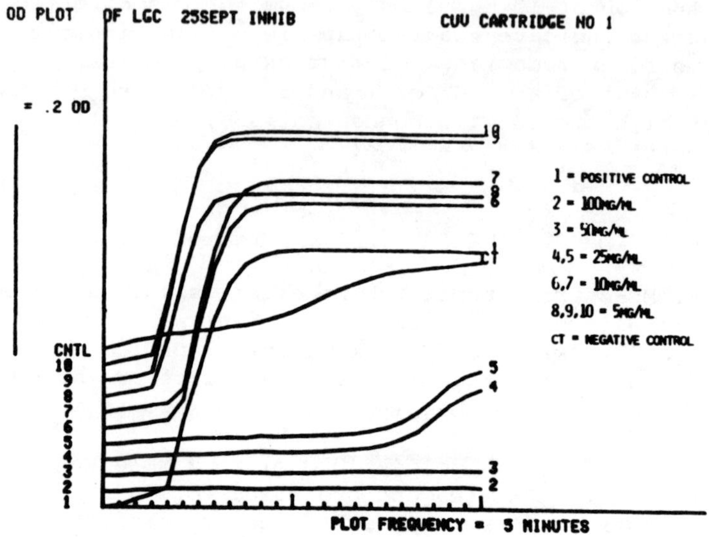

Fig. 2. Expanded MS-2 graph of the kinetics of the LAL reaction. Concentrations of disodium ticarcillin ranging from 5 mg/ml to 100 mg/ml were each spiked with 0.1 ng/ml of a control standard *E. coli* preparation.

Table 2

Inhibition of LAL by Disodium Ticarcillin Using the MS-2

Concentration of Disodium Ticarcillin Spiked with 0.1 ng/ml E. coli Endotoxin	Incubation Time to Reach 0.1 O.D. Units
100 mg	> 2 hr.
50 mg	> 2 hr.
25 mg	> 2 hr.
10 mg	31-32 minutes
5 mg	25 minutes
Control-Water	29 minutes

incubation. In contrast, the 10 mg/ml and 5 mg/ml concentrations each attained an 0.1 unit change in 31-32 minutes and 25 minutes, respectively. Both times compare favorably with the endotoxin-water control, which required 29 minutes of incubation to change 0.1 unit.

FDA PROPOSED GUIDELINES FOR LAL VALIDATION

We have thus far demonstrated two important concepts in the LAL validation process:

1. The product being tested does inhibit the LAL reaction, but that inhibition is readily overcome by dilution.

2. The rabbit colony reacts equally to either the injections of endotoxin-water or the injections of endotoxin-ticarcillin.

Our next step in the validation study was to determine if the conditions under which we perform the test comply with the proposed (1980) FDA Guidelines for LAL validation.

These Guidelines state that the test sample concentration be the "rabbit or maximum human dose of product per kilogram body weight,which ever is more stringent. Use 70 kg as the weight of the average human when calculating the maximum dose/kg." In the case of disodium ticarcillin, the 100 mg/kg dose specified in 21 CFR sections 436.32 and 440.90a for the Rabbit Pyrogen Test is the more stringent dose. We therefore used this concentration as the basis for our minimum valid dose (MVD) calculation.

$$\text{MVD} = \frac{\lambda\theta}{k}$$

Where λ = Sensitivity of the LAL reagent in Endotoxin Units per ml.

θ = Rabbit or maximum human dose per kg.

k = Tolerance limit of 2.5 E.U./kg (0.5 ng EC-2/kg)

Since we require a 1:10 dilution in the product to overcome

inhibition, we considered 10 mg/ml our MVD and solved for λ, the lysate sensitivity we would need to detect the published tolerance limit.

Substituting values, the equation becomes:

$$10 \frac{mg}{ml} = \frac{(\lambda)\ (100 \frac{mg}{kg})}{0.5 \frac{ng}{kg}}$$

$$\lambda = 0.05 \text{ ng/ml}$$

Solving the equation for λ shows that we would require lysate with a minimum sensitivity of 0.05 ng/ml to conform with the FDA proposed Guidelines.

LAL VS. RABBIT PYROGEN TEST

Gel-Clot Method

To date, we have tested a total of sixty-four (64) batches of disodium ticarcillin at a concentration of 10 mg/ml by the gel-clot method, using lysate validated at sensitivities of either 0.05 ng/ml or 0.025 ng/ml. Of these sixty-four (64) batches, forty-five (45) were "spiked" with 0.05 ng/ml of a Control Standard Endotoxin (Lot 6, obtained from Associates of Cape Cod). All 45 spiked samples had a positive response with the gel-clot method.

Turbidimetric Method

Fifty-two (52) batches of disodium ticarcillin have been tested using turbidimetric methodology. Each group of samples tested was accompanied by a ticarcillin standard curve from which the endotoxin concentration for each test batch was extrapolated. During the course of our validation studies, we have not found a level greater than 0.16 ng/100 mg in our product.

In total, we have tested 116 batches of disodium ticarcillin using LAL. All of these batches were also tested using the CFR Rabbit Pyrogen Test with negative results.

CONCLUSION

The data we have presented validate the methodology employing LAL as a substitute for the CFR Rabbit Pyrogen Test for the semisynthetic penicillin, disodium ticarcillin. Each of our products is currently being screened for compatibility with the LAL test. Preliminary data indicate that all are inhibitory, with the degree of inhibition depending upon the chemical composition of the side chain and/or formulation of the filled drug product. We are continuing to research the use of LAL as a substitute pyrogen test methodology for our other penicillin products, and we will be happy to exchange technical information and methodology with others doing similar studies.

AUTHOR'S ADDENDUM

Additional statistical analyses of raw rabbit data (Table 1) and reanalyses of these data corrected for diluent and antibiotic "effect" substantiate that the slopes of the regression lines for the two test substances are not different.

ACKNOWLEDGMENTS

The author wishes to thank Mrs. Doris Conrad, Quality Assurance Director, Beecham Laboratories, for her careful review of the manuscript. In addition, I would like to thank Dorothy Anderson, Joan Wilson and Vivian Gardner for typing the manuscript; Mr. Andrew S. Casella, Computer Systems Analyst, for assistance with statistical analyses; Dr. Peter Newsome of Beecham Research Laboratories, Surrey, England, for sharing unpublished results; and Drs. Stanley Watson and Tom Novitsky of Associates of Cape Cod, Inc. for many helpful discussions and for allowing me to use the MS-2.

Bondar RJ, Teller JD, Rowanko A, Kelly KM (1979). Properties of Limulus amebocyte lysate and the turbidimetric assay for the quantitative determination of gram negative bacterial endotoxin. In Cohen E, Bang F, Levin J, Marchalonis JJ, Pistole TG, Prendergast RA, Shuster CJr, Watson SW (eds): "Biomedical Applications of the Horseshoe Crab (Limulidae)," New York: Alan R. Liss, p 435.

Harrison SJ, Tsuji K, Enzinger RM (1979). Application of LAL for detection of endotoxin in antibiotic preparations. In Cohen E, Bang F, Levin J, Marchalonis JJ, Pistole TG, Prendergast RA, Schuster CJr, Watson SW (eds): "Biomedical Applications of the Horseshoe Crab (Limulidae)," New York: Alan R. Liss, p 435.

Health Industry Manufacturers Association (1979). HIMA Collaborative Study for the Pyrogenicity Evaluation of a Reference Endotoxin by the USP Rabbit Test.

McCullough KZ, Scolnick SA (1976). Effect of semisynthetic penicillins on the Limulus lysate test. Antimicrobial Agents and Chemotherapy 9:856-858.

Newsome PM (1977). Penicillins and the Limulus amebocyte lysate test for endotoxin. J Pharm Pharmac 29:704-706.

Endotoxins and Their Detection With the Limulus Amebocyte Lysate Test, pages 101–104

A METHOD FOR DETERMINING PYROGEN BURDEN ON PACKAGING COMPONENTS

Elsie J. Helme

Schering-Plough Corporation

1011 Morris Ave.
Union, NJ 07083

Earlier in the year a project was proposed to develop a procedure for determing pyrogen burden on 21 packaging components (for parenteral products), as they are received from the supplier, in order to determine the challenge levels of the depyrogenation systems.

In order to establish a protocol for such a project, two questions had to be resolved: 1) the test system to be employed, i.e., rabbit pyrogen test, LAL, or both and 2) the quantity of rinse water to be used for rinsing each component. The answer to the first question was easy to answer. There is ample data in the literature that list the advantages of the LAL test over the rabbit pyrogen test. To list a few:

1. It has a greater reliability and is more extensive than the rabbit test.
2. The results can be quantitated in terms of ng/ml of endotoxin.
3. It is faster and less expensive.
4. Cases of nonendotoxin pyrogens are rarely seen, and all of the above mentioned advantages outweigh the chance of encountering one.

The second question was not as easy to answer. We felt that it might be appropriate to use the pyrogen test for transfusion and infusion assemblies that appears in USP XX as a guide. Upon reading through the test, however, we discovered that the same amount of rinse solution is used for each assembly tested, regardless of the size of the assembly. Since the components that were to be included in this study ranged from stoppers to 1 ml ampuls to 250 ml vials, it was felt that some attempt should be made during procedure development to equate volume of rinse water to the surface area of the component.

PROCEDURE DEVELOPMENT

Twenty samples from the 21 packaging components to be included in the study were delivered to the laboratory for procedure development. The samples were divided into two groups of ten. One group was used to determine appropriate container and rinse water volume for each component and the second group was used in an inhibition test to demonstrate that there was nothing present in any of the components that would inhibit the reaction between LAL and endotoxin, thereby preventing gel formation.

Container and Soaking Solution Volume Determination

To determine the appropriate container and soaking solution volume to be used for each component, component specification drawings were obtained. The total exposed surface area, inside and out for vials and ampuls, of each component was calculated from the dimensions of the component. After experimenting with several different stopper components, a surface area to volume ratio of 0.6 used the least volume of soaking solution to sufficiently cover the components. This ratio was also applied to vials and ampuls. Using the proposed volumes, the containers to be used for soaking were selected based on their ability to hold 10 components and allow those components to be sufficiently covered with rinse solution.

Component Compatibility with LAL

To test for compatibility with LAL, ten samples of each component were placed in appropriate depyrogenated containers and an amount of endotoxin solution, concentration 1 ng/ml, equal to the previously determined soaking solution, was added to each. The samples were soaked for 1 hour, removed and air dried. They were then placed in another depyrogenated container and the same volume of pyrogen-free water was added. The samples were soaked for 1 hour. LAL testing was performed on the soaking solution using Mallinckrodt lysate lot 0 GR and endotoxin lot 9 HI. An endotoxin dilution series was used as a positive control. A positive response was indicated by a gel formation that did not collapse upon 180 degree inversion of the test tube. No component tested inhibited gel formation. After procedure development was completed, the following test protocol was written:

PROTOCOL

Place ten samples of each component under test in the appropriate depyrogenated container. Add the previously determined volume of pyrogen-free water. Set aside an aliquot of pyrogen-free water to serve as a negative control. Soak the component samples for 1 hour. At the end of this time, agitate the container and withdraw enough of the soaking solution to perform LAL testing.

In the case of ampuls and vials, pool aliquots of rinse water from inside as well as outside the components. Spike a portion of the soaking solution with concentrations of endotoxin that bracket the lysate label sensitivity by one two-fold dilution. This will serve as a positive component control. Prepare an endotoxin dilution series that brackets the lysate sensitivity by one 2 fold dilution to serve as a positive control. Perform LAL testing on:

- test component soaking solution
- positive component control soaking solution
- endotoxin dilution series
- negative control

If a positive response is observed for the component soaking solution, perform LAL tests on a dilution series of this solution until an end-point is reached. Calculate the amount of endotoxin in the soaking solution from the end-point of the endotoxin dilution series.

In summary, this paper briefly describes a method for determining pyrogen burden on packaging components used in sterile manufacturing. During procedure development, a surface area to volume ratio of 0.6 was used to determine the amount of soaking solution to be used for each component. Compatibility with LAL was demonstrated for all components. A procedure was then written that included a step to quantitate the amount of endotoxin found in any positive test component soaking solution.

Acknowledgements

The author wishes to extend sincere appreciation to Ms. Carolyn Dobbins for her contribution in preparing this paper.

References

Mascoli, Carmine C., and Weary, Marlys E. "Limulus Amebocyte Lysate (LAL) Test for Detecting Pyrogens in Parenteral Injectable Products and Medical Devices: Advantages to Manufacturers and Regulatory Officials", J. Parenteral Drug Assoc., 33, 81-95 (1979).

U.S. Pharmacopeia XX (1980).

Endotoxins and Their Detection With the Limulus Amebocyte Lysate Test, pages 105-113

PYROGENICITY VERSUS ENDOTOXICITY: WHICH IS MORE SIGNIFICANT?

Edward Shanbrom, M.D., F.A.C.P.

Mallinckrodt, Inc.

St. Louis, Missouri 63134

Somewhere between Pasteur and the antibiotic era the concept of how bacteria injure the host got lost. It seems the idea is we need only kill bacteria and we will make the patient well. Rarely in clinical literature is it suggested that by destroying bacteria we may also injure the patient by liberating excessive toxins. Yet even before the antibiotic era tremendous progress had been made by the advent of antitoxins, e.g., diphtheria, tetanus, and clostridia. But these were exotoxins and once these diseases were controlled, interest in such toxins waned. Perhaps the newly recognized "Toxic Shock Syndrome" is a blessing in disguise since it has provoked renewed interest in toxins in general, and an associated interest in the mechanism by which bacteria induce pathology. In recent years a great deal of work has been done to establish that the diarrheal disorders induced by various bacteria (cholera, salmonella, shigella, *E. coli*, etc.) are a result of enterotoxins elaborated by these organisms.

For reasons that are not entirely clear the significance of the most potent bacterial toxin, endotoxin, is still not well recognized in clinical medicine, although this is beginning to change. I believe one reason for the problem is that the most emphasized pathogenetic effect of gram-negative disease is the production of fever, and of all the effects of endotoxin, fever is probably the least important biologically and clinically. In fact there is a growing body of evidence which suggests that fever may produce biologic benefit.[1] Endotoxin contamination may well be one of the most common causes of so-called "anaphylactoid" or "idiosyncratic" reactions to injected therapeutic agents. Yet rarely, if ever, is this discussed in industry or clinical medicine. A recent review in the Annals of Internal Medicine[2] from the prestigious National Cancer Institute epitomizes the problem. The article is entitled "Hypersensitivity Reactions to Cancer Chemotherapeutic Agents" and in spite of the fact that fever was an associated side effect in many instances, not a single mention of the words "endotoxin" or "lipopolysaccharide" (LPS) appears in the article. One of the drugs cited is the antibiotic bleomycin and here is what the authors write: "Bleomycin causes what might be termed a 'pseudoanaphylactic' reaction in a few patients. It induces fevers in 20% to 25% of patients 1 to 4 hours after injection. In a few instances it has induced fever over 40°C that was associated with severe shaking, chills, diaphoresis, wheezing, mental confusion, marked hypotension, and anuria.....The cause of these intense hypotensive and hyperpyrexic episodes is uncertain. It is not an antibody-mediated reaction because some patients have developed the problem with administration of the first dose of the drug. Studies in rabbits have shown that intravenous bleomycin causes release of a circulating pyrogen." The authors describe various "allergic" reactions in a dozen drugs, fever being a frequent but not omnipresent manifestation. Of special importance is the description of reactions to *L. asparaginase* (ELSPAR -- Merck) which "is responsible for the highest prevalence of hypersensitivity reactions"... Table 1 lists these reactions, presumably in order of frequency.

TABLE 1

"Hypersensitivity" Reactions to *L. asparaginase* (ELSPAR)
Urticaria (65% of cases studied)
Agitation
Dyspnea
Hypotension
Facial Edema
Abdominal Cramps
Laryngospasm and stridor
Rash (localized or generalized)
Pruritus of hands and feet
Loss of consciousness
Fever
Chronic Urticaria
Arthralgias
Proteinuria

One or more symptoms developed in 6-43% of patients.

25-43% of reactions occurred in patients receiving no other chemotherapeutic agent.

L. asparaginase is derived from *E. coli* and as can be imagined, it contains significant amounts of LPS. When I telephoned one of the authors of the study and asked why they did not discuss endotoxin as a potential cause of drug reactions, I was told they hadn't thought of it. When I asked if they had tested the drugs for LPS, he said no but that in all cases the drugs had passed the rabbit test. I have subsequently learned that another research unit within the NIH had performed Limulus Amebocyte Lysate (LAL) tests on these chemotherapeutic agents utilizing various procedures and did find "significant" amounts of endotoxin.

Cancer chemotherapeutic drugs are rarely used as single agents. Often 3 or more drugs are administered concomitantly often in very high doses and the additive effect of small amounts of endotoxin in immunocompromised patients could be expected to produce endotoxicity. The suppression of the reticuloendothelial system by these same drugs limits leukocyte production and the elaboration of endogenous pyrogen, so fever may not develop. In my own experience treating large numbers of patients with antitumor drugs, certain individuals developed a rapid pulse and decreased blood pressure, but no fever, several hours before clinical recognition of gram-negative sepsis.

The antifungal antibiotic, Amphotericin B, has a high incidence of adverse reactions: chills, sweating, fever, muscle or joint pains, general malaise and various gastrointestinal symptoms. A recent report in The New England Journal of Medicine has described lethal pulmonary reactions when the drug had been administered together with leukocyte transfusions.[3] Eighty-two percent of the patients had gram-negative septicemia and were febrile. Only 6% of patients receiving leukocyte infusions alone and no patients receiving Amphotericin alone developed the lung disease. The authors concluded therefore that a third factor, endotoxin, could play a role in producing the lung problems. There is other evidence that endotoxin may have destructive effects on lung tissue both in experimental models and in clinical syndromes. The so called "allergic pneumonias" (bysinossis, bagassosis, humidifier fever, etc.) are associated with inhalation of air heavily laden with endotoxin and the clinical manifestation of these diseases is similar to that seen following intravenous injection of Amphotericin and/or endotoxin.

The bacteria which causes Legionnaires' Disease seems unusually lethal when one compares the relatively moderate degree of lung involvement seen in X-ray to that produced by other more common causative organisms. Studies by groups at CDC and BoB[4] have shown that the endotoxin of the Legionnaires' bacillus has a different spectrum of toxicity than the more common gram-negative organisms. In particular, the Legionnaires' endotoxin was not very pyrogenic but was exquisitively active in the LAL --- 1,000 fold difference between the two tests. Surely in this situation the rabbit test is inadequate for determining the potency of the toxins present.

What I have said so far leads to my theme that monitoring the amount of endotoxin injected, infused, or inhaled is critical and it should no longer suffice to simply say a substance "passes the rabbit test." Other clinical situations would seem to strengthen this idea:

1. Injection of LPS into cerebrospinal fluid is about 1000x more toxic than other routes of administration.

2. Renal dialysis is frequently associated with "hypotensive episodes" with or without chills and fever.

3. Clinical trials with human growth hormone produced by recombinant DNA techniques in *E. coli* have produced "endotoxic" side effects in the majority of patients. The material is said to have passed rabbit tests.

4. Virtually all patients treated with Interferon so far reported in the English literature develop pyrogenic reactions. In this regard, the putative activities of Interferon are of interest when compared to the biologic activities of endotoxin.

TABLE 2

	BIOLOGICAL ACTIVITIES OF	ENDOTOXIN	INTERFERON
1.	Fever	+	+
2.	Activation of the complement system	+	
3.	Activation of coagulation	+	
4.	Activation of the kinin-forming system	+	
5.	Activation of the fibrinolytic system	+	
6.	Elaboration of tissue factor from monocytes	+	
7.	Activation and depression of platelets	+	
8.	Alteration of neutrophil function and number	+	+
9.	Stimulation of mononuclear macrophage activities	+	+
10.	Stimulation of cytotoxic and cytolytic activity	+	+
11.	Release of prostaglandins	+	
12.	Production of "lethal factor"	+	
13.	Injury of endothelium	+	
14.	Shock	+	
15.	Depression of myocardial contractility	+	
16.	Stimulation of B lymphocytes	+	+
17.	Depression of T lymphocyte activity	+	+
18.	Lung toxicity	+	

Not all the biologic effects of endotoxin are exhibited by all bacterial species but with such profound undesirable effects and the well-recognized ubiquity of gram-negative bacteria, it should be obvious that there must be various mechanisms which modify endotoxicity. Table 3 shows some examples of such mechanisms:

TABLE 3

FACTORS MODIFYING ENDOTOXICITY

Dose of endotoxin
Route of administration
Granulocytes
Platelets
Plasma proteins
Potentiating agents

1. Dose of endotoxin -- a total dose of 1.0 nanogram per Kg of body weight is the estimated threshold for pyrogenicity. The amount required to produce other biologic effects are not established but in the age of polypharmacy it is not difficult to imagine that multiple drugs which pass rabbit tests may total significantly high amounts of endotoxin.

2. Route of administration -- injection into cerebrospinal fluid is highly toxic while inhalation of LPS can resemble intravenous infusions.

3. Granulocytes -- no cells, no fever, since endogenous pyrogen could be inadequate.

4. Platelets -- platelets are one mechanism of clearing endotoxin and thrombocytopenia may therefore result in increased sensitivity to LPS.

5. Plasma proteins -- many plasma proteins bind LPS and lipoproteins may specifically detoxify. Patients constantly challenged by bacteria or deficient in these proteins, e.g., cirrhotics and burn patients, may have these detoxifying mechanisms overwhelmed.

6. Endotoxin potentiation -- a number of compounds are known to enhance the adverse effects of LPS, e.g., lead acetate, dactinomycin, carageenan. This is an area for fruitful research and undoubtedly a number of other potentiating agents exist.

While there seems no question that substances other than endotoxin may produce fever, it remains to be shown that these substances have toxic effects more onerous than fever. In the case of bacterial LPS it is clear that most if not all of its biologic effects are of much greater significance than fever. While a great deal of effort has been made relating LAL testing to rabbit testing (i.e., validation), it would seem equally prudent and probably more rewarding to attempt to quantitate endotoxin burden as accurately as possible. It is not the fever that harms the patient, but rather the other toxic effects of LPS. For this reason emphasis in language and thought should be placed on the concept of "endotoxicity" rather than "pyrogenicity."

REFERENCES

1. Donaldson JF (1981). Therapy of acute fever: a comparative approach. Hosp Practice 16:125.
2. Weiss RB, Bruno S (1981). Hypersensitivity reactions to cancer chemotherapy agents. Ann Int Med 94:66-72.
3. Wright DG, Robichaud KJ, Pizzo PA, Deisseroth AB (1981). Lethal pulmonary reactions associated with the combined use of Amphotericin B and leukocyte transfusions. N E J of Med 304(20):1185.
4. Wong KH, Moss CW, Hochstein DH, Arko RJ, Schalla WO (1979). "Endotoxicity" of the Legionnaires' disease bacterium. Ann Int Med 90:624-627.

SHANBROM SPEAKER
TITLE: PYROGENICITY VS. ENDOTOXICITY - WHICH IS MORE SIGNIFICANT?

FINK: You mentioned that B cells were stimulated by LPS, what system are you referring to?
SHANBROM: I can not answer that, maybe Dr. Morrison can.
MORRISON: It appears that under appropriate culture conditions, high concentrations of LPS in prolonged contact with the culture will stimulate B cell proliferation. It appears that the protein associated with LPS will also stimulate B cell proliferation.
FINK: What kind of killer cells are activated by LPS?
MORRISON: Both LPS and LPS associated protein (LAP) will react directly with macrophages, leading to their activation to a state in which they can kill tumor cells.
FINK: Have you evidence that the lipid portion of LPS can activate macrophages?
MORRISON: Experiments in our laboratories and others, which have been published, indicate that Lipid A will activate macrophages. I think Dr. Shanbrom is pointing out that LPS interacts with many relevant biological systems which we should consider. One is pyrogenicity, one is LAL activity, only the latter of which activates a precursor protease activity.
SHANBROM: We just seem to always equate LPS activity with pyrogenicity.
FINK: What kind of leucocytes will produce endogenous pyrogen?
SHANBROM: Only granulocytes have been shown to do this so far. The classic experiments using nitrogen mustard to knock out granulocytes, which I am sure you are aware of, have shown that granulocytes are required for fever.
ENZINGER: Referring to your slide which shows about 18 clinical manifestations of endotoxin, getting to fever which manifests itself at about 1 ng/kg. What about the other 17, what dose is required for these other manifestations?
SHANBROM: As far as I know no one has published this information since these are quantitative clinical observations.

MUNSON: Dr. Elin from N.I.H. has studied some of the effects of endotoxins discussed by Dr. Shanbron. From the preliminary data which I have seen the white blood cell counts are affected starting at about 0.5 ng/kg. Quite high concentrations are needed to change the platelet count. Growth hormone was stimulated with about 4 ng/kg.
SHANBRON: It is a common practice for manufacturers of albumin to blend lots of non-pyrogenic albumin with lots of pyrogenic albumin. From time to time we see symptoms such as tight chest etc., all symptoms of a quasi-antiphylactic shock. This is really not a rare happening.
PEARSON: Did you suggest that endotoxin can be responsible for hypersensitivity and if so is this published information?
SHANBRON: Yes I am suggesting this.
COOPER: When L-asparaginase was initially marketed, it usually contained 1 to 100 ng/ml concentrations of endotoxin by LAL assay. Administration of these products to rabbits did not always produce pyrogenicity. In fact, some rabbits became hypothermic, and within 24 to 36 hours, they were dead. In this case, pyrogenicity did not accompany endotoxicity.

Endotoxins and Their Detection With the
Limulus Amebocyte Lysate Test, pages 115–119

CHARACTERIZATION OF THE USP REFERENCE STANDARD ENDOTOXIN

Aubrey S. Outschoorn, L.M.S., Ph.D.
Senior Scientist, Drug Standards Division
U. S. Pharmacopeial Convention, Inc.
12601 Twinbrook Parkway
Rockville, Maryland 20852

There was, and unfortunately still is, considerable confusion from neglect of certain elementary considerations in measuring the biological potency of endotoxins. The Bureau of Biologics, with long experience in the measurement of potency of various biologics, early defined an Endotoxin Unit. This can only mean, by definition, the activity contained in a stated amount of a particular preparation designated the U. S. Standard. It is amazing that in this day and age there are still endeavors to express endotoxin activities in weight of endotoxin. It is inescapable that equal weights of any two endotoxins are not necessarily possessed of equal potency, i.e., endotoxin activity, whether on lysates or in producing fever in rabbits or humans. Unfortunately, and this is owing to the great haste with which certain interested parties wanted indiscriminately to use the lysate test, different types of articles have been evaluated with different endotoxin references. This is of little value unless the potency of these references is known in relation to their effects on humans. The least that might have been done was for any candidate references to have been compared with different lysates. The most logical way to express their relative potencies would be in relation to a single reference, i.e., in terms of the legal, defined U. S. Endotoxin Unit. The contention that valid comparisons of these different endotoxin references cannot be done with different lysates is not acceptable. If that is so, then the results of estimating endotoxin levels of pharmaceutical articles with lysate and in comparison with any endotoxin references is also not acceptable. This means that the test would then not be a valid alternative to the pyrogen test with rabbits. But those who want to use the test and ignore the doctrine of "similarity" in biological assays are quite prepared to insist that such measurements do in fact evaluate pyrogenic potency in rabbits or humans. The prime requirement is that each candidate article for the lysate test must be individually validated.

The essential requirement of the official USP Bacterial Endotoxins Test for the application of the procedure for determining endotoxin limits was the provision of a characterized Reference Standard Endotoxin (RSE). It was fortunate that the FDA Bureau of Biologics had a fairly large quantity of the parent or master material (designated lot EC), from which they were preparing successive lots of the U. S. RSE. In view of the potential value in the control of medicaments for human administration of the use of a preparation derived from a single source, USP endeavored to procure material similar to the current U. S. RSE. Recognizing this, the Bureau of Biologics was prepared to donate, for compendial use, a quantity of their parent material. The preparation and properties of this national reference endotoxin had already been published (Rudbach, Akiya, Elin, Hochstein, Luoma, Milner, Milner, and Thomas 1976). It was made from *Escherichia coli* (Braude strain) 0:113 which had been serotyped by the Bacteriology Division, Bureau of Laboratories, Center for Disease Control. The quantity supplied to USP would suffice for putting up large numbers of containers, each containing sufficient material for several assays by current lysate tests.

It was then necessary to solicit the collaboration of a laboratory which had experience in processing such endotoxin and to explore the possibility of their putting up a number of vials of it, each containing a suitable amount. The laboratories of Mallinckrodt, Inc. were willing to undertake this work. They made a few pilot lots, which were studied for both adequate potency in lysate gelation as well as for stability. The procedure finally selected did, in fact, achieve a preparation with these properties.

While these investigations and arrangements were being developed, it happened that stocks of the current lot of the U. S. RSE established by the FDA Bureau of Biologics (then designated EC-4), were running low and it would have to be replaced. This turned out to be a great opportunity for a collaborative effort by the Bureau of Biologics and USP, in that we could both aim at setting up and establishing for our respective purposes, a single reference preparation. Further, the information obtained from the studies on processing, and on stability of the processed material by our collaborating laboratory were also available. It was finally agreed by the Bureau of Biologics and USP that the preparation would be made in that way from the master material (EC, referred to previously). The plan was to prepare sufficient final bulk for many thousands of vials, to fill those vials from a single outlet, and if possible to freeze-dry the entire number as one lot. This was largely fulfilled, except that one lyophilizer with a large enough capacity to take the entire lot in one load was not available. The lot was therefore freeze-dried in five sublimators. The vials thus obtained from each sublimator have been kept identifiable. In fact, our collaborators went to the extent of keeping identifiable each tray in each sublimator. Whether or not any significant differences will be found (using significant in the non-statistical sense), remains to be seen. However, some information is already available to show that any differences, if detectable, are likely to be small, and well within the limits of variations detectable by the lysate tests currently available.

A preliminary evaluation was made with the collaboration of the Bureau of Biologics and a few other collaborating laboratories, using lysate tests and tests with rabbits. In these tests both the proposed new RSE and the U. S. RSE Lot EC-2 (not EC-4) were compared. The results were encouraging and showed many useful properties of the proposed RSE. In view of the immense interest in obtaining the new RSE, it was then arranged to carry out a joint collaborative assay of the preparation, comparing it with the U. S. RSE Lot EC-2, which had been used first to define the Endotoxin Unit. Sixteen laboratories were invited to participate, and fifteen agreed to do so. A detailed description, consisting of a protocol and instructions, was drawn up, and the collaborative assay is now in progress.

The instructions were formulated between the Bureau of Biologics and USP, and the design was made according to statistical principles that would facilitate and be likely to yield good information on subsequent analysis of results. Invited participants were invited to make any comments or suggestions. However, no proposals were received for any substantial changes and the assay is therefore being carried out essentially as designed. The actual procedure for each participant making the assay was requested to be "in the usual manner". The participants were, however, required to make replicate assays, using the U. S. Reference Lysate (Lot 8) which was distributed, as well as a lysate preparation of their own choosing. The assay design was such that each set of dilutions compared (i.e., concomitantly prepared dilutions of the proposed new reference and the U. S. RSE Lot EC-2) constituted an independent assay. This was replicated in parallel four times on each occasion, with each lysate. Three sets of such parallel assays were required, in each of which each vial of the proposed new reference was assayed twice. This amounted to each participant being expected to submit finally 60 result sheets, 30 with the Reference Lysate and 30 with their own lysate, each sheet providing 4 independent assays. This provides a total of 240 assays (or replicates, whatever term one may prefer). To minimize variables, each participating laboratory was required to have all the dilutions and assays on the nine working days done by one assayist/technician. An important requirement was that participants should not try to add to the number of procedures outlined by including additional endotoxins or additional lysates once a particular lot of lysate had been selected, other than the distributed U. S. Reference Lysate (Lot 8), or by introducing any other variables.

The results from all the participants are now expected to be returned within days. The statistical analysis is being done by the Bureau of Biologics. If the assay design is strictly adhered to, the actual calculations should not pose any problems and could be completed in short order. Although no information on any assay results can be given at this time, it is expected that the new RSE could be established by the FDA Bureau of Biologics, and by the USP Reference Standards Committee, each according to routine procedure, in a few weeks.

The success of this collaborative effort could result in there being a single preparation of a reference endotoxin for use in the U. S., according to requirements for the FDA, for manufacturers, and for compendial purposes. The USP has also offered a quantity of the new preparation to the World Health Organization, so that the same identical preparation would also be available for international use. In fact, the WHO is itself in the process of carrying out a large international collaborative study of different lysates (and similar preparations) and many endotoxins, including the International Pyrogen Reference Preparation which was established decades ago: (WHO 1959). The advantages of having such a widely available single reference would be many, but one of the most important would be in relation to the defined unit of potency. For quite some time there would not arise the question of different units of endotoxin activity based on different Reference Standard Endotoxins, and any difference would appertain only to Control Standard Endotoxins calibrated with the RSE. If this does not occur, in all cases, it would be to the loss of any who do not choose to use the RSE. What will remain to be determined would be whether or not different kinds of endotoxins, i.e., from different microorganism, purified and unpurified, naturally occurring in manufactured products or artificially prepared in the laboratory, can all be validly assayed with the single reference standard endotoxin. Also much remains to be assessed of the behavior of such different endotoxins in rabbits, and presumably in humans. The single RSE will therefore, while useful, need to be used in many further studies to evaluate these problems.

Rudbach RJ, Akiya FI, Elin RJ, Hochstein HD, Luoma MK, Milner ECB, Milner KC, Thomas KR (1976). Preparation and properties of a national reference endotoxin. J Clin Microbiol 3:21.

Wld Hlth Org Techn Rep Ser (1959). Twelfth report of the expert committee on biological standardization 172:11.

Endotoxins and Their Detection With the Limulus Amebocyte Lysate Test, pages 121–130

REFERENCE ENDOTOXIN: A PRACTICAL RATIONALE

Joseph R. Firca and Jon A. Rudbach
Abbott Diagnostics Div., Abbott Laboratories, North Chicago, Illinois, 60064 and The Stella Duncan Memorial Research Institute, Dept. of Microbiology, University of Montana, Missoula, Montana 59812

Shortly after the discovery of the phenomenon that bacterial endotoxin induces clot formation in Limulus amebocyte lysate (LAL), the utility of this reaction for the pharmaceutical industry was recognized. Here was a simple test for pyrogenic endotoxins that was a thousand-fold more sensitive than the rabbit pyrogen test, could be performed in a few hours, avoided the expense and variability of testing living rabbits, and could be automated. However, the test, as originally conceived, had several drawbacks. There were materials in blood and in other biological products that interfered with the LAL test for endotoxin. Some of these prevented clot formation and others caused "false-positive" reactions. However, once the conditions under which these phenomena occurred were defined, interpretation of the LAL test for endotoxin in these materials could be qualified. The other problem with the LAL was related to its use for quantification of endotoxin. Batches of LAL differed in their sensitivity for detecting endotoxin. Furthermore, a preparation of LAL might change in its sensitivity, with time and conditions of storage. Finally, the conditions of individual assays (times, temperatures, volumes, etc.), performed in different laboratories could influence the concentrations of endotoxin that were detectable. Thus, for quantification of endotoxin by the LAL test, it was recognized that an endotoxin control had to be run in each assay. This need for the availability of a standard endotoxin that could be used as a positive control for establishing the sensitivity (end point) of LAL in a given test situation was accentuated when the

LAL test was contemplated as a replacement for the rabbit pyrogen test for limiting the levels of endotoxin in injectable solutions.

The first reference standard endotoxin that was employed by the Bureau of Biologics, FDA, was a purified lipopolysaccharide from Klebsiella pneumoniae. However, it became apparent that samples of a reference endotoxin had to be made available to pharmaceutical manufacturers and to qualified biomedical investigators so that the LAL tests and other bioassays could be standardized. Therefore, a new, large batch of endotoxin had to be prepared in order to meet the anticipated requests for a national reference standard preparation.

Considerable thought and discussion were put into the rationale for the properties of the reference endotoxin that would be most desirable. These were never before articulated, publicly, and the national reference endotoxin that was prepared, although it met the stated criteria, has been criticized on several bases. In the following, the rationale for the desired properties of reference endotoxin EC will be outlined, so that they might form the bases for future discussions of desirable properties for other reference standard endotoxins which might be prepared.

First to be recognized must be that this was a biological standard. The reference endotoxin EC has been criticized because it was not the "ultimately refined endotoxin", it was not "pure" lipid A, etc. Apparently, in some quarters, it was felt that a chemical formula should have been presented for the reference endotoxin. Many highly chemically refined and defined endotoxins, or preparations of lipid A, have low biological activity or undesirable properties of solubility and stability. Therefore, the targeted chemical goal for the reference endotoxin was to prepare a lipopolysaccharide that was free of potentially biologically active proteins, peptides, polynucleotides, or polysaccharides.

Another criticism that has been directed at the reference endotoxin EC, was that other types of endotoxin could have been prepared that have higher specific activities, on weight bases. Certainly, this was the case. However, the goal was to produce a reference standard of a "typical" endotoxin with _average_ biological potency. Such a preparation would serve as the best reference against which endotoxins with a wide range of potencies could be tested. For this reason an extract of one of the most "typical" gram negative bacteria, _Escherichia coli_, was chosen. The particular strain of _E. coli_ was also selected with care. Lipopolysaccharide extracted from _E. coli_ 0113 (Braude strain) does not contain the dideoxyhexoses that are immunodominant sugars in so many endotoxins from enteric bacteria. Therefore, if it ever became necessary to quantify the endotoxin by a chemical test (for example, this might be necessary because of changes in water content of the hygroscopic material), ketodeoxyoctulosonic acid could be quantified directly with the periodate-thiobarbituric acid test, in the absence of the interfering chromogen generated from dideoxyhexoses. Furthermore, this strain of _E. coli_ is K negative; it is genetically incapable of making a capsular polysaccharide that would contaminate the extracted endotoxin and which could prove difficult to remove from the reference endotoxin. Finally lyophilized preparations and solutions of endotoxin from this strain of _E. coli_ have proved to be stable over decades of work, under average storage conditions.

Thus, by design, the reference endotoxin subsequently produced represented a typical preparation, of average biological potency, with sufficient chemical purity, and excellent properties of stability and solubility.

As a final note, apparently some problems have been encountered during preparation of the individual lots of reference endotoxin EC. Some of these difficulties might be due to the addition of albumin as a "filler" before the minute amounts of endotoxin are freeze-dried. For years, the pharmaceutical industry has employed mannitol as a "filler"; it lyophilizes well, dissolves readily, and should have no interaction with endotoxin. Certainly, mannitol should be tested for use with the reference endotoxin.

In order to assess the value of _Escherichia coli_ 0113 as a representative endotoxin for a national reference standard for the LAL test, a spectrum of endotoxins were examined for their reactivity with LAL. All of the endotoxins employed in this study, with the exception of one, were prepared and purified according to the procedures established for preparing the national reference endotoxin (Rudbach, 1976). The exception is Novo-Pyrexal[R], an endotoxin produced by Hermal-Chemie. Preparation and properties of Novo-Pyrexal[R] have been reported (Galanos, 1979). LAL was purchased from Associates of Cape Cod, Inc. Woods Hole, Mass. The sensitivity of the lot of LAL was 0.05 ng/ml as determined by the clot method using the U.S. standard endotoxin, (RE-EC). LAL was reconstituted according to the manufacturer's instructions. Reactivity of LAL to endotoxin was monitored with Abbott's MS-2[R] Research System. Kinetic measurements were recorded at a wavelength of 670 nanometers at 1 min intervals for a total reaction time of 96 min. Drs. James H. Jorgensen and Gary A. Alexander developed the automatic methodology and described its performance with various LAL preparations (Jorgensen, 1981). Stock solutions of endotoxin were prepared in pyrogen-free Abbott 0.9% sodium chloride solution. Standard solutions of endotoxins to be examined consisted of a series of 2-fold dilutions from 1000 pg/ml to 1.95 pg/ml in pyrogen-free Abbott water. A single endotoxin-LAL reactivity experiment consisted of reacting all ten dilutions of the endotoxin plus the water control with a standard volume of LAL. The procedure consisted of adding 400 ul of each endotoxin concentration to a single pyrogen-free glass Ampvette[TM]. Monitoring began when 100 ul of LAL was added to each Ampvette. Results are displayed as in the graphic representations of Figures 1 through 4. Reactivity for the various concentrations of each endotoxin was determined by the time of onset of the initial rapid rise in optical density. Figures 1 through 4 are illustrated because they represent rates of reactivity that varied from very fast to very slow, respectively. Because these experiments analyzed the kinetics of the LAL reaction, the results could be quantified. An interesting observation is the initial steep slope followed by a second shallow slope. The lower the concentration of endotoxin, the more discernable are the two slopes. The time of onset of the first slope was concentration-dependent while the second slope appeared to be dependent on endotoxin concentration as well as on the rate of reaction of the first slope. Employing these kinetic parameters, all seventeen endotoxins were examined and assigned a position with respect to LAL reactivity (Table 1). Although it was easy to assign positions to the most rapid reactors and

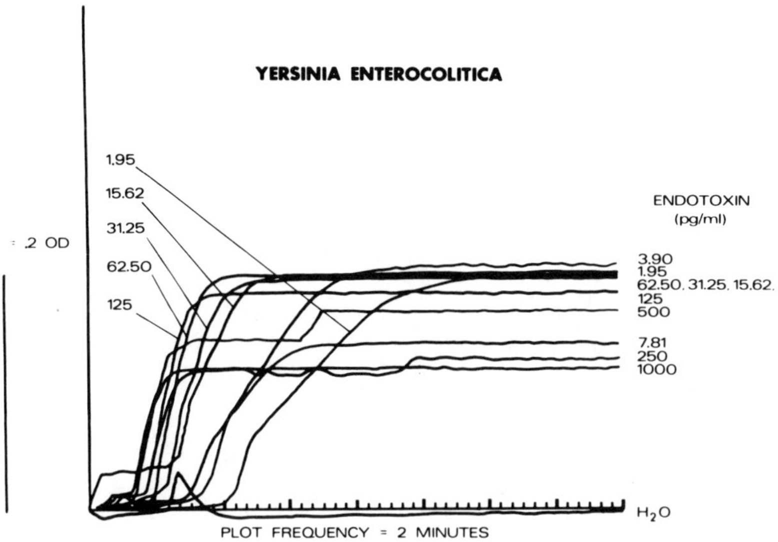

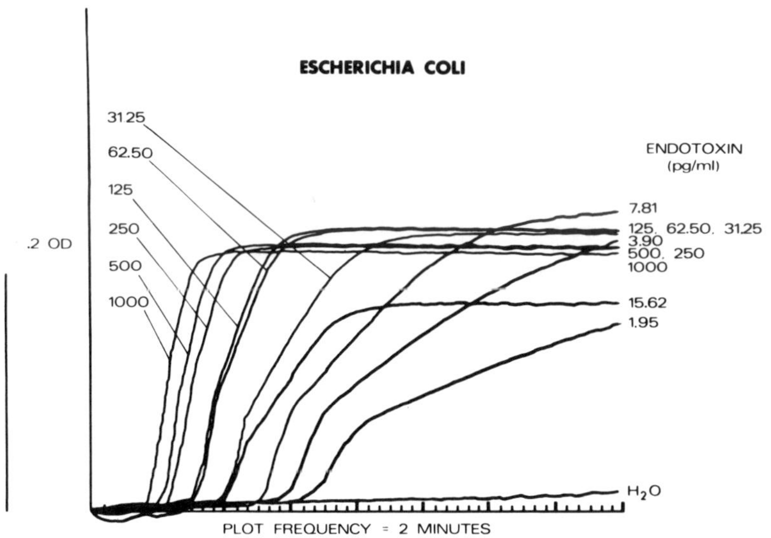

Fig. 1. (Top) and Fig. 2. (Bottom). MS-2 graphic representation of LAL reactivity with endotoxin from *Yersinia enterocolitica* and *Escherichia coli*, respectively.

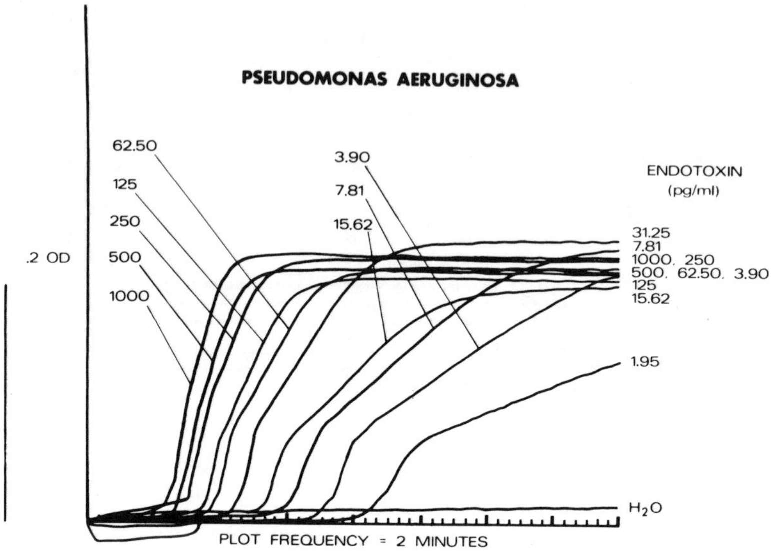

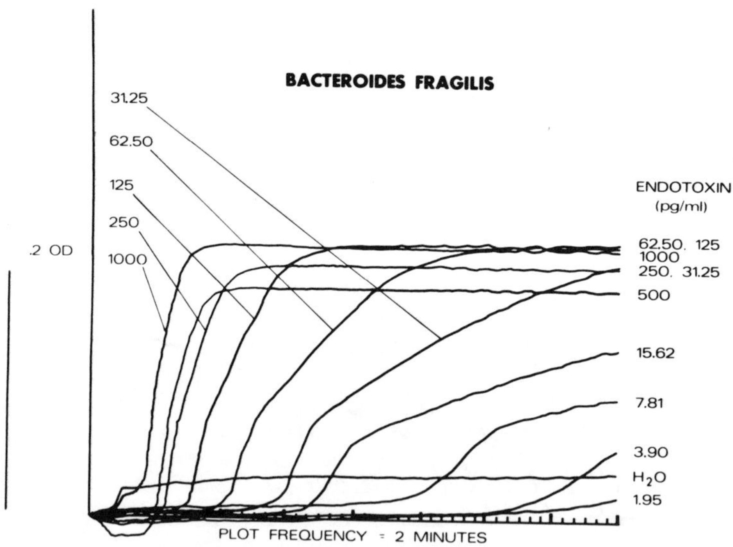

Fig. 3. (Top) and Fig. 4. (Bottom). MS-2 graphic representation of LAL reactivity with endotoxin from Pseudomonas aeruginosa and Bacteroides fragilis, respectively.

Table 1. Ranking of endotoxins by time of onset of reaction with LAL.

	ENDOTOXIN	RELATIVE REACTIVITY WITH LIMULUS AMEBOCYTE LYSATE
A	1. YERSINIA ENTEROCOLITICA	FAST
	2. PROTEUS VULGARIS	
B	3. NEISSERIA LACTAMICA	
	4. PROTEUS MIRABILIS	
	5. NOVO-PYREXAL®	
	6. PROTEUS MORGANII	
	7. SERRATIA MARCESCENS	
	8. ESCHERICHIA COLI O113	
	9. ENTEROBACTER CLOACAE	
	10. SALMONELLA ENTERITIDIS	
C	11. PROVIDENCIA STUARTII	
	12. PSEUDOMONAS AERUGINOSA	
	13. KLEBSIELLA OXYTOCA	
D	14. KLEBSIELLA PNEUMONIAE	
	15. SHIGELLA DYSENTERIAE	
	16. BACTEROIDES FRAGILIS	
	17. PSEUDOMONAS FLUORESCENS	SLOW

to the poor reactors, caution was exercised in assigning the intermediate positions. Due to the sensitivity of the instrument, LAL-reactivities at all concentrations for each endotoxin was considered before assignment to a position in the Table. It is apparent from Table 1 that *Escherichia coli* 0113 endotoxin represented an average endotoxin. Table 1 could be further subdivided into four groups, based upon similar LAL reactivities of the endotoxins. Rank within the above sub-groups was not absolute but appeared to vary with changing environmental influences. Age of the reconstituted endotoxin solution and adsorption onto the surfaces of the vessel containing the endotoxin solution are likely explanations, although others are possible. The rabbit pyrogen study of the seventeen endotoxins has not been completed, but experimental data by other investigators tends to substantiate the ranking in Table 1. Comparisons by the LAL clot test of various lipopolysaccharide (LPS) preparations demonstrate high LAL reactivity for a variety of *Escherichia coli* preparations when compared with LPS from *Klebsiella*, *Salmonella* and *Veillonella* (Elin, 1976).

A recent independent study has demonstrated that lipopolysaccharides purified from Veillonella and Fusobacterium were more reactive with LAL, on a weight basis, than was LPS from Bacteroides (Sween, 1977). Pyrogenicity values in both studies paralleled the LAL reactivity of the endotoxins. The endotoxin of Neisseria gonorrhea is so LAL-reactive that a test for the detection of these organisms in cervical secretions has been suggested (Spagna, 1980).

In summary, the rationale behind the design and preparation of national reference standard endotoxin EC was presented. This endotoxin preparation was shown to perform in a reproducible and predictable manner in the LAL assay, as quantified by the sensitive Abbott MS-2 instrumentation methodology.

Galanos C, Luderitz O, Westphal O (1979). Preparation and properties of a standardized lipopolysaccharide from Salmonella abortus equi (Novo-Pyrexal). Zbl Bakt Hyg, I Abt Orig A 243:226.

Jorgensen JH, Alexander GA (1981). Automation of the Limulus amoebocyte lysate test by using the Abbott MS-2 Microbiology System. Appl Environ Microbiol 41:1316.

Rudbach JA, Akiya FI, Elin RJ, Hochstein HD, Luoma MK, Milner ECB, Milner KC, Thomas KR (1976). Preparation and properties of a national reference endotoxin. J Clin Microbiol 3:21.

Elin RJ, Sandberg AL, Rosenstreich DL (1976). Comparison of the pyrogenicity, Limulus activity, mitogenicity and complement reactivity of several bacteria endotoxins and related compounds. J Immunol 117:1238.

Sween K, Hofstad T, Milner KC (1977). Lethality for mice and chick embryos, pyrogenicity in rabbits and ability to gelate lysate from amoebocytes of Limulus polyphemus by lipopolysaccharides from Bacteroides, Fusobacterium and Veillonella. Acta Path Microbiol Scand Sect B 85:388.

Spagna VA, Prior RB, Perkins RL (1980). Rapid presumptive diagnosis of gonococcal cervicites by the limulus lysate assay. Am J Obstet Gynecol 137:595.

FIRCA SPEAKER
TITLE: DESIGN OF A NATIONAL REFERENCE ENDOTOXIN STANDARD OF THE PHARMACEUTICAL INDUSTRY.

LEVIN: At the high concentrations, one would have expected the reaction to have gone to completion and therefore the maximum optical density should have been found. Yet I noticed tremendous variations which seemed to be unrelated to the concentration. One would have predicted they would have gone to the same optical density.
FIRCA: Two possible reasons for the variations observed in optical density may be 1) the physical state of the endotoxin in solution relative to its concentration, i.e. micelle formation and 2) the degree of lattice assembly formed as influenced by endotoxin concentration, i.e. multiple foci.
MORRISON: What is the maximum and minimum range, in terms of the concentration of endotoxins, required to elicit responses to LAL in reference to your graph where you showed the fastest and slowest response? Does the concentration vary by an order of magnitude, two orders of magnitude or by a factor of two?
FIRCA: In the present study, the figures show a range of LAL sensitivity of two orders of magnitude. However, in this particular assay system, the concentration range for *Yersinia enterocolitica* (fastest response) can extend up to five orders of magnitude.
BANG: Did your results with LAL correlate with pyrogenicity?
FIRCA: We are completing the pyrogen testing for all seventeen endotoxins. Other investigators have observed a direct relationship between LAL reactivity and rabbit pyrogenicity for some of these endotoxins.
BANG: Does the character of an endotoxin manifest itself in the same way in the two tests?
FIRCA: We do not have enough information at this time to answer your question.
BANG: From what you say we can conclude that we don't know what the correlation is between tests based on the species of bacteria and the method of extraction.
FIRCA: That is correct.

TSUJI: We know that some brands of LAL do not increase in turbidity when reacted with endotoxins. What are we measuring with the turbidimetric method? Obviously we are not measuring the same enzymatic endpoint. Could you speculate the modes of action between turbidity vs gelation? We may be measuring the endotoxin by two entirely different enzymatic schemes.

FIRCA: The MS-2 methodology is measuring the development of turbidity which may be due to the activation and activity of one or more enzymes within the LAL reagent. Dr. Watson, as a manufacturer and supplier of LAL, would you care to comment on this question?

WATSON: There are probably two different cascades: one responsible for the gel and one for turbidity. A given brand of LAL may contain the constituents for both or only one of these cascades.

Endotoxins and Their Detection With the Limulus Amebocyte Lysate Test, pages 131–140

A COMPARISON OF PYROGENICITY OF BACTERIAL ENDOTOXINS FROM A VARIETY OF GRAM-NEGATIVE BACTERIA AS DETERMINED BY THE LAL TEST

Jeffrey Fleishman and Freddie Fowlkes

Difco Laboratories, Inc.
P. O. Box 1058A
Detroit, Michigan 48232

INTRODUCTION

This study was designed to evaluate the relative potencies of various endotoxin preparations in the Limulus Amebocyte Lysate (LAL) assay, using the gel clot endpoint technique. Initially, we were interested in determining the batch-to-batch uniformity of *E. coli* 055:B5 and 0127:B8 endotoxins prepared as secondary standards. This study followed the HIMA collaborative study conducted in 1980 on the lot-to-lot and vial-to-vial variation of standard endotoxin preparations. The second phase of our study was to determine the relative potencies of representative samples of all the lipopolysaccharides prepared at Difco. In all, 29 lots of endotoxins, including both Westphal and Boivin preparations from 6 strains of *E. coli*; 6 species of *Salmonella* and one species of *Shigella* were tested. Finally, the potencies of the more highly purified FDA's EC2 (*E. coli* 0113 Braude), Novo Pyrexal's *S. abortus equi* and List Biological's *E. coli* 0111:B4, *E. coli* 055:B5 and *S. minnesota*, were compared to the Difco equivalent preparations. This report summarizes the results of these studies.

MATERIALS AND METHODS

Limulus Amebocyte Lysate

Six different lots of LAL were used throughout these

experiments. These included 4 lots of Pyrotest™ 1 Test (Difco), 1 lot of Pyrotest™ 50 Test (Difco), and the FDA reference lysate Lot #8. See Table 1 for the lot numbers and sensitivities of the lysates used in the individual studies.

Table 1. Sensitivities and lot numbers of LAL preparations

Lot Number	Sensitivity[a] (ng/ml)	Table Number (Study)
Pyrotest™ 1 Test		
HJ6846A	0.125	2
HJ6848	0.25	3, 4
HK5897	0.05	3, 4
HK6782	0.025	3, 4
Pyrotest™ 50 Test		
HJ3673	0.25	2
FDA Reference Lysate		
8	0.025	2, 3, 4

[a] Sensitivity of lysate are relative to the FDA standard endotoxin EC2.

Preparation of Endotoxin Dilution

The endotoxin preparations used in these studies were supplied either as bulk powders or lyophilized in vials. Stock solutions containing 1 mg endotoxin per ml pyrogen-free distilled water were prepared from the bulk powders. The stock solutions were mixed well and allowed to remain overnight in the cold room prior to preparing the final dilutions. Four series of dilutions were made on 4 days within a two week period from each stock solution. The dilutions were all made in pyrogen-free distilled water resulting in final concentrations of endotoxin 10 ug/ml, 100 ng/ml, 1 ng/ml, 0.4 ng/ml, 0.2 ng/ml, 0.1 ng/ml thru 0.003 ng/ml. Endotoxins supplied in vials were rehydrated in pyrogen-free distilled water according to the label directions and dilutions were prepared so that the same titration sequence beginning at 0.4 ng/ml was achieved. Three sets of dilutions were prepared for the vials in the batch-to-batch variation study, and four sets of dilutions were prepared for comparison studies. (See Experimental Model below.)

LAL Assay Method

The standard LAL gel clot endpoint method was employed throughout. This consisted of either the rehydration of the Pyrotest™ 1 Test lysate with 0.2 ml of each endotoxin dilution, or the addition of 0.1 ml of each endotoxin dilution to 0.1 ml of the properly rehydrated Pyrotest™ 50 Test or FDA reference lysate in depyrogenated 12X75 mm tubes (the same size tubes used to package the Pyrotest™ 1 Test). After initial gentle shaking, the tubes were left undisturbed for 60 minutes in a 37°C water bath. At the end of the incubation period, the tubes were individually inverted through 180°. The gel clot endpoints were determined as the dilution of endotoxin in the last tube in which a firm clot remained intact upon inversion.

Experimental Model

The first study, analyzing the batch-to-batch uniformity of the secondary standard endotoxins *E. coli* 055:B5 and *E. coli* 0127:B8, consisted of 9 observations. Three different lots of LAL, including the FDA reference lot 8, were used on each of 3 days. Dilutions of each endotoxin stock solution were freshly prepared on each day and tested in duplicate. The second study, comparing the potencies of the different endotoxin preparations, consisted of 16 observations. Four different lots of LAL, including the FDA reference lot 8, were used on each of 4 days. As before, dilutions of each endotoxin stock solution were freshly prepared on each day and tested in duplicate. The EC2 standard endotoxin was used on every test day as a reference and control to calculate the relative potencies. The following formula was used to obtain relative potencies:

$$\text{Potency of Unknown} = 5\ \text{EU/NG}^{*} \div \frac{\text{Gel Clot Endpoint of Unknown}}{\text{Gel Clot Endpoint of EC2}}$$

*5 EU/NG is the potency assigned to the FDA Standard EC2.

Table 2. Batch-To-Batch Uniformity of *E. coli* 055:B5 and *E. coli* 0127:B8 Secondary Reference Standards

Endotoxin	Bulk Batch	Lot#	Potency[b] Number of Observations At Each Dilution (EU/NG)	Average Potency (EU/NG)	Potency[c] Ratio
E. coli 055:B5	CC7817A		NA[e]	-	-
		18090	9 @ 10 EU/NG	10	2
Boivin Extraction		18090C	9 @ 10 EU/NG	10	2
(4/74)		18229B	2 @ 20 EU/NG; 7 @ 10 EU/NG	12.2	2.44
LD_{50}[a] = 0.25 mg		18694	9 @ 10 EU/NG;	10	2
Lipid A = 11.9%		18981	8 @ 10 EU/NG; 1 @ 5 EU/NG	9.4	1.88
% N_2 = 0.31		19037	9 @ 10 EU/NG	10	2
		18090B[d]	9 @ 10 EU/NG	10	2
E. coli 055:B5	HJ2710B		9 @ 10 EU/NG	10	2
		591321	9 @ 10 EU/NG	10	2
Boivin Extraction		19301B[d]	9 @ 10 EU/NG	10	2
(1/80)		HH3855	1 @ 20 EU/NG; 8 @ 10 EU/NG	11.1	2.22
LD_{50} = 0.20 mg		HH8696A	1 @ 20 EU/NG; 8 @ 10 EU/NG	11.1	2.22
Lipid A = 6.8%		HH8696C	9 @ 10 EU/NG	10	2
OD_{260nm} = 9.0		HJ6028A	3 @ 20 EU/NG; 6 @ 10 EU/NG	13.3	2.66
% N_2 = 3.53		HK4842	1 @ 20 EU/NG; 8 @ 10 EU/NG	11.1	2.22
		HK6235	9 @ 10 EU/NG	10	2

Table 2 (Con't)

Endotoxin	Bulk Batch	Lot	Potency[b] Number of Observations At Each Dilution (EU/NG)	Average Potency EU/NG	Potency[c] Ratio
E. coli 055:B5	A02700		3 @ 20 EU/NG; 6 @ 10 EU/NG	13.3	2.66
		GG9851	2 @ 20 EU/NG; 7 @ 10 EU/NG	12.2	2.44
Westphal Extraction		HJ1807A	9 @ 10 EU/NG	10	2
(3/70)		HJ1808A	3 @ 20 EU/NG; 6 @ 10 EU/NG	13.3	2.66
LD_{50} = 0.20 mg		HJ4899A	9 @ 10 EU/NG	10	2
% N_2 = 3.85		HJ4899B	9 @ 10 EU/NG	10	2
		HJ6783B	2 @ 20 EU/NG; 7 @ 10 EU/NG	12.22	2.44
E. coli 0127:B8	GG8036		9 @ 5 EU/NG	5	1
(6/76)		HH8726	8 @ 5 EU/NG; 1 @ 2.5 EU/NG	4.7	0.94
LD_{50} = 0.11 mg		HH4697B	6 @ 5 EU/NG; 3 @ 2.5 EU/NG	4.1	0.83
% N_2 = 2.9		HH0101	9 @ 5 EU/NG	5	1
		633353	9 @ 5 EU/NG	5	1

(a) LD_{50} in 15 gm mice.
(b) Potency in EU/NG is determined relative to EC2 having a potency of 5 EU/NG using the formula:

$$\text{Potency of Unknown} = 5\ \text{EU/NG} \div \frac{\text{Gel endpoint of unknown}}{\text{Gel endpoint of EC2}}$$

(c) Potency Ratio is derived by dividing the potency of the unknown by 5 (potency of the EC2 reference endotoxin) so that relative comparison can easily be made.
(d) 18090B = HIMA Study 504089, 19301B = HIMA Study 675514
(e) NA = Material not available.

The averages of all observations were calculated based on EU/NG to facilitate comparisons. The potency ratios were derived to compare the potencies of all the endotoxins relative to EC2, by dividing each average potency by 5.

RESULTS

Sixteen lots of E. coli 055:B5 (Boivin, trichloroacetic acid extraction) were tested. (See Table 2.) Seven of the lots were derived from a single bulk batch prepared in 1974; the remaining 9 were derived from a second bulk batch prepared in 1980. There was no apparent difference in sensitivity between the two bulk batches, nor was there any difference between the different lots prepared from each bulk batch. Most observations placed the potency of E. coli 055:B5 at about twice that of the standard EC2 endotoxin (i.e., 10 EU/NG).

Four lots of E. coli 0127:B8 (Westphal, phenol extraction) endotoxin prepared from one bulk batch, GG8036, were tested and observed to have the same potency as the standard endotoxin EC2 (i.e., 5 EU/NG). (See Table 2.)

Seven lots of E. coli 055:B5 (Westphal, phenol extraction) were also tested for uniformity. These were all found to be of equivalent potency to the Boivin extracted lots (i.e., 10 EU/NG).

Table 3 compares the endotoxin potencies of 12 E. coli preparations relative to EC2. As can be seen in this table, the potency ratios range from 1.25 to 2.88 for the Difco preparations and were slightly higher, 3.25 and 3.5, for the chromatography purified List preparations. There was no real difference observed in Boivin vs. Westphal preparations of the same strain of E. coli endotoxin. However, the endotoxin preparations prepared from E. coli 0127:B8 were observed to be about half as potent as the other strains of E. coli endotoxins tested.

Table 4 compares the potencies of endotoxins extracted from 6 Salmonella and 1 Shigella species. As in Table 3, all potencies are relative to the standard endotoxin EC2. Again minimal potency differences were observed between the Boivin and Westphal extractions of the same organism, except for S. abortus equi. However, up to a four-fold difference

Table 3. Comparison of Potencies of Various *E. coli* Endotoxins

Endotoxin	Preparation	LD_{50} in 15 gm Mice	% Lipid A Content	O.D. 260nm	% N_2	Potency[a] EU/NG	Potency[b] Ratio
FDA EC2	Westphal					5.0	1
FDA EC4	Westphal					1.95	0.39
E. coli 0113	Boivin (D)[c]	0.2 mg	6.8	9.0	2.98	6.25	1.25
E. coli 0127:B8	Boivin (D)	0.31 mg	NA	NA	3.97	6.56	1.31
E. coli 0127:B8	Westphal (D)	0.11 mg	NA	NA	2.90	6.25	1.25
E. coli 0128:B6	Boivin (D)	0.28 mg	15.3	25.75	3.77	10.31	2.06
E. coli 0128:B12	Boivin (D)	0.20 mg	NA	NA	4.07	10.31	2.06
E. coli 0128:B12	Westphal (D)	0.20 mg	9.3	11.0	2.45	8.25	1.65
E. coli 0111:B4	Boivin (D)	0.16 mg	15.68	NA	6.19	11.88	2.35
E. coli 0111:B4	Westphal (D)	0.20 mg	20.6	13.25	3.00	12.8	2.56
E. coli 0111:B4	Westphal (L)[d]	Chromatography purified				17.5	3.5
E. coli 055:B5	Boivin (D)	0.25 mg	11.9	NA	NA	10	2
E. coli 055:B5	Westphal (D)	0.20 mg	NA	NA	3.85	14.38	2.88
E. coli 055:B5	Westphal (L)	Chromatography purified				16.25	3.25

a. Potency in EU/NG is determined relative to EC2 having a potency of 5 EU/NG using the formula:

$$\text{Potency of Unknown} = 5\ \text{EU/NG} \div \frac{\text{Gel endpoint of unknown}}{\text{Gel endpoint of EC2}}$$

These data represent the average of 16 observations.

b. Potency ratio is derived by dividing the potency of the unknown by 5 (potency of the EC2 reference endotoxin) so that a relative comparison can easily be made.

c. (D) Difco Laboratories preparation

d. (L) List Biological Laboratories preparation.

TABLE 4. Comparison of Potencies of 6 Salmonella and 1 Shigella Endotoxins

Endotoxin	Preparation	LD50 in 15 gm Mice	% Lipid A Content	O.D. 260nm	% N_2	Potency[a] EU/NG	Potency[b] Ratio
S. abortus equi	Boivin (D)[c]	0.2	NA[f]	NA	2.78	6.25	1.25
S. abortus equi	Boivin (D)	0.27	NA	10.7	2.96	5.93	1.19
S. abortus equi	Westphal (D)	0.25	NA	NA	3.46	10.0	2.0
S. abortus equi	Westphal (NP)[d]					6.25	1.25
S. enteritidis	Boivin (D)	0.45	12.92	24.0	4.71	2.96	0.59
S. enteritidis	Westphal (D)	0.24	12.90	30.0	1.71	2.58	0.52
S. marcescens	Boivin (D)	0.41	12.0	12.0	3.8	5.3	1.06
S. marcescens	Westphal (D)	0.3	8.28	9.0	1.71	6.25	1.25
S. minnesota 9700	Westphal (D)	0.4	NA	NA	5.55	10.31	2.06
S. minnesota Re595	(L)[e]	Morrison and Leive Lipid A preparation				19.38	3.88
S. typhosa 0901	Boivin (D)	0.25	10.4	8.8	3.0	6.95	1.39
S. typhosa 0901	Westphal (D)	0.22	11.0	3.3	3.21	9.38	1.88
S. typhimurium	Boivin (D)	0.11	13.0	34.7	3.6	8.13	1.63
S. typhimurium	Westphal (D)	0.15	16.2	44.35	5.53	10.63	2.13
Shigella flexneri	Westphal (D)	0.49	NA	NA	2.2	4.68	0.94

a Potency in EU/NG is determined relative to EC2 having a potency of 5 EU/NG using the formula:

$$\text{Potency of Unknown} = 5\ \text{EU/NG} \div \frac{\text{Gel endpoint of unknown}}{\text{Gel endpoint of EC2}}$$

These data represent the average of 16 observations.

b Potency ratio is derived by dividing the potency of the unknown by 5 (potency of the EC2 reference) so that relative comparisons can easily be made.

c (D) Difco Laboratories preparation

d (NP) Novo Pyrexal preparation

e (L) List Biologicals preparation

in potency was observed between Difco extracts from different species (i.e., *S. enteritidis* (W) 0.52 (2.58 EU/NG) and *S. typhimurium* (W) 2.13 (10.63 EU/NG). The List chromatography purified endotoxin was twice as potent as the comparable Difco preparation.

The Difco *S. abortus equi* (Westphal) endotoxin was found to be twice as potent as the comparable, more highly purified Novo Pyrexal (Westphal), whereas the Difco Boivin endotoxin was equivalent to the Novo Pyrexal preparation. In addition, the Difco preparation of *E. coli* 0113 (Braude) (Boivin) was comparable to the more highly purified FDA EC2, which is a Westphal extraction of the same *E. coli* strain. Furthermore, the FDA EC4 standard endotoxin was found to be about one-third weaker than the EC2 preparation.

DISCUSSION

Although additional lots of *E. coli* 055:B5 were prepared between 1974 and 1980, it was decided to use the oldest available preparation to compare with the most recent one to determine if there was a potency loss due to stability of the bulk endotoxin. It should also be noted that the Westphal extraction of *E. coli* 055:B5 was prepared in 1970 and still maintained a potency equivalent to the Boivin extractions. In addition, human albumin was used as a filler in all the *E. coli* 055:B5 (Boivin) and *E. coli* 0127:B8 (Westphal) lots without any noticeable effect on their potencies.

The Difco preparations of *E. coli* 055:B5 were found to be uniform in potency throughout all the lots tested. The solubility of this material was excellent and probably aided in the reproducibility of the titration dilutions. We feel that these attributes of this preparation would ideally suit it as a secondary standard. In contrast, we feel that the need for a highly purified, well characterized endotoxin standard is over-stressed. We found the potencies of the Difco preparations to be equal to or greater than those of the more highly purified comparable endotoxins and exhibited greater vial-to-vial uniformity than the EC2 and EC4 vials.

FLEISHMAN SPEAKER
TITLE: A COMPARISON OF THE PYROGENICITY OF BACTERIAL ENDOTOXINS FROM A VARIETY OF GRAM-NEGATIVE BACTERIA AS DETERMINED BY THE LAL TEST.

TSUJI: I'm wondering if you compared your LAL test to the rabbit test?
FLEISHMAN: We do not have a rabbit colony at Difco. If anyone would like to try it out I can supply the samples.
MORRISON: In at least two cases the *S. typhosa* Westphal preparation appeared to be twice as reactive as the Boivin extraction. Is this a general finding?.
FLEISHMAN: In general the Westphal extraction is more potent than the Boivin extraction as judged by the LAL test. But the significance isn't that great as you are only talking about a twofold difference.
JEFFRIES: I wonder if you have any information on the different sensitivities of the endotoxin you studied. In a meeting in London several days ago the point was made that endotoxins varied greatly in their heat sensitivities. Some are destroyed by boiling for 10 minutes while others lose no activity when autoclaved for 5 hours. Is it bad to heat pharmaceutical preparations in order to overcome inhibition?
FLEISHMAN: I can't answer that question. We haven't done any heat inactivation studies.
JEFFRIES From the literature it appears that *E. coli* endotoxin is one of the most stable endotoxins. Now *E. coli* is very unlikely to be present in pharmaceutical preparations; one would expect to find endotoxins from Pseudomonas and other organisms. How many of these environmental endotoxins are not detected by the LAL test?
FLEISHMAN: I can't answer your question..

Endotoxins and Their Detection With the
Limulus Amebocyte Lysate Test, pages 141-151

REVIEW OF THE BUREAU OF BIOLOGIC'S EXPERIENCE WITH LIMULUS AMEBOCYTE LYSATE AND ENDOTOXIN

H. Donald Hochstein

Food and Drug Administration
Bureau of Biologics
8800 Rockville Pike
Bethesda, Maryland 20205

In 1970, a graduate student at the Johns Hopkins University, working for the Bureau of Radiological Health, asked if he could use our laboratory during off-hours for his research with horseshoe crabs. The request was approved and, before long, Mr. James Cooper convinced most of us in the laboratory that the lysate made from the blood cells of the *Limulus polyphemus* had potential for detecting bacterial endotoxins in biological products. It was not long thereafter that we installed several holding tanks, and spent one week each summer at Chincoteaque, VA bleeding and collecting crabs. We would usually bleed about 1,000 crabs per week which yielded enough lysate for one year.

Mr. Cooper completed his M.S. degree and left the government but we continued to monitor for endotoxin in as many biological products as possible with the Limulus Amebocyte Lysate (LAL) test, noting any inhibition or enhancement that a product might exhibit.

From time to time we would purchase endotoxin from Difco, Inc. to check the sensitivity of our lysate. It did not take long to realize that different endotoxins gave different potencies with the same lysate, indicating that on a per weight basis, endotoxin prepared from *E. coli* could vary in potency as much as 100-1000 times that prepared from *Pseudomonas spp.* This meant that each LAL value had to be prefaced with the endotoxin and lysate lot so another laboratory could repeat the results.

It was about this time that the LAL test was beginning to be widely used in the medical community and the Food and Drug Administration received complaints that some people were using lysate to diagnose septicemia, meningitis, and urinary tract infections. The FDA decided that some regulations had to be written to control the sale and use of this product. Since LAL was a blood product, the Bureau of Biologics (BoB) assumed responsibility for its control. In 1973 the first Federal Register notice (2) was published concerning the LAL test. Essentially this notice stated that: a. LAL was a blood product and subject to licensure, b. all drugs subject to a pyrogen test are required to be tested by the rabbit pyrogen test, c. the LAL test cannot be used as a final pyrogen test.

Shortly thereafter a manufacturer submitted the first license application for LAL. The Bureau determined that, to control lots of lysate, both a standard endotoxin and a standard lysate were needed. The Bureau purchased a large lot of LAL from the manufacturer applying for a license and labeled it Reference Lysate Lot 1. However, the endotoxin was a different situation. What were we looking for as a Reference Standard Endotoxin (RSE)? Table I lists eight requirements for an RSE. It would be difficult to meet all the listed requirements but the more that are met the better the reference preparation.

Dr. D.B. Castor (1) of the Bureau of Drugs had just completed a collaborative study using a *Klebsiella pneumoniae* endotoxin. He was kind enough to supply us with a small quantity of this material which he had prepared and tested in rabbits. This met most of our requirements for an RSE. The two requirements not met were availability in large quantities and preparation with a common organism. We freeze-dried the Klebsiella endotoxin (Lot 1b) in 1972 (Table II) and used it for the next several years. This preparation was dried with 0.1% Normal Serum Albumin (Human) and was quite stable.

In 1974 we decided to replace Lot 1b with an RSE that met all of our requirements. The Bureau subsequently purchased 30 grams of endotoxin made from *E. coli* 0113 from Dr. J.A. Rudbach at the University of Montana. The details of this EC preparation were outlined in a publication by Rudbach et al, 1976 (4). This quantity of material should meet our demands for about 20 years if used only as a reference preparation. We currently store the bulk EC endotoxin (as recommended by Dr. Rudbach) in a desiccator at room temperature.

In 1976 we prepared our first lot of endotoxin from the EC bulk and designated this as Lot EC-1. Before freeze drying, 0.1% Normal Serum Albumin (Human) was added as a stabilizer. Lot EC-1 was a small pilot lot of approximately 200 vials which lasted less than a year.

We then decided to prepare a large lot in the same manner in which EC-1 was prepared. This lot, designated EC-2, consisted of approximately 1500 vials, each vial containing 1.0 μg of EC endotoxin. After accumulating extensive data on this very stable reference preparation, we decided to assign a unit value to this preparation. Most control authorities agree that is better to use a unit of activity rather than a concentration based on dry weight in expressing the strength of reference preparations. Based on a collaborative study involving BoB and licensed LAL manufacturers, we assigned a value of 5.0 endotoxin units/vial to Lot EC-2 (3).

The only criticism of lot EC-2 was that it contained Normal Serum Albumin (Human) which might bind endotoxin in an unknown product. To eliminate this hypothetical problem, lot EC-3 was prepared in 1980 as a pilot lot with no fillers or stabilizers. Following several months use this lot appeared satisfactory and, therefore, Lot EC-4 was prepared in the same way. After freeze drying, the potency of EC-4 was just half that of EC-2. This low potency value was remedied by reconstituting EC-4 with half as much diluent as EC-2. Some laboratories have experienced problems with EC-4 because, with no filler or stabilizer, the vials not only appeared empty but it was also quite difficult to dissolve all the endotoxin into water. This lot is now depleted.

Early this year (1981) it was decided to have a large lot of EC endotoxin prepared that would match as closely as possible lot EC-2. This was prepared by a licensed LAL manufacturer in accordance with a protocol prepared by the United States Pharmacopeia (USP) and Bureau of Biologics using lactose and polyethylene glycol as stabilizers. Between 28,000 and 30,000 vials were filled from a single bulk and freeze-dried. Five freeze-dryers had to be used to accommodate the large number of vials. The final residual moisture of this new proposed RSE was 0.5%. A pilot sensitivity study indicates that both EC-2 and the proposed replacement, designated EC-5, have almost an identical potency of between 10 and 15 picograms/ml with U.S. Reference Lysate, Lot 8.

The label unitage on the proposed RSE will be in EU per vial which will be in accordance with the USP and new FDA guidelines. Also, within the next several months, all lysate labels will have the sensitivity in EU/ml so the user can verify the labeled sensitivity in his own laboratory.

Acknowledgements

The author expresses his sincere thanks to Dr. Edward A. Fitzgerald for his critical review and to Ms. Renee C. Burt for her help with the preparation of this manuscript.

Table I

GENERAL REQUIREMENTS FOR A REFERENCE STANDARD ENDOTOXIN:

a. a dry preparation

b. available in substantial quantities

c. stable room temperature

d. only slightly hygroscopic

e. goes readily into aqueous solution

f. clear solution in water and saline

g. potency tested in rabbit and man

h. prepared from a common bacterial species

Table II

BOB EXPERIENCE WITH REFERENCE ENDOTOXIN PREPARATIONS:

Lot Designation	Organism	Year Freeze-dried
1b	*Klebsiella pneumoniae*	1972
EC	*E. coli* 0113	1975
EC-1	" "	1976
EC-2	" "	1977
EC-3	" "	1980
EC-4	" "	1980
EC-5*	" "	1981

* Proposed U.S. Reference Standard Endotoxin now being tested in a collaborative study.

Castor, G.B., N. Kantor, E, E. Knoll, J. Blakely, J.K Nielsen, J.O. Randolph, and A. Kirshbaum, 1971. Characteristics of a highly purified pyrogenic Lipopolysaccharide, J. Pharm. Sci. 60: 1578-1580.

Federal Register (January 12, 1973) Vol. 38, No. 8, p. 1404.

Hochstein, H.D., 1981. The LAL test versus the rabbit pyrogen test for endotoxin detection. Pharm. Tech., Aug. 36-42.

Rudbach, J.A., Akiya, E.I., Elin, R.J., Hochstein, H.D., Luoma, M.K., Milner, C.B., Milner, K.C. and Thomas, K.R. (1976). Preparation and properties of a national reference endotoxin. J. Clin. Micro., 3: 21-25.

DON HOCHSTEIN SPEAKER

TITLE: PREPARATION OF A NEW U.S. REFERENCE ENDOTOXIN STANDARD.

FAULKNER: I have two questions. 1) Can you tell us when the information will be published on the pyrogen study? 2) Can you give us advice on how we should handle the reference endotoxin solution? If we get one or two vials from you, presumably we would need to resuspend it, then store it in some way until we use it.

HOCHSTEIN: We recommend that the protocol for reconstitution, storage and use of EC-5 should be identical to the protocol recommended for EC-2. That is, you would rehydrate it and then make dilutions from this rehydrated standard. You should freeze the remaining concentrated endotoxin in the vial, once frozen it can maintain its potency for months to years. However, the frozen standard should be thawed only once; after thawing throw away the material you do not use; do not refreeze it. Calibrate your own standard against EC-5. Use EC-5 for calibration of your own in-house standard rather than for daily use. You should calibrate your own standard occasionally or as the need arises. We don't know where the data on EC-5 will be published; it could be published in Biological Standardization or some other similar journal. EC has been tested in humans by Dr. Elin. This study was completed a year or so ago. It will be published in the October Journal of Infectious Diseases. He found that the minimum pyrogenic dose in humans is about 0.5 ng/kg. There is some controversy about the controls he used.

WATSON: You brought up an important point about freezing. There always has been a big question concerning whether you can or cannot freeze endotoxins. In our laboratory we normally observe a loss in endotoxin when we freeze at -20C. I know of no reference concerning the effect of freezing on endotoxins but the general belief is that freezing an endotoxin will decrease its potency. Is this belief based on fact or fiction? Does anyone have any information on the loss of potency due to freezing?

ALBAUGH: We tried spiking solutions for validation procedures and we found that freezing did in fact reduce the apparent concentration of endotoxins. This was stored at -70 over a few days; we lost potency. The initial concentration was 150 pg/ml and we could recover only 85 pg/ml after freezing.

WATSON: Certainly it is not advisable to freeze such low concentrations. The question is whether you can freeze anything at 100ng level without loss. I think this should be carefully studied by several laboratories. There may be some trouble in freezing changing the aggregation state. I don't have any actual data except we observe a loss of potency when we freeze.

LEVIN: I think the point about not freezing at too low a concentration is a very important one. We use Difco material and we freeze at 100 mg/ml. This will last four months to a year. I'm not implying that a year is the limit but that seems to be a good concentration at which to freeze for a long period of time. What does Difco add to their endotoxin?

FLEISHMAN: We added 0.5% albumin to the 055:B5 standard we prepared for Mallinckrodt. We do not add any filler to the rest of our endotoxins. I have frozen 055:B5 at concentrations of 25 ng/ml for over a year and had no loss in potency.

MORRISON: In the 1973 Jour. of Infect. Diseases Supplement there is a paper by Erwin Neter which reports that freezing of endotoxins causes them to lose their immunogenicity.

STOTT: I have used endotoxins as a stimulant in lymphocyte stimulation assay. I found that when I froze endotoxins in tissue culture media, the endotoxin decreased in potency if the material was frozen and thawed twice. The greatest loss in activity was found after the second freezing and thawing.

DITTER: We found a significant loss of endotoxin activity in biological samples when stored at -18C.

TSUJI: We froze Difco 0127B at a concentration of 1,000 ng/ml. After one freezing and thawing we lost potency as judged by LAL.

WATSON: Would it be possible to have a study to answer this question?

HOCHSTEIN: Yes, we could have a study where the freeze-thawing of endotoxin would be evaluated.

DILEO: What is the concentration of lactose and polyethylene glycol (PEG) in the new endotoxin standard.

MILLS: 1% lactose and PEG is 1%.

HOCHSTEIN: One percent of each preparation.

FLEISHMAN: One more thing; from the conversation so far, I gather that those people who find endotoxin is stable when frozen are refering to endotoxin prepared with albumin. Maybe albumin has some stabilizing effect. Is the discrepancy between those who report a loss of activity and those who report no loss the effect of albumin? Are they all talking about stability with respect to the same initial concentration of endotoxin. Except for the 055:B5 prepared for Mallinckrodt, the endotoxins from Difco do not contain albumin.

SWEET: Are we sure that this is not an effect of adsorption on the glass rather than inactivation?

HOCHSTEIN: You mean you can't get it out of the vial. I think you have to vortex a long time to get it into solution. It goes in a lot easier if you have a sugar added.

SWEET: Albumin also prevents adsorption to the glass.

HOCHSTEIN: I am not sure if albumin helps prevent adsorption or if it just aids in rehydrating the endotoxin.

URBASCHECK: After liquification of our endotoxin measured in biological parameters, we observe a very rapid loss of activity after one week. I think we should look at this time period after freezing.

TSUJI: Difco 1278 at a concentration of 100 pg/ml is stable for one week.

PEARSON: I wonder if any of those who have found a loss in activity have tried a mild sonication or have added a surfactant such as Tween to see whether or not you can recover the activity.

NEELY: We have done just that and we always see a loss in activity after freezing. However you can recover this activity by long and vigorous vortexing or, at least in one case, by sonication. Since Pyrosperse reagent works as well as anything else and has the effect of liberating "lost" endotoxin activity, I think the loss of activity is due to aggregation.

HOCHSTEIN: A number of people have asked when they will be hearing about their license ammendments to use the LAL test. These applications have been in the Bureau for a year or more waiting until the guidelines were approved. Now that the guidelines are out you should be hearing from the Bureau in the near future.

COOPER: With respect to labeling, Dr. Hochstein, will the BOB require the suppliers of endotoxin to label their CSE in endotoxin units?

HOCHSTEIN: This will be strictly voluntary. Even if the endotoxin is supplied in an LAL kit, the manufacturers will not have to label in endotoxin units.
WATSON: Does the BOB have any plans to license endotoxin standards?
HOCHSTEIN: We have been asked to do so but don't plan to in the near future.

Endotoxins and Their Detection With the Limulus Amebocyte Lysate Test, pages 153–160

INTERNATIONAL ACCEPTANCE OF REFERENCE ENDOTOXINS

G. Alan Sabey

The Wellcome Foundation Ltd.,
Dartford, Kent,
England

In view of the papers and comments already presented at this conference, some of the statements in this address have already become superfluous. Some, I will make again for emphasis, but three important points need to be clearly stated which we all may realise, but which have so far not been overtly recognised:

1. Both the pharmacopoeial Rabbit Pyrogen Test and the Limulus Amoebocyte Lysate (LAL) test for Bacterial Endotoxins are intended as Limit tests to ensure the quality of pharmaceutical products and their safety to the user.

2. For a pharmacopoeial end-product control test, the method should be Simple, Reliable and Reproducible.

3. Because of the twofold dilution ratio used in the LAL gelation test, the apparent difference between sensitivities of lysates or potencies of endotoxins may not be a real measure of the differences between claimed and estimated values. For example, a lysate with a true sensitivity of 35 pg/ml, and a labelled sensitivity of 30 pg/ml could have an estimated sensitivity of 60 pg/ml.

The LAL test has been universally-received as being of tremendous importance for the detection of gram-negative endotoxins in parenteral pharmaceuticals and infusion fluids. However, before crossing the barrier between acceptance as a useful quality assurance tool and world-wide pharmacopoeial recognition as an official quality control standard, two vital variables have to be reckoned with. Both of these are essential to the LAL reaction, for they are the LYSATE itself and the ENDOTOXIN to be detected. The lysate may vary with the source, the condition of the Limulus when the blood is taken, the method of preparation and purification employed, the manufacturer who supplies it, the particular endotoxin against which the sensitivity is determined or the conditions under which it is stored (Cooper 1977). Before the lysate can be standardised and defined, there must be a standard endotoxin against which any lysate can have its sensitivity measured and clearly stated. The FDA has defined a test that lysate manufacturers are now using: the reference standard endotoxin chosen by the FDA has been a preparation of Escherichia coli, EC2 first and now EC4, although this agency has also used at least one other reference endotoxin - Lot 1B of a purified lipopolysaccharide of Klebsiella pneumoniae.

So then is there a need to look further? Should not other organizations national and international just accept EC4 as the one definitive reference preparation? Unfortunately, the chosen standard is no longer considered ideal - because results obtained by manufacturers and users alike, have suggested that lysates whose sensitivities have been determined against EC4 have not given consistent results in either validation tests or routine use.

The World Health Organisation set up a collaborative trial, the results of which are being examined at a meeting in Geneva. At least two of the collaborating laboratories found that the preparation EC4 distributed under a code number only gave very variable results. My own laboratory ran into problems of reproducibility, a colleague independently encountered difficulties and a call to the lysate supplier revealed they too were experiencing similar anxieties.

If there is to be international acceptance of a reference endotoxin, I suggest there are seven characteristics which must be demonstrated for that preparation.

I hope these proposals will go some way to answering those questions posed by Dr. Perkins, of the WHO at the Symposium in Budapest five years ago (Perkins 1977).

1. <u>It Should be a Typical Pyrogenic Endotoxin</u>.

Several endotoxins have already been used by different manufacturers and agencies as reference endotoxins for pyrogenic effects in rabbits or for positive controls in LAL testing.

In the rabbit pyrogen test, the febrile response varies considerably between different endotoxins and between the responses demonstrated for the same endotoxin in different strains, colonies, individual rabbits and even on different occasions. Recent work in Italy has also shown a difference between responses at different times of the day. None of this is surprising, but it has made it difficult for regulatory authorities to define a standard pyrogen test - hence the many variations seen in the pyrogen test requirements of the different pharmacopoeias - and impossible to define a standard response to a specific dose of any partic - ular endotoxin.

Nevertheless, such endotoxins as <u>E. coli</u>, <u>Shigella dynsenteriae</u>, <u>Klebsiella pneumoniae</u>, <u>Proteus vulgaris</u> have been used in both rabbit and LAL tests to demonstrate positive responses.

The more well known it is and the more widely it is found as a contaminating endotoxin, the better it should serve as a universal standard, providing the other criteria are met.

2. <u>It Must be Well Defined and Characterised.</u>

Initially, the potency has been defined in terms of the weight of endotoxin required to give a positive response under specified conditions of time and temperature in a LAL gelation test. In view of the variations in lysate sensitivities, the aim must now be to define a Standard International Endotoxin Unit (IEU). There need not be too much concern that such a unit may need redefinition when experience has been gained in its use. The nearest anyone has come to this so far has been the FDA, who defined what weight of EC2, and later EC4, should bring about gelation in 1 hour at 37°C with a stated volume of lysate. The purpose of this has been to quantify the sensitivity of a lysate preparation.

3. <u>It Should Give a Measured Febrile Response in Rabbits.</u>

In the many collaborative trails and individual laboratory tests vast data have been obtained and accumulated which would enable a statistical mean dose response to a particular endotoxin to be calculated. From limited data in my own laboratory, we consider, for example, that 1 ng/kg of <u>E. coli</u> endotoxin as distributed by one LAL supplier will elicit a threshold pyrogenic response in our colony of Dutch rabbits. We define threshold response as approximately 0.55^{o}C per rabbit, which is,of course, the mean response above which a substance fails the European Pharmacopoeia test for freedom from pyrogens in the full 12 rabbit cumulative test.

It should be possible to define the potency of the reference endotoxin with greater precision in terms of the LAL test than in terms of the response in rabbits.

However, a statement of the quantity of endotoxin expected to produce a given temperature rise in rabbits would be a useful bonus, even if covered by a typical pharmacopoeial caveat that 'variations may be expected for local populations of rabbits'!

4. It Should be Capable of Quantitative Cross-relation with other Endotoxins.

Such a characteristic may not prove to be as formidable as it sounds. Only experience with different lysates and different endotoxins will confirm this. It may not be being too optimistic to hope that the relative gelling activity of a particular endotoxin to the chosen reference endotoxin will be reasonably constant when used with different lysates. It may even turn out that relative gelling activity will reflect relative pyrogenicity of different endotoxins. If this proves to be so, it will go a long way to the international acceptance of the reference endotoxin; it would also invalidate any demand that there should be a range of reference endotoxins as implied by the title of this paper.

5. It Should be Stable over a Number of Years under Suitable Conditions of Storage.

Once established and a potency has been assigned, the reference endotoxin must retain its potency over the whole period of its use as the standard preparation. Ideally the greater the stability and the longer the useful life of the preparation, the better. Standardization and establishment of any international reference preparation is time consuming and costly; any deterioration in the reference preparation will cause confusion, concern and the need for its rapid replacement. Under such conditions the old reference would not be suitable for standardization of its successor and we would be back to the position we are in today.

6. It Must be Available in Sufficient Quantity.

For international use, the demands on the reference preparation will be extensive and widespread. In order to last for as many years as possible and in this context, thinking of decades, one should decide upon a realistic amount of a single adequately purified preparation. In this way it should be available on demand in sufficiently large amounts to meet the needs of regulatory agencies, manufacturers and other users. To this end, care must be taken that the individual amount per ampoule or vial is sensible for the intended use.

I have deliberately used the phrase 'adequately purified' because it is debatable just how pure that should be. If it is too well purified it may well be unrepresentative of the species we are expecting to detect; if it is insufficiently purified it could be less homogeneous in constitution and response.

7. It Must be Reliable and Reproducible in Use.

This is perhaps the most fundamental and vital of all the characteristics of an internationally acceptable reference preparation. Especially knowing the variability in the lysates to be used, it is essential that the reference endotoxin will give consistent results - or, to express it in another way, that the potency of the reference material will be truly a constant and not a varible. To achieve this it must to:

(a) Homogeneous across every container.

(b) Readily soluble when reconstituted in the approved way.

(c) Not absorbed on the container or on the vessels used for dilution.

Some of these characteristics may be ideal but not absolutely essential. Without doubt, for a reference endotoxin to be internationally acceptable, authorities such as WHO, FDA and many other national and multinational control organizations will demand that it be well defined and extensively tested; that it is available in sufficient quantity and be of adequate stability to last for at least twenty years; and that above all it be reproducible under defined test conditions so that it will provide a reliable standard for a reliable assay. For the pharmaceutical manufacturer, the reference endotoxin with such excellent credentials will be exprected to contribute significantly to the acceptance of the LAL test, not only for quality assurance in production, but also for quality control of those products for which its use has been validated.

References

1. Cooper JF (1977). International Symposium on pyrogenicity, innocuity and toxicity test systems for biological products, Budapest 1976. Develop biol Standard Vol 34, pp 33-36 (Karger S, Basel 1977). Joint Discussion on pyrogenicity test systems.

2. Perkins FT ibid.

SABEY SPEAKER
TITLE: INTERNATIONAL ACCEPTANCE OF REFERENCE ENDOTOXINS.

HANKE: Novo-Pyrexal is manufactured in a large frozen dried lot. Frozen dried the endotoxin is completely stable. Once reconstituted the standard is stable for two years which is sufficient for a CSE.
SABEY: I recognize this. I want to refute what Dr. Watson said that ideally a standard should be in solution rather than freeze-dried. I just want to emphasize that it is impractical to have an endotoxin standard kept indefinitely in liquid solution.
OUTSCHOORN: It is difficult to come to any conclusion concerning the stability of any endotoxin simply because the stability studies have not, in general, been done properly. A statistically valid stability study must be done by withdrawing samples periodically, placing them under the conditions of the test, and testing all the samples at once. It is very important to test all the samples at one time with the same LAL. If this isn't done, there can be no comparison. The USP cannot set up endotoxin requirements until stability studies are conducted properly.
HOCHSTEIN: The stability of the EC-2 seems quite good. EC-2 was prepared six years ago. In the spring of 1981 we tested the original frozen dried material supplied to us by Dr. Rudbach and found no loss of potency.

Endotoxins and Their Detection With the
Limulus Amebocyte Lysate Test, pages 161–171

COMPARISON OF SEVERAL ENDOTOXIN STANDARDS WITH A KINETIC LAL TEST

Marilyn J. Case and Thomas J. Novitsky

Associates of Cape Cod, Inc.
Woods Hole, Massachusetts 02543

INTRODUCTION

The chemical composition of an endotoxin preparation varies (Galanos et al, 1979b) with the source of the organism, with the nature of the extraction procedure, and with the level of purification. Several endotoxin preparations have been used or have been proposed for use as reference standard endotoxins in the *Limulus* Amebocyte Lysate (LAL) test. These preparations were derived from several different bacterial strains, were extracted and purified by various procedures, and were bottled for storage in several different forms.

The reactivities of these endotoxin preparations were compared in order to evaluate their suitability as reference standard endotoxins (RSE). A modified kinetic turbidimetric assay (Jorgensen and Alexander, 1981; Levin and Bang, 1968; Levin et al, 1970) was chosen rather than the endpoint assay because we have found the kinetic test to be more sensitive to minor differences in endotoxin reactivity.

MATERIALS AND METHODS

Several of the endotoxin preparations were derived from a phenol-water (Westphal et al, 1952) extraction of *Escherichia coli* (Braude) strain 0113 (Rudbach et al, 1976). These included U. S. reference standards EC-2, EC-4 and the pending USP reference standard (EC-5); and two lots of 0113 endotoxin bottled by Associates of Cape Cod, Inc. (ACC) lots 7 and 13. One endotoxin preparation was a trichloroacetic acid (Boivin et al, 1933) extraction of E. coli strain 055:B5, control #657846, from Difco Laboratories. The last preparation, Novo-Pyrexal (Hermal-Chemie Kurt Herrmann), was a phenol-water extraction of Salmonella abortus-equi purified to the uniform sodium salt (Galanos et al, 1979a). Lyophilized preparations of E. coli 0113 were reconstituted according to manufacturers' instructions with pyrogen-free sterile water for irrigation (Cutter Laboratories) to give stock solutions of 100 ng/ml; or for EC-5, 200 ng/ml. The E. coli 055:B5 preparation was a lyophilized bulk lot that was weighed out in the laboratory and reconstituted to give a stock solution of 1000 ug/ml. Novo-Pyrexal was a liquid preparation in ampules. All stock preparations were transferred to pyrogen-free test tubes, sealed with parafilm (American Can Co.), and stored at 5 C.

Sterile water for irrigation (SWI) was tested and used for dilutions (SWID) only if it contained less endotoxin than was detectable within the limits of the kinetic assay.

LAL was obtained from Associates of Cape Cod, Inc., Pyrotell #52-76-245 (LAL lot 245) and #96-02-264 (LAL lot 264). These were reconstituted with SWID.

For the kinetic assay, four parts sample were mixed with one part LAL in disposable cuvettes, Ampvettes (Abbott Laboratories). The Ampvettes were placed into cartridges which were inserted into the incubation chamber of an MS-2 Microbiology System research analysis module (Abbott Laboratories). Optical density, OD(670), measurements were made and recorded automatically every minute throughout the LAL reaction period. At the end of an incubation period, the MS-2 graphics program was utilized to display the data as OD vs time.

Glassware was depyrogenated by dry heat incubation at 180 C for 4 hr or 250 C for 1 hr. Sterile disposable borosilicate glass pipettes (Corning) and plastic pipette tips for automatic pipettors (Brinkmann Instruments, Inc.) were found to be free of detectable endotoxin as determined by testing rinse water.

RESULTS AND DISCUSSION

When endotoxin and LAL were incubated together, a flocculant suspension was formed in the reaction vessel. The rate of formation of this material was measured in the kinetic turbidimetric test. Figure 1 shows a typical set of curves generated by concentrations of endotoxin ranging from 100 pg/ml to 0.2 pg/ml in two-fold serial dilution.

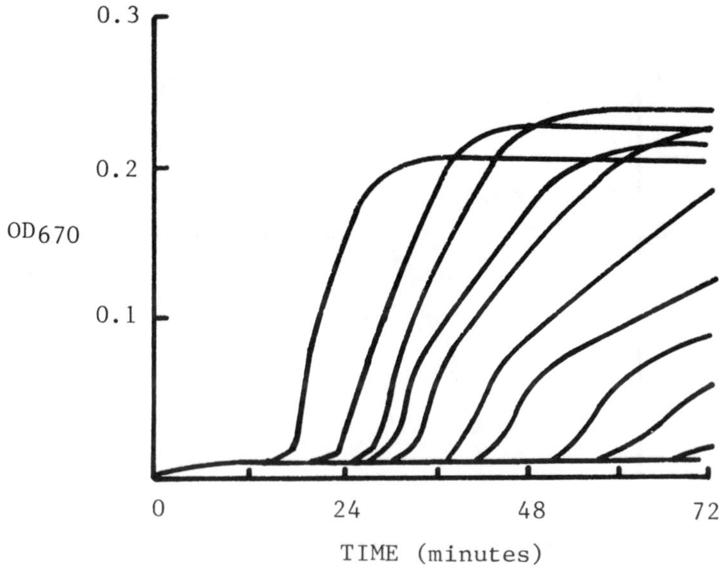

Fig. 1. Kinetic response of a series of two-fold dilutions of endotoxin in water. The curves are displayed from left to right, from the greatest (100 pg/ml) concentration of endotoxin to the least (0.2 pg/ml).

The kinetics of increase in turbidity were complex, but several characteristics of the curves were found to be functions of endotoxin concentration: the lag time prior to onset of an increase in turbidity increased with decreasing concentration of endotoxin, the overall rate of the reaction decreased with decreasing endotoxin concentration, and the maximum OD (not shown for lower endotoxin dilutions) seemed to increase slightly with decreasing endotoxin concentration. The negative control, or blank, was LAL mixed with SWID. The negative control showed no response over the course of the incubation period.

ACC lots 7 and 13 were prepared from the same bulk preparation of *E. coli* O113 endotoxin. These lots had similar titers by the gel-clot test. Figure 2 shows overlays of the curves generated by 100 pg/ml of lots 7 and 13 when tested in the kinetic assay. The kinetics demonstrated by each lot of endotoxin were closely similar as was seen by the identical lag times and the shape of the curves.

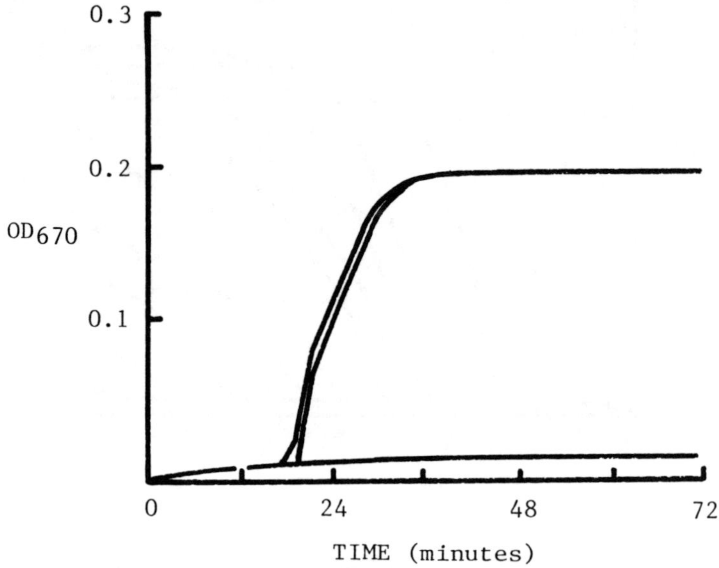

Fig. 2. Overlay of curves generated by 100 pg/ml of ACC lots 7 and 13 (LAL lot 264).

U. S. reference lots EC-2 and EC-4 were from the same bulk preparation of *E. coli* O113 endotoxin. EC-2 differed from EC-4 in that EC-2 was lyophilized with human serum albumin as a filler. EC-4 was intended to be as potent as EC-2, but for unknown reasons was less potent in the gel-clot test. Figure 3 shows overlays of the curves generated by 100 pg/ml of EC-2 and EC-4 (LAL lot 264). EC-4 was clearly less reactive than EC-2 as seen by the longer lag time and decreased rates of increase in turbidity. When

EC-4 was compared to a set of curves generated by a dilution series of EC-2 in the kinetic assay, EC-4 was determined to be 4 to 8-fold less reactive than EC-2.

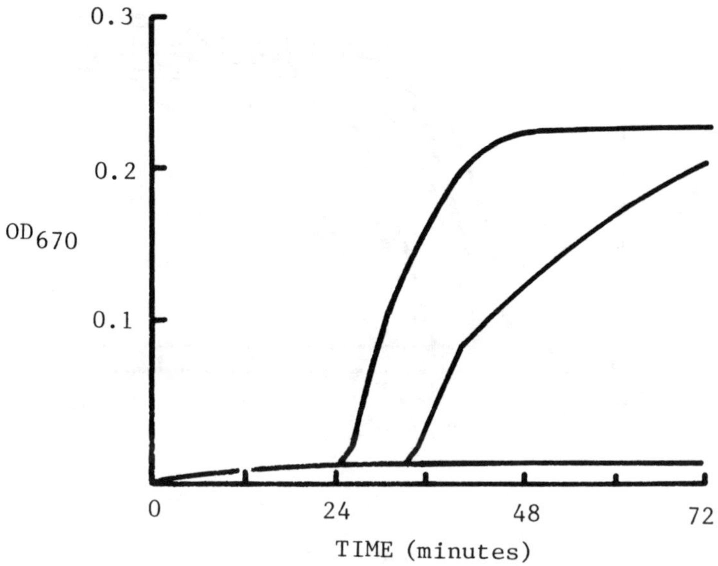

Fig. 3. Overlay of curves generated by 100 pg/ml of EC-2 on the left and EC-4 on the right (LAL lot 264).

Comparison of these first four endotoxin preparations indicated that the kinetic assay was a sensitive measure of the relative reactivities of the endotoxin preparations. Figure 4 shows the results of all the endotoxin preparations tested by the kinetic assay. Solid lines indicate those preparations tested with LAL lot 264 and the dashed lines indicate three, Novo-Pyrexal, EC-2, and lot 7, that were also tested with LAL lot 245. With LAL lot 264, these three endotoxin preparations practically overlaid each other in the cluster of curves having lag times of 18 min. The comparison of the three endotoxin preparations tested with the two LAL lots, indicated that lag time and, to a lesser extent, rate were influenced by the lot of LAL as well as by the concentration of endotoxin. There was also the suggestion in figure 4 that the type of endotoxin preparation also influenced lag time; compare O55:B5 with

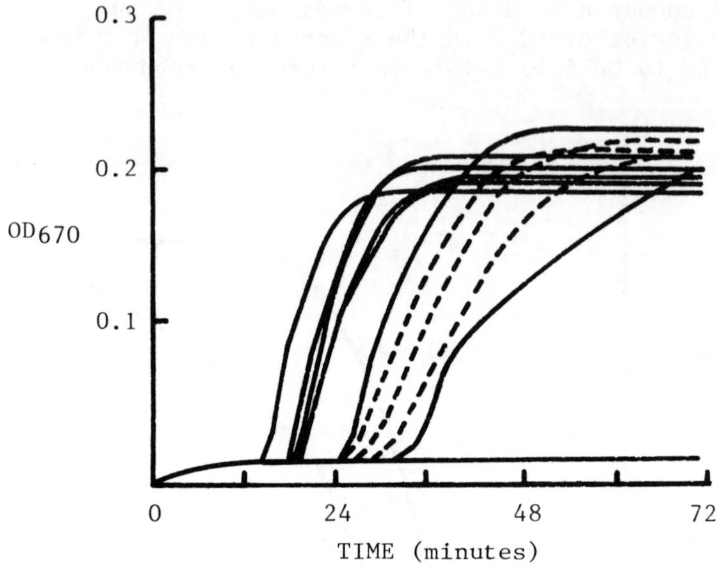

Fig. 4. Overlay of curves generated by 100 pg/ml of all the endotoxin preparations tested. Solid lines from left to right are 055:B5, EC-5, Novo-Pyrexal, ACC lot 7, ACC lot 13, EC-2, and EC-4 (LAL lot 264). Dashed lines from left to right are Novo-Pyrexal, ACC lot 7, and EC-2 (LAL lot 245).

lot 13 or EC-2. Just how important lag time is in judging endotoxin potency is unclear. It would seem that this may be a reflection of the relative availability of the endotoxin in the various preparations to the LAL substrate enzyme system. The variation in lag times between the endotoxin preparations may also lead to detected differences in potency when the same preparations are assayed by an end point LAL method.

Assuming that lag time was not a necessary parameter in measuring reactivity, the curves for the various endotoxin preparations were overlaid (figure 5) without regard to lag time in order to compare the shapes of the curves. All the preparations except EC-4 were seen to respond in similar manner.

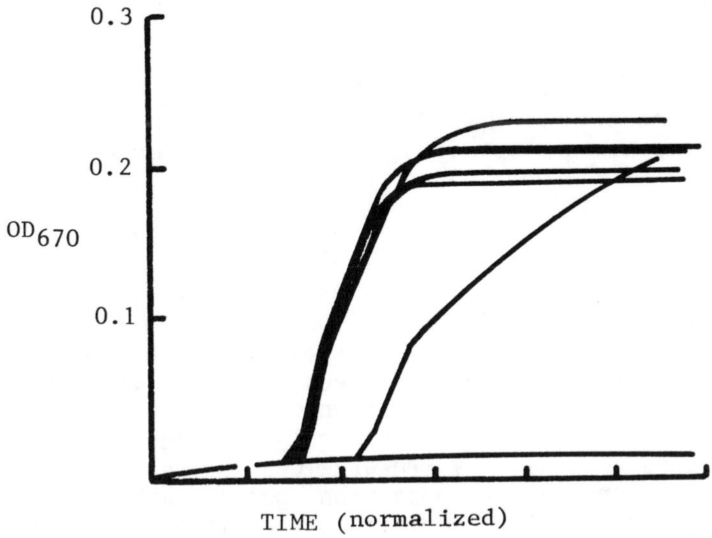

Fig. 5. Overlay of solid lines from fig. 4 (LAL lot 254). All curves except that for EC-4 have been superimposed without regard to lag time.

When the kinetics of the turbidimetric reaction of the various endotoxin preparations with LAL were compared, and the differences in lag times normalized, all preparations except EC-4 were found to have similar potencies. The same conclusion was reached for the three preparations tested with a second lot of LAL, although some variability was detected between lots of LAL against the same endotoxin preparation.

It was suggested (Dr. H. D. Hochstein, personal communication) that the difference between EC-2 and EC-4 (fig. 2) was the result of a filler (human serum albumin) in EC-2 that was not in EC-4. Yet, when we compared ACC lots 7 and 13 (without a filler) to EC-2 and EC-4, these preparations more closely resembled EC-2. EC-5, containing lactose and polyethylene glycol fillers, was different again, at least with respect to lag time, from either EC-2 or EC-4, so the presence or absence of a filler was not sufficient to predict the relative reactivities between these endotoxin preparations. In fact, there was no obvious basis of comparison between these preparations in spite of their bulk similarity.

We would like to stress that all these preparations with the exception of Novo-Pyrexal may contain undefined non-LPS components (i.e. ions, proteins, polyamines). The composition of the non-LPS components is influenced by the growth conditions of the organism, the species, the extraction procedure, and the level of purification of the endotoxin (Galanos et al, 1979b). The non-LPS components influence the aggregation state of the LPS and possibly other characteristics of the LPS that are known to alter its activity in bioassays (Galanos et al, 1979b) and may alter its reactivity with LAL. Because of the variability that is possible between similar endotoxin preparations, endotoxin preparations that are highly purified and chemically defined are the only preparations that can truly serve as a reference for comparison against each other and, ultimately, against less purified preparations especially when the detection and quantitation of the endotoxin depends on activities in biological assays.

REFERENCES

Boivin A, Mesrobeanu I, Mesrobeanu L (1933). Technique pour la preparation des polysaccharides microbiens specifiques. Seances Soc Biol Filiales 113:490.

Galanos C, Luderitz O, Westphal O (1979a). Preparation and properties of a standardized lipopolysaccharide from Salmonella abortus-equi (Novo-Pyrexal). Zbl Bakt Hyg, I Abt Orig A 243:226.

Galanos C, Freudenberg MA, Luderitz O, Rietschel, ET, Westphal O (1979b). Chemical, physicochemical and biological properties of bacterial lipopolysaccharides. In Cohen E (ed): "Biomedical Applications of the Horseshoe Crab (Limulidae)," New York: Alan R. Liss, p 321.

Jorgensen JH, Alexander GA (1981). Automation of the Limulus amoebocyte lysate test by using the Abbott MS-2 microbiology system. Appl Environ Microbiol 41:1316.

Levin J, Bang FB (1968). Clottable protein in Limulus: its localization and kinetics of its coagulation by endotoxin. Thromb Diath Haemorrh 19:186.

Levin J, Tomasulo PA, Oser RS (1970). Detection of endotoxin in human blood and demonstration of an inhibitor. J Lab Clin Med 75:903.

Rudbach JA, Akiya FI, Elin RJ, Hochstein HD, Luoma MK, Milner ECB, Milner KC, Thomas KR (1976). Preparation and properties of a national reference endotoxin. J Clin Microbiol 3:21.

United States Pharmacopeia (1980). Bacterial endotoxins test. In "United States Pharmacopeia," 20th revision, Mack Publishing Co., Easton, PA, p 888.

Westphal O, Luderitz O, Bister F (1952). Uber die extraktion von bacterien mit phenol/wasser. Naturforsch 7b:148.

CASE SPEAKER
TITLE: COMPARISON OF SEVERAL ENDOTOXIN STANDARDS WITH KINETIC LAL TEST.

LEVIN: I hope everyone noticed that the final optical densities were approximately the same for each concentration tested, which is what I would have expected. I don't understand the differences between your results and Dr. Firca's results since you both use the same LAL and type of equipment.
CASE: I am not sure what you were referring to in Dr. Firca's graphs, but he did have some erroneous values (glitches) that are not handled in the graphics routine and which artificially raise or lower the individual plots with respect to the axes. There is also the observation that the maximum O.D. seems to increase as the concentration of endotoxin tested decreases. The absolute values for the maximum O.D. vary somewhat. As you go above 100 pg/ml, there is very little variation in maximum O.D.
LEVIN: Are those previous data not reproducible?
CASE: Yes, they are reproducible with respect to onset and rates. The maxium O.D. is usually, but not always reproducible.
WATSON: In the higher concentrations aren't larger aggregates formed resulting in a lower O.D. than found when slightly lesser concentrations are employed?
CASE: Yes that is true.
REICHLER: Have you ever taken material out of the ampvettes after the experiment, mixed and read them in a spectrophotometer; because we are in some degree reading a slighly different aggregate than is published in the classic turbidimeteric LAL method. I think that is what Dr. Watson is referring to.
CASE: No, we haven't tried this.
NEELY: How many replicates were done to determine each curve?
CASE: Different numbers. Most of our replicates were done by filling 11 ampvettes with the same concentration of endotoxin and running them simultaneoulsy. The plots overlaid very well.

NEELY: So the assay variation is small; it seems that the most striking thing that you showed was not the activation time but the slope and the maximum O.D.. I was wondering if the intra-assay variability may explain the small differences in the activation time.
CASE: I can't answer that at the present time.
SWATON: Do you think part of this is just due to the time it takes to get the sample in the ampvette?
CASE: Yes, that may be part of it.
MILLS: Are the kinetics influenced by the formation of the clot?.
CASE: No, at the ratio of LAL to sample that is used, a clot will not form unless the endotoxin concentration is extremely high.
FINK: What happens at very high concentrations of endotoxin?
CASE: The reaction rate is very fast and the time of onset becomes negligible.
BANG: Has anyone analyzed the role of NEM on the enzymatic activity of LAL?
WATSON: Not everyone is using NEM.
LEVIN: NEM does not inhibit the LAL enzyme(s).
BANG: I would still like to see some quantitative data on the effect of NEM on the enzyme.

Endotoxins and Their Detection With the
Limulus Amebocyte Lysate Test, pages 173–182

PREPARATION AND APPLICATION OF A NEW ENDOTOXIN DETERMINATION KIT, PYRODICK®, USING A CHROMOGENIC SUBSTRATE

Yoshimasa Fujita and Chizuko Nakahara

Seikagaku Kogyo Co. Ltd.
1253, Tateno-3-chome, Higashiyamato
Tokyo, Japan

Since Levin and Bang (Levin and Bang, 1964) discovered that gram-negative bacterial endotoxin could rapidly induce gelation of horseshoe crab amoebocyte lysate in 1964, the amoebocyte lysate has been used widely to detect endotoxin in various fields. Because of the extremely high specificity and sensitivity of the amoebocyte lysate to endotoxin, elucidation of gelation mechanism and its application to the micro-determination of endotoxin became an interesting subject for many investigators.

Among these studies, several groups of investigators (Solum, 1973; Nakamura et al., 1976a, 1976b; Tai et al., 1977; Mosesson et al., 1979) separated and highly purified a clotting protein, named coagulogen, from the amoebocyte lysate, and showed that the gelation of coagulogen was induced through limited cleavage of coagulogen by an endotoxin activated clotting enzyme.

In 1976, Iwanaga et al. (Nakamura et al., 1976a, 1976c)

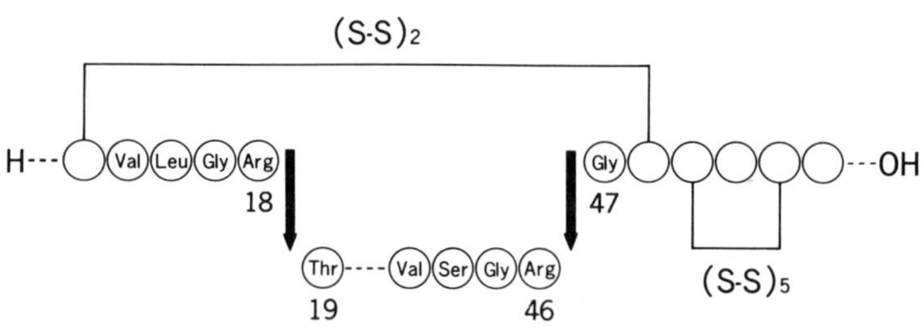

Fig. 1. Peptide linkages of coagulogen cleaved by Tachypleus clotting enzyme. (Nakamura et al., 1976c)

found that the gelation of the coagulogen isolated from the Japanese horseshoe crab, Tachypleus tridentatus, involved the cleavage of peptide bond associated with limited proteolysis by Tachypleus clotting enzyme at two peptide linkages with concomitant release of peptide C. One of these linkages was Arg-Gly at the 46th and 47th position, and the other was Arg-Thr at the 18th and 19th position on the covalent structure of coagulogen (Fig.1). This indicated the clotting enzyme must have a proteolytic or esterolytic activity like α-thrombin.

Following these experimental results, they further investigated the substrate specificity of the clotting enzyme using various chromogenic peptides (Table 1). They found that certain peptide amides having a sequence, Gly-Arg at the carboxyl terminal were suitable substrates for the clotting enzyme. According to these facts, they developed a new sensitive colorimetric assay method for bacterial endotoxin using horseshoe crab amoebocyte lysate and chromogenic substrates (Harada et al., 1978).

Table 1. The Km and Vmax values of horseshoe crab clotting enzyme toward peptide-pNAs and peptide-MCAs. (Narada et al., 1978)

Substrate	Peptide-pNAs			Peptide-MCAs		
	Km (M)	Vmax μmoles/min/ A 280 = 1.0	Vmax/Km	Km (M)	Vmax μmoles/min/ A 280 = 1.0	Vmax/Km
Tos-Ile-Glu-Gly-Arg-	2.1×10^{-4}	2.7	13	—*	—*	—*
Boc-Ile-Glu-Gly-Arg-	—*	—*	—*	1.3×10^{-4}	4.0	31
Boc-Val-Leu-Gly-Arg-	5.4×10^{-5}	3.2	59	6.0×10^{-5}	4.3	72
Boc-Leu-Gly-Arg-	6.5×10^{-5}	4.2	65	2.7×10^{-5}	4.8	178
Z-Leu-Gly-Arg-	—*	—*	—*	5.8×10^{-5}	1.4	24
Boc-Val-Ser-Gly-Arg-	1.4×10^{-4}	3.3	24	1.1×10^{-4}	3.1	28
Boc-Ser-Gly-Arg-	—*	—*	—*	2.4×10^{-5}	4.3	179
Boc-Ser(OBz)-Gly-Arg-	4.9×10^{-5}	3.3	67	5.1×10^{-5}	6.0	118
BZ-Val-Gly-Arg-	4.6×10^{-5}	3.3	72	—*	—*	—*

* not determined

pNA : p-Nitroanilide
MCA : 4-Methyl-coumarin amide

Based on their studies, we prepared a new simple and rapid colorimetric endotoxin determination reagent kit, named PYRODICK. The kit is composed of the PYRODICK reagent, a buffer solution and a reference endotoxin. Since the PYRODICK reagent contains Tachypleus amoebocyte lysate and a

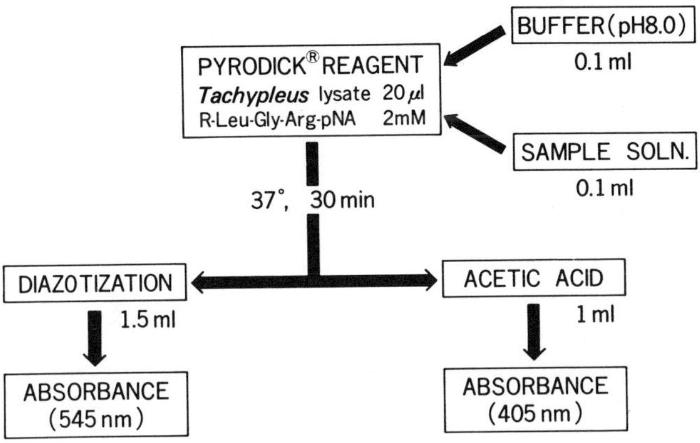

Fig. 2. Endotoxin assay procedure with PYRODICK kit.

chromogenic substrate in the same tube, it makes the assay procedure simple and convenient.

Figure 2 shows standard procedure for the endotoxin determination using PYRODICK kit. PYRODICK reagent is composed of 20 μl of Tachypleus tridentatus lysate and 2 mM of chromogenic substrate which are contained in one tube and lyophilized. The reagent is stable when stored below 5 °C in the dark.

For the assay of endotoxin, dissolve the reagent with 0.1 ml of a buffer solution attached with reagent, and add 0.1 ml of a sample solution. These procedures must be done in an ice bath. Then the mixture is incubated at 37 °C for 30 min. To the colored mixture, 1 ml of 0.6 N acetic acid solution is added to stop the enzyme reaction, and the optical density is measured at 405 nm. When higher sensitivity is required, the incubation mixture is subjected to diazotization and the measurement of red color developed is done at 545 nm.

Unless otherwise stated, the quantity of endotoxin is expressed by the amount of E. coli 0111:B4 endotoxin distributed by Difco throughout the present study.

Figure 3 shows calibration curves at different incubation time. Basically, PYRODICK kit is designed to be capable of determining endotoxin most suitably at around the concentration of 1 ng/ml of sample solution at 30 min incubation, it is also possible to enhance the sensitivity by varying the incubation time as needed. In this way, an

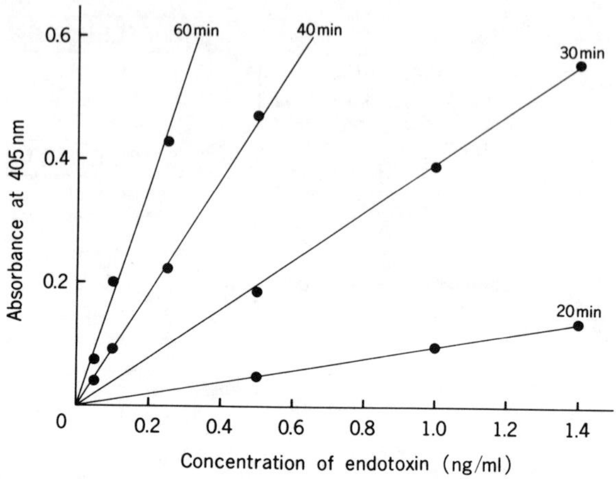

Fig. 3. Calibration curves at different incubation time. The reaction mixture containing the PYRODICK reagent dissolved in 0.1 ml of attached buffer solution pH 8.0 and 0.1 ml of an endotoxin solution (0.02 - 1.4 ng/ml), was incubated at 37 °C.

amount of endotoxin from as little as 20 pg/ml to 30 ng/ml is quantitatively determined by selecting the incubation time between 20 min and 60 min without dilution or concentration of sample solutions. Further elongation of incubation time to 90 min makes it possible to determine as little as 5 pg/ml of endotoxin.

With the latest progress on the fractionation and and purification of horseshoe crab amoebocyte lysate (Morita et al., 1981a, 1981b), it will soon be possible to prepare an improved reagent kit having higher sensitivity and specificity by using more purified lysate.

It is well known that the sensitivity of various endotoxin preparations to the *Limulus* test are considerably different with the difference of bacterial species as well as their grades of purification, solubility and so on. Figure 4 shows a comparison of calibration curves for 4 different endotoxin preparations. Preparation D is used as a reference endotoxin in our present study. Preparation A possesses more than 8 times higher reactivity to PYRODICK reagent as compared to preparation D. Preparation A is a kind gift from Dr. Kanoh of National Institute of Hygienic Science, Osaka, who extracted and purified extensively the endotoxin preparation from *E. coli* UKT-B. Chemical analysis

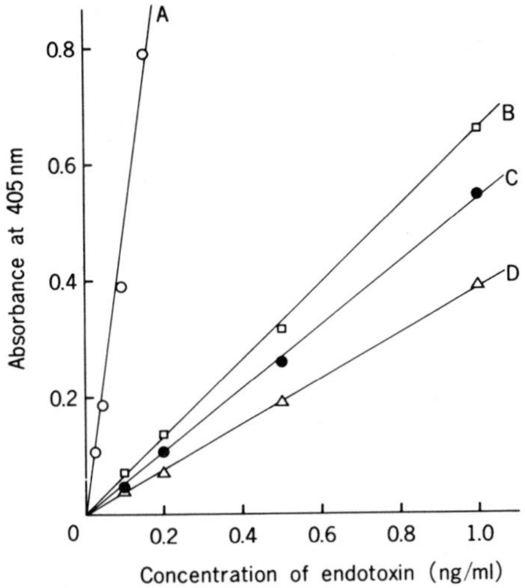

Fig. 4. Calibration curves with various endotoxins. The reaction mixture containing the PYRODICK reagent dissolved in 0.1 ml of attached buffer solution pH 8.0 and 0.1 ml of each endotoxin solution (0.025 - 1.0 ng/ml), was incubated 30 min at 37 °C. A : *E. coli* UKT-B (Kanoh S), B : *Shigella flexneri* (Difco), C : *Salmonella minnesota* (Difco), D : *E. coli* O111:B4 (Difco).

of the purified preparation showed that it was more homogeneous than crude products, and the content of nucleic acid and protein decreased considerably. The preparation possessed high pyrogenic potency and showed elevated reactivity to *Limulus* test (Kanoh et al., 1981).

Since the mechanism of endotoxin determination with PYRODICK reagent is based on an enzymatic process, it is easily presumed that the reaction will be influenced by substances coexisting in the assay system. Therefore, similar to the gelation method (Kobayashi et al., 1975a, 1975b), it should be necessary to ascertain the degree of influences with coexisting substances prior to the determination of endotoxin with PYRODICK reagent.

Table 2 shows the comparison of three different methods applied on the endotoxin determination in the process check of sodium chondroitin sulfate preparation. Samples number 1 to 4 passed rabbit pyrogen test, and their 2 % solution gave negative reaction to the LAL test. The endotoxin content of

Table 2. Endotoxin determination in the process check of sodium chondroitin sulfate preparation.

Sample	Pyrogen test*	LAL-test** (37°, 60 min)		PYRODICK® (37°, 20 min)		
		Concentration of sample (%)	Result	Concentration of sample (%)	Endotoxin (ng/ml)	Endotoxin (ng/g)
1	passed	2	−	2	0.086	4
2	passed	2	−	2	0.089	4
3	passed	2	−	2	0.080	4
4	passed	2	±	2	0.105	5
5	passed	2	+	2	0.203	10
6	failed	1	+	1	0.272	27
7	failed	0.5	+	0.5	0.451	90
8	failed	0.5	±	0.5	0.559	112

* Pyrogen test : J. P. X

** LAL-test : PREGEL®

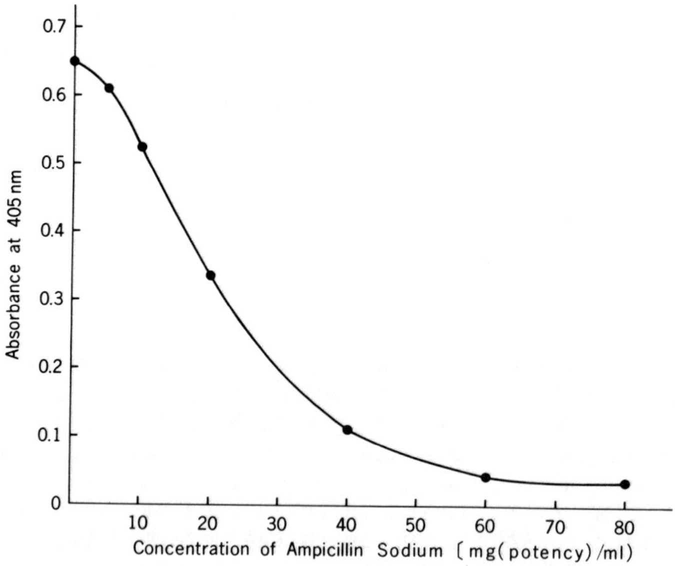

Fig. 5. Influence of the concentration of ampicillin sodium on the determination of endotoxin. The reaction mixture containing the PYRODICK reagent dissolved in 0.1 ml of attached buffer solution pH 8.0 and 0.1 ml of a sample solution (endotoxin 0.4 ng/ml, ampicillin 10 - 80 mg/ml), was incubated at 37 °C for 50 min. (Takahashi et al., 1981)

these samples determined with PYRODICK reagent were below 5 ng/g sample. Sample number 5 which passed pyrogen test reacted positively with LAL test and the endotoxin content determined with PYRODICK reagent also increased. Samples 6 to 8 which failed to pass pyrogen test gave positive reaction to the LAL test as well as high endotoxin contents determined with PYRODICK reagent. The correlation between these three methods were satisfactory, and the colorimetric method using PYRODICK kit seemed to be superior to the other two in that it offered not only a simple, rapid and sensitvie means for the endotoxin determination but gave exact numerical data.

Takahashi et al. examined the influences of ampicillin sodium upon the determination of endotoxin with PYRODICK reagent (Takahashi et al., 1981). Figure 5 shows an inhibition

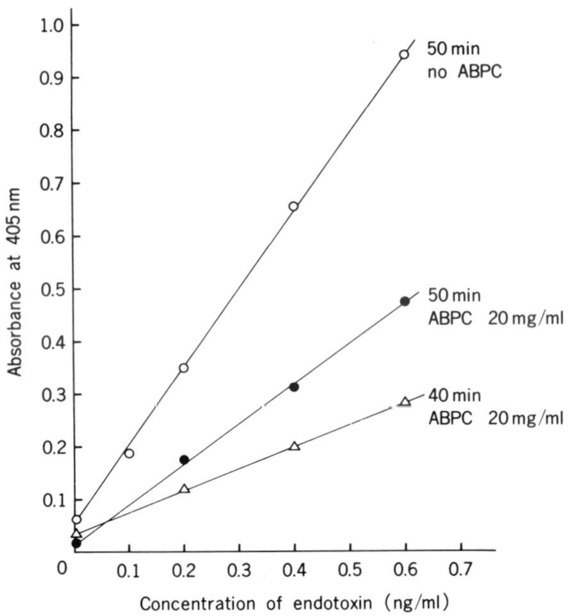

Fig. 6. Influence of ampicillin sodium and incubation time on the standard curve. To the PYRODICK reagent dissolved in 0.1 ml of attached buffer solution pH 8.0 was added 0.1 ml of an endotoxin solution (0.1 - 0.6 ng/ml) containing 0 (○) or 20 mg/ml of ampicillin (● and △). The reaction mixtures were incubated at 37 °C for 50 min (○ and ●) and 40 min (△). ABPC : Ampicillin sodium. (Takahashi et al., 1981)

curve of ampicillin on the endotoxin determination at different concentration. The assay system containing an endotoxin solution at a concentration of 0.4 ng/ml gave an absorbance value of 0.65 at 405 nm when there was no coexistence of ampicillin. When ampicillin was added to the system at a concentration of 20 mg/ml, the absorbance value decreased to 0.35 and the value decreased in proportion as the antibiotics concentration increased. However, at a fixed concentration of ampicillin, the standard curve for the determination of endotoxin maintained excellent linearity (Fig. 6). Thus the PYRODICK reagent enabled them to determine endotoxin in the ampicillin even though there was up to 50 % inhibition of the

Table 3. Influence of the concentration of ampicillin sodium on the LAL test. (Takahashi et al., 1981)

Concentration of endotoxin (ng/ml) (*E. coli* 0111:B4)	Concentration of Ampicillin Sodium [mg(potency)/ml]							
	0	1	5	10	20	40	60	100
1	+	+	+	+	+~±	±	−	−
10	++	++	++	++	++	+	−	−

LAL-test : PREGEL®, 37°, 60 min

Table 4. Pyrogen concentration in intravenous solutions. (Shimohira et al., 1981)

Sample number	Formulation	Manufacturer	PYRODICK® (ng/ml)	LAL-test*	Pyrogen test** (rabbit)
1	Xylitol (5%)	A	0.0	−	−
2	Xylitol (5%)	B	0.1	−	−
3	Xylitol (5%)	C	0.0	−	−
4	Xylitol (20%)	D	0.0	−	−
5	Fructose (5%)	B	0.2	−	−
6	Maltose (5%)	A	0.0	−	−
7	Maltose (5%)	E	0.0	−	−
8	Glycerin (10%) Fructose (5%) NaCl (0.9%)	F	0.0	−	−

* LAL-test : PREGEL®

** Pyrogen test : J. P. X

reaction by the antibiotics. Table 3 shows the results of endotoxin detection by ordinal gelation method in the co-existence of ampicillin. The gelation method showed merely an ampicillin concentration for almost complete inhibition, but did not give any information on the degree of inhibition.

Table 4 shows the comparison of three different endotoxin detection methods applied to various intravenous solutions (Shimohira et al., 1981). There were two preparations which gave positive reaction with PYRODICK reagent, although all samples including these two passed rabbit pyrogen test and gave negative reaction with the LAL test.

In conclusion, the PYRODICK kit should become a useful means for the determination of endotoxin, since it gives not only more objective and quantitative results as compared to the ordinal gelation method, but also enables a numerical quantitation of endotoxin accurately. It is expected that the new reagent kit can be applied in many fields such as the in-process checks of parenteral drugs, examination of water pollution, improvement of food hygiene and clinical investigations.

Acknowledgement

The authors wish to express their thanks to the investigators who gave us generous permissions to quote their experimental results in the present paper.

REFERENCES

Harada T, Morita T, Iwanaga S (1978). A new assay method for bacterial endotoxins using horseshoe crab clotting enzyme. J Med Enzymol 3:43.

Kanoh S, Ogawa Y, Komuro T, Kawasaki K, Niji K (1981). Studies on the extraction process of E. coli LPS and concomitant changes in physical properties, Limulus activity and pyrogenicity. Japan J Med Sci Biol (in press).

Kobayashi M, Morimoto T, Yamamoto M (1975a). Studies on the gelation reaction of Limulus lysate (Pre-gel). III. Influences of various substances on the gelation reaction of Limulus lysate. Yakugaku Zasshi 95:720.

Kobayashi M, Yamamoto M (1975b). Studies on the gelation reaction of Limulus lysate (Pre-gel). IV. Effect of alkaline earth metals and chelating agents on the gelation reaction of Limulus lysate. Yakugaku Zasshi 95:959.

Levin J, Bang FB (1964). The role of endotoxin in the extra-

cellular coagulation of *Limulus* blood. Bull Johns Hopkins Hosp 115:265.
Morita T, Tanaka S, Nakamura T, Iwanaga S (1981a). A new (1→3)-β-D-glucan-mediated coagulation pathway found in *Limulus* amoebocytes. FEBS Letter 129:318.
Morita T, Tanaka S, Nakamura T, Ohki M, Iwanaga S (1981b). The reaction mechanism of "*Limulus* test". Thrombosis Haemostasis 46:44.
Mosesson MW, Wolfenstein-Todel, Levin J, Bertrand O (1979). Characterization of amebocyte coagulogen from the horseshoe crab (*Limulus polyphemus*). Thrombosis Res 14:765.
Nakamura S, Takagi T, Iwanaga S, Niwa M, Takahashi K (1976a). A clottable protein (coagulogen) of horseshoe crab hemocytes: Its structural change of its polypeptide chain during gel formation. J Biochem 80:649.
Nakamura S, Iwanaga S, Harada T, Niwa M (1976b). A clottable protein (coagulogen) from amoebocyte lysate of Japanese horseshoe crab (*Tachypleus tridentatus*): Its isolation and biochemical properties. J Biochem 80:1011.
Nakamura S, Takagi T, Iwanaga S, Niwa M, Takahashi K (1976c). Amino acid sequence studies on the fragments produced from horseshoe crab coagulogen during gel formation: Homologies with primate fibrinopeptide B. Biochem Biophys Res Commun 72:902.
Shimohira A, Usami H (1981). Personal communications.
Solum NO (1973). The coagulogen of *Limulus polyphemus* hemocytes. A comparison of the clotted and non-clotted forms of the molecule. Thrombosis Res 2:55.
Tai JY, Seid RC JR, Huhn RD, Liu TY (1977). Studies on *Limulus* amoebocyte lysate II. Purification of the coagulogen and the mechanism of clotting. J Biol Chem 252:4773.
Takahashi S, Yano S, Nagaoka Y, Kawamura K, Minami S (1981). High sensitive pyrogen test for antibiotics I. Detection of a trace amount of endotoxin in *Ampicillin Sodium* for injection. J Pharmaceut Sci (submitted, 8.18.1981).

Endotoxins and Their Detection With the Limulus Amebocyte Lysate Test, pages 183–194

AUTOMATED METHODOLOGY FOR THE LIMULUS AMEBOCYTE LYSATE (LAL) ASSAY USING THE MULTISKAN MICROPLATE READER

Bruce R. Albaugh and Charles B. Chandler
Flow Laboratories, Inc.
McLean, Virginia 22102
Associates of Cape Cod, Inc.
Woods Hole, Massachusetts 02543

INTRODUCTION

The endotoxin activated Limulus reaction has proven to be an important quality control tool for in-process pyrogen testing of: raw materials, components in various stages of manufacture and finished products. (Federal Register, Volume 45, No. 13, 1980). The most commonly utilized method employs the unique ability of endotoxin to induce gelation of Limulus Amebocyte Lysate and form a firm clot when incubated in the presence of minute quantities of bacterial endotoxins. This method is accurate and highly reproducible (Musiani, et al., 1976), however it is only semi-quantitative, and is not easily adapted to testing large numbers of samples. Levin and Bang (1968) observed that the reaction between purified lysate and endotoxin can be easily measured by recording alterations in optical density. This ability to quantitatively measure turbidity and interpolate the endotoxin concentration from a standard curve (optical density vs. known endotoxin concentrations) form the basis of the LAL-microplate assay. Other turbidimetric methods have been utilized (Bondar, et al., 1979). The purpose of this paper is to present validation for an alternate photometric assay which offers the advantages of reduced reagent volumes, rapidity, ease of use and automation potential inherent in the micro procedure.

METHODS AND MATERIALS

All glassware used in the following procedures were depyrogenated by exposure to heat at 180°C for a minimum of 4 hours; all other materials were pre-tested and determined to be endotoxin-free. Water used to reconstitute lysate, endotoxin and used as a diluent was purchased from Abbott Laboratories, and labeled as sterile and non-pyrogenic. All lysate and endotoxin (E.coli) standards used in this work were supplied by Associates of Cape Cod Inc.

The assay is performed as follows:

1) Test samples or endotoxin standards (0.1 ml) are placed in appropriate wells of a microplate (96 round bottom wells with a capacity of 0.25 ml) (Flow Laboratories Cat. No. 76-012-05).

2) Reconstituted lysate (0.1 ml) is added to each well containing samples or standards and the plate sealed with tape (Flow Laboratories Cat. No. 76-401-05). This plate sealer is recommended to retard evaporation during incubation.

3) The plate is placed on a shaking apparatus for approximately 60 seconds to ensure adequate mixing.

4) The plate is incubated in a forced air incubator at 37°C ± 1° for 60 minutes and then read at 380 nm on a high speed photometer (Flow Laboratories Multiskan™), which was developed for through the plate reading of microplates.

To reduce errors in the assay created by time lag between addition of lysate to various samples, an eight channel micro pipetter (Flow Laboratories Cat. No. 77-844-00) was utilized. This instrument uses disposable tips (pre-tested prior to use and found to be endotoxin-free) and is capable of accurately dispensing 0.1 ml aliquots of reagent into 96 wells in less than a minute. Delays in addition of reagents may produce significant errors in this type of kinetic reaction.

The Multiskan has computer interface capabilities and would adapt easily to large scale LAL pyrogen testing. It would be conceivable to run 80 samples through the photometer and obtain a linear regression analysis of the standard curve with all standard values registered and computed in several minutes.

RESULTS AND DISCUSSION

Although the performance of the Multiskan has been tested extensively by studying the extinction values obtained by experimental variation of the optical density and reaction volume of a stable test dye (Ruitenberg, et al., 1980), little work has been documented using optical density measurements of a stable turbid protein solution. The sensitivity and reproducibility of the instrument was measured by reading a plate containing 4 solutions of Bovine Serum Albumin - Fraction V. The turbidity readings chosen represent values commonly encountered in running the assay. Each well was read 4 times at 1 minute intervals and the mean values, standard deviation and the coefficient of variation are presented in Table 1.

Table 1

Reproducibility of Turbidity Readings [1,2]

$\bar{X}$	S.D.	% C.V.
.310	.004	0.14
.185	.0005	0.30
.144	.001	0.91
.042	.0005	1.26

1. $\bar{X}$ is an mean of 4 replicate readings
2. Bovine Serum Albumin - Fraction V Solution

Analysis revealed no systematic differences among the optical densities read, and no internal drift was noted. Channel-to-channel variation was tested by exposing all 8 photodiodes to a homogeneous neutral density filter. The average coefficient of variation never exceeded 1%. Test data indicates that the total contribution of photometric inaccuracy was negligible at the ranges measured and that the instrument is capable of giving highly reproducible results.

The incubation time for the reaction is critical as is demonstrated in Figure 1. The turbidity will continue to increase after the 60 minute incubation when allowed to remain at room temperature (approximately 22°C). It can be demonstrated that an approximate 10% error in O.D. results can be expected at lower endotoxin concentrations when incubated longer then 10 minutes at 22°C. It is therefore, extremely important that the total incubation time be the same for all samples in a given assay. This time represents not only the time of incubation at 37°C but also the time prior to and post incubation.

Sensitivity and linearity will vary from lot to lot of lysate (Valois F.W., 1979). It is extremely important that candidate lots be examined very closely for acceptable response before being incorporated into the micro procedure. Figure 2 represents a typical turbidimetric LAL response when used in the micro procedure against endotoxin standards ranging from 8 to 1000 picograms/ml. The linear range will vary from lot to lot and once past the linear portion, a plateau will occur followed by a drop off in O.D. A candidate lot should show acceptable linearity (measured by regression analysis) with approximately a 1 log range coverage. The differences in O.D. between lowest standard values and upper values should be sufficient to afford good resolution in interpolation of samples off the standard curve. Figure 3 shows an expanded version of lot 264 showing the linear portion of the response curve (10 - 100 picograms/ml), a linear regression analysis of this line between 20 and 100 picograms shows acceptable linearity.

To compare the results of the conventional spectrophotometric assay with the microplate procedure, a single lot of lysate was tested against endotoxin standards ranging from 10 --> 100 picograms/ml. The linear regression analysis is presented in Figure 4. It

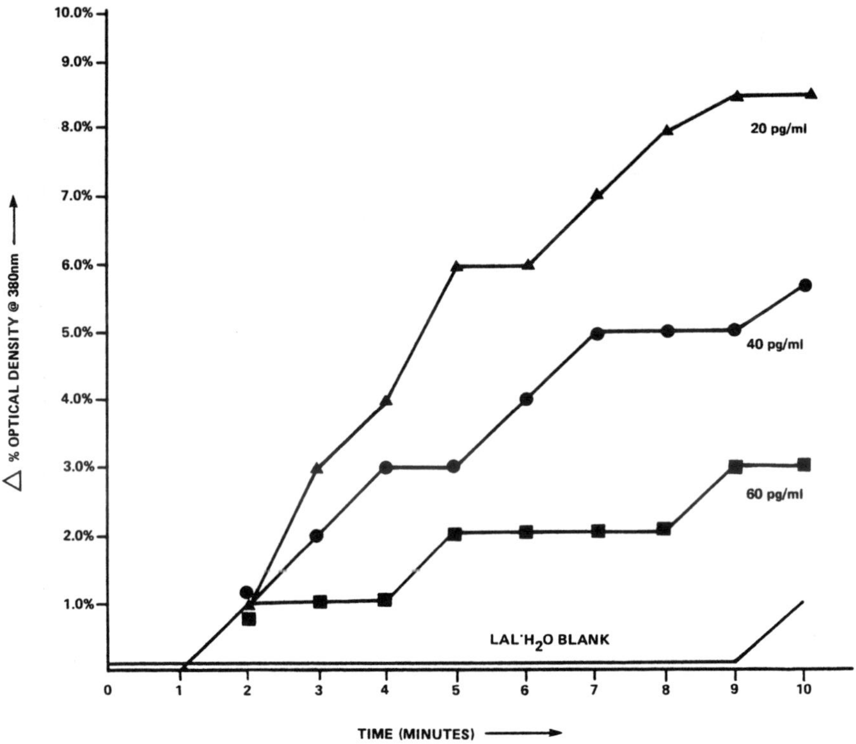

FIGURE 1 INCREASE IN % OPTICAL DENSITY AT 22° C.

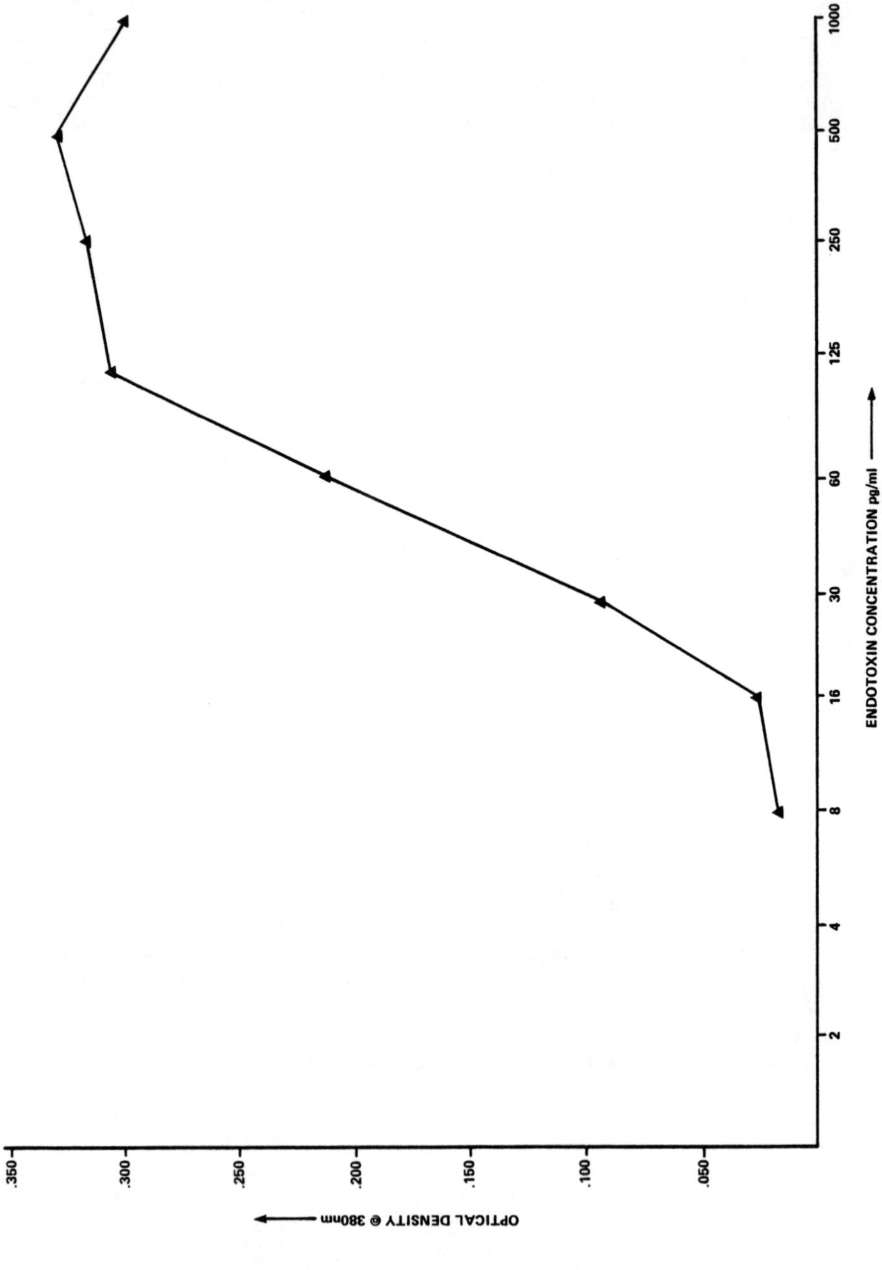

LAL TURBIDIMETRIC RESPONSE LOT No. 264

FIGURE 2

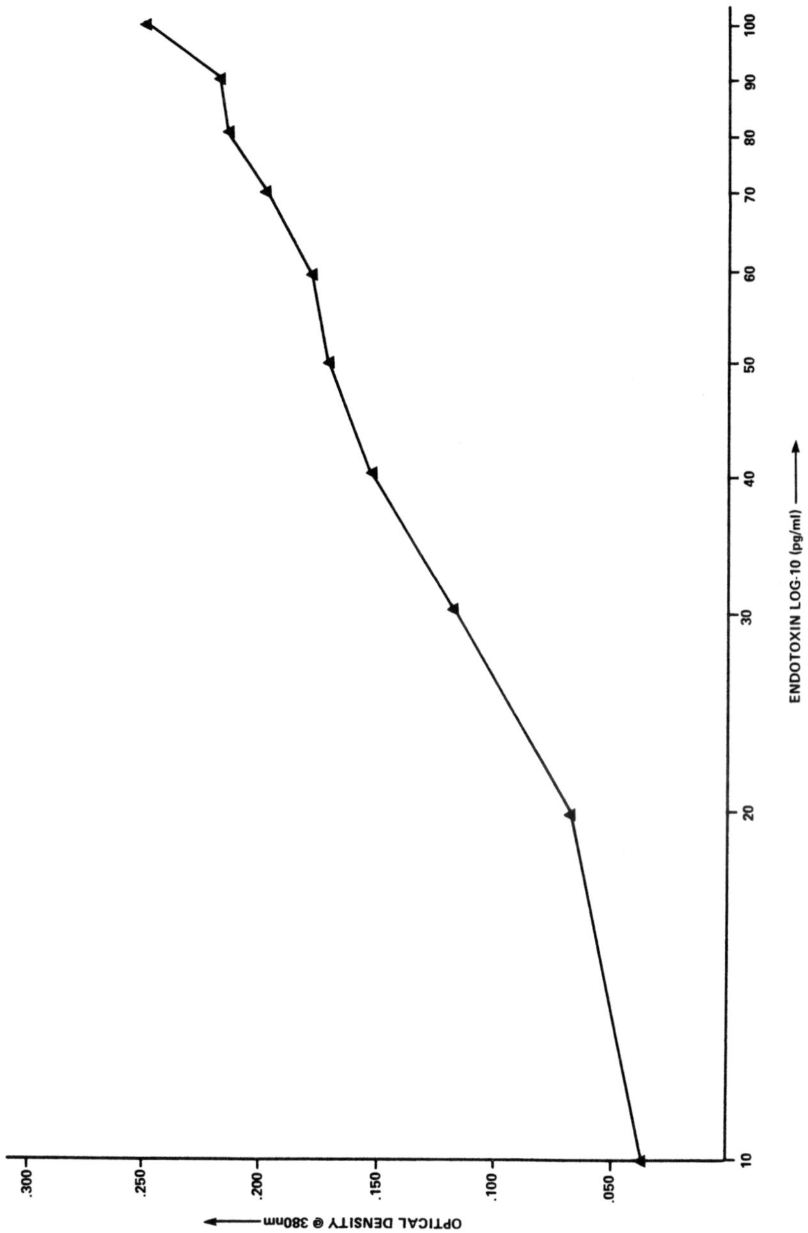

LAL TURBIDIMETRIC RESPONSE LOT No. 264

FIGURE 3

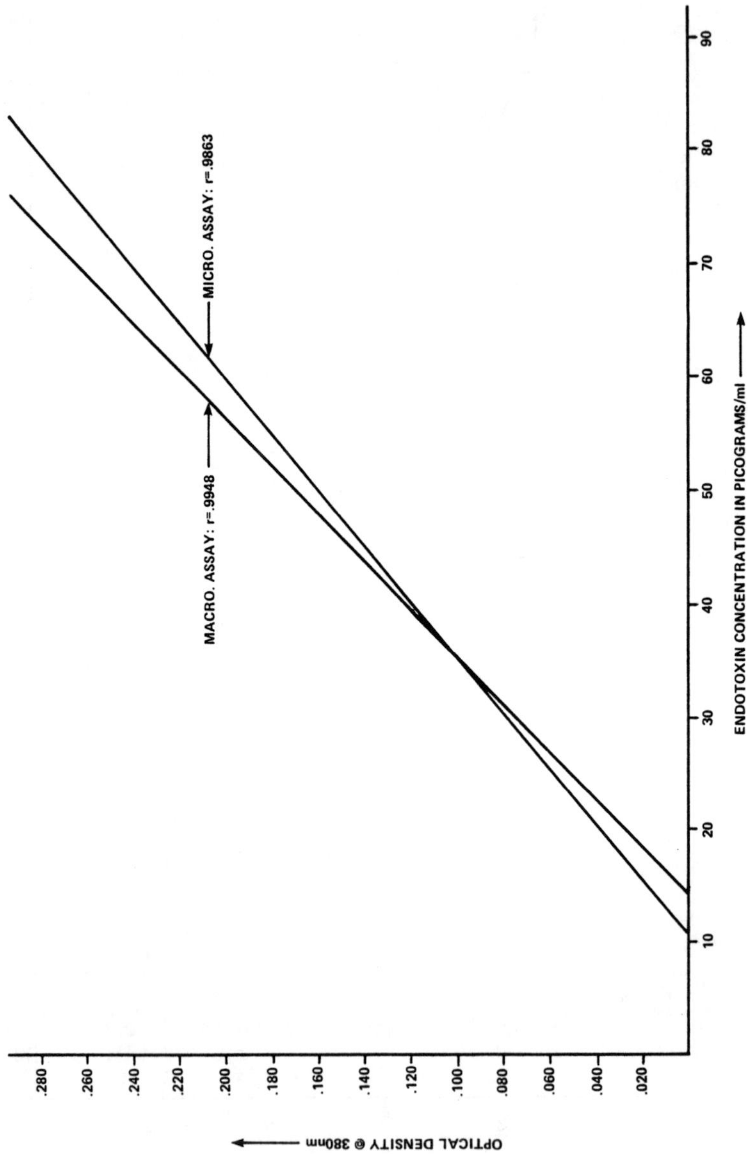

FIGURE 4

LINEAR REGRESSION ANALYSIS OF LAL LOT No.259

can be demonstrated that the lines show approximately the same response with slightly different slopes. Linearity analysis for both procedures represented by the coefficient of correlation, show acceptable values.

Table 2

Coefficient of Variation of Standard Endotoxin
(4 Replicate Samples)

	10	20	30	40	50	60	70	80	90	100
$\bar{X}$ (A.O.D.)	.028	.034	.079	.135	.178	.240	.271	.310	.314	.322
S.D.	.002	.002	.004	.006	.004	.008	.007	.012	.010	.013
% C.V.	6.2	5.9	5.1	4.4	2.2	3.3	2.6	3.2	3.2	4.0

The accuracy in determining the O.D. at different levels of endotoxin standards is shown in Table 2. The mean value for 4 replicate samples is presented along with the standard deviation of each endotoxin concentration, and its coefficient of variation. All values show acceptable levels of error and compare very well with the error inherent in the macro spectrophotometric method (Bondar, et al., 1979). The largest amount of error is found in the lower endotoxin concentrations since any slight amount of variation would show the greatest C.V.

Table 3

Assay of Spiked Water Samples

Sample	$\bar{X}$ of 4 Replicates*	S.D.	% C.V.
A	85 pg/ml	4.2	4.9
B	81 pg/ml	3.1	3.8
C	89 pg/ml	2.0	2.2
D	80 pg/ml	4.4	5.5
E	78 pg/ml	0.8	1.0
F	82 pg/ml	3.0	3.7
G	77 pg/ml	2.7	3.5

A spiked water solution was prepared and aliquoted into seven pyrogen-free vials and stored at -70°C. Single aliquots were removed at random and assayed for endotoxin content. This data is presented in Table 3. The ability to accurately measure endotoxin varied from 1.0% C.V. to 5.5% C.V. The overall average C.V. is 3.51%.

The spectrophotometric assay has proven to be an invaluable asset in quantitation of bacterial endotoxin. The assay is more complex than the standard clot method but offers numerous advantages. An alternate photometric assay which overcomes many of the inherent problems associated with conventional photometric techniques has been demonstrated. The use of state of the art photometric techniques coupled with semi-automation and computer interface capabilities make this procedure extremely advantageous to large scale use. However, as with all complex kinetic reactions, extreme user caution and rigid quality control (Rastogi, et al., 1977) must be employed to ensure full benefit.

Bondar, R.J.L., Teller, J.D., Bowanko, A., Kelly, K.M. (1979) "Biomedical Applications of the Horseshoe Crab (Limulidae)" Alan R. Liss, Inc. pp. 435-451.

Levin, J., Bang, F.B. (1968) Clottable Protein in Limulus: Its localization and kinetics of coagulation by endotoxin. Thromb. Diath. Haemorrh. 19:186.

Musiani M., Gatti M., Cevenini, R., LaPlace, M. (1976) Study of the Reliability of the Limulus Amebocyte Test. Quad Sclavo Diag. 12 (2) p. 123.

Rastogi , S.C., Hochstein, H.D., and Seliymann, E.B. (1977) Quality Control Procedures for testing limulus amebocyte lysate sensitivity Using Reference Endotoxin. Ib. Bull. Parenteral Drug Assoc., 31 120-127.

Ruitenberg, E.J., Sekhuis, J.M., Brosi, B.J.M. (1980). Some characteristics of a New Multiple-Channel Photometer for Through-the-Plate Reading of Microplates to be Used in Enzyme-Linked Immunosorbent Assay. J. of Clin. Microbiol., Vol. II, No. 2: 122-134.

Valois, F.W. (1979) "Biomedical Applications of the Horseshoe Crab (Limulidae)" Alan R. Liss, Inc. pp. 415-422.

ALBAUGH SPEAKER
TITLE: AUTOMATED METHODOLOGY FOR THE LAL ASSAY USING THE MULTISKAN MICROPLATE READER.

FINK - I wonder about the precision data performing solely replicate assays. According to the recommendation of clinical chemistry towards the coefficient of variation at least 10 replicate assays must be done. If you look at your slides, performing a 4 replicate assay at 77 pgs, you have a coefficient of variation of .9 and doing 78 pgs you have 3.5. This will make a difference, but as I said at least 10 replicate assays should be done, otherwise the coefficient of variations is not objective.
ALBAUGH: Your point is well taken, I agree with you 100%. Please try to bear in mind that we tried to do 4 assays 7 times, so if you look at the total number of independent assays performed you have 28 replicate assays which is in excess of the minimum of 10 recommended. Performing the statistical analysis you can determine that the coefficient of variation, or the inherent error of the Multiskan procedure is quite low. Another point I feel needs to be addressed is if you discuss a 5% coefficient of variation at 70 - 80 pg we are talking about a very small % error (3-4 pg). When you compare this type of error inherent in the quantitative procedure with the twofold dilution error in the gel-clot assay, the amount of error is absolutely insignificant.

**Endotoxins and Their Detection With the
Limulus Amebocyte Lysate Test, pages 195–206**

A QUANTITATIVE ENDOTOXIN ASSAY UTILIZING LAL AND A CHROMOGENIC SUBSTRATE

P. Friberger, M. Knös and L. Mellstam

Kabi AB, Peptide Research
Mölndal
Sweden

Gram-negative bacterial endotoxin catalyzes the activation of a proenzyme in the Limulus Amebocyte Lysate (LAL) (1, 2). It has recently been demonstrated that at least two proteins are involved in the activation of the clotting enzyme (3). The amount of enzyme needed to obtain a suitable hydrolysis of a chromogenic substrate is less than the amount needed to form a clot. This makes it suitable to change the detection of activation from a clot formation to the formation of a coloured compound (4).

By using an excess of LAL, the amount of enzyme formed is directly proportional to the amount of endotoxin present. It is the initial rate of activation that is determined. The enzyme formed after a certain incubation time then catalyzes the splitting of p-nitroaniline (pNA) from the colourless substrate Ac-Ile-Glu-Gly-Arg-pNA. The rate at which the pNA is released is measured photometrically at 405 nm after stopping the reaction with acetic acid. The correlation between the absorbance (A) and the endotoxin concentration is linear in the 0.01-0.1 ng/ml range.

MATERIALS AND METHODS

The equipment used is: a spectro- or filterphotometer with 405 nm filter, semi-micro cuvettes, heat block or water bath ($37 \pm 0.2^{\circ}C$), stop watch, disposable pyrogen-free glass test tubes and measuring pipettes. Automatic hand pipettes with sterile tips (eg Eppendorf or MLA).

Substrate: 2.2 mmol/l (1.5 mg/ml) S-2423 (Ac-Ile-Glu--Gly-Arg-pNA from Kabi) which is stable at +2 to +8°C for one month if made up with pyrogen-free water (E. Ph.).

Buffer for the substrate reaction: 0.05 mol/l Tris buffer pH 9.0 I 0.20M (NaCl). An autoclaved solution is stable at +2 to +8°C for one month once opened.

Solvent buffer for the LAL: 0.05 mol/l Tris buffer pH 7.4, 0.08 mol/l NaCl, 0.01 mol/l $MgCl_2$. Once opened, the autoclaved solution is stable for one month at +2 to +8°C.

LAL: Lyophilized LAL is available from several licensed distributors (see eg Table 5) and is reconstituted in a predetermined amount of solvent buffer. If more than one vial is needed, pool before use. If less than one vial is needed, dispense 100 µl each in a number of pyrogen-free test tubes and freeze. Thaw at 37°C immediately before use. Reconstituted LAL is kept on ice.

Endotoxin standard (from Kabi or M A Bioproducts, Md): 2 ng of lyophilized endotoxin from E.coli 0111:B4 is reconstituted by adding 1.00 ml of pyrogen-free water of room temperature and shaken vigorously during at least three minutes, preferably with a Vortex mixer. This stock solution is stable for one week at +2 to +8°C. Before subsequent use, the solution must be warmed to room temperature and vigorously mixed for one minute. This is important because endotoxin tends to attach to glass.

Sterile pyrogen-free water: shall be used to dissolve all the reagents and as a negative control (blank). It shall also be used for the preparation of standards and predilution of samples.

Substrate-buffer solution: is made each day by mixing equal volumes of 2.2 mol/l substrate and buffer pH 9.0.

Precautions! All materials coming in contact with specimen or test material must be pyrogen-free. Clean glassware may be rendered pyrogen-free by heating at 180°C for four hours. A septic technique must be used. Strict adherence to time and temperature of the test procedure is required.

Standardization: In each series of determinations, use

the standards 0.025 and 0.075 ng/ml according to Table 1.

Table 1. Preparation of standards

Endotoxin conc ng/ml	Stock solution 2 ng/ml µl	Standard solution 0.100 ng/ml µl	Pyrogen-free water or sample ref[x)] µl
0.100	100		1900
0.025		500	1500
0.075		Remaining 1500	500

[x)] Samples other than water

The standards can be saved for one week if they are kept at +2 to +8°C and warmed to room temperature and vigorously shaken for one minute before use.

Preparation of samples: Samples must be collected and stored in such a way that bacterial contamination is avoided. Store the samples at +2 to +8°C. Before measuring, the test sample must be warmed to room temperature and shaken. Adjust the pH of the sample to within the range of 7.0-8.0, using pyrogen-free sodium hydroxide (eg 0.1 mol/l) or hydrochloric acid (eg 0.1 mol/l). Unbuffered solutions are automatically adjusted by the LAL solution.

Table 2. Method

End point method		Sample Tube No 1	Blank Tube No 2
LAL (on ice)	µl	100	100
Thermostate (37°C) 3-5 minutes		X	X
Test sample or standard (20°C)	µl	100	
Sterile pyrogen-free water	µl		100
Mix and incubate (37°C) exactly	min	10[x)]	10[x)]
Substrate-buffer solution (37°C)	µl	200	200
Mix and incubate (37°C) exactly	min	3	3
Acetic acid 50%	µl	200	200
Mix immediately		X	X

[x)] The incubation time varies with the LAL (producer and batch). See Table 5.

Timing of the test must begin at the addition of the first reagent to each tube and must continue uninterrupted to the last step. It is suggested that in a series of tests, a new test starts every 15 seconds (12 determinations in each series).

Read the absorbance of the samples and standards (Tube 1) against the blank (Tube 2) in a photometer at 405 nm and calculate the endotoxin concentration of the test sample from the standards. The colour is stable for at least four hours.

Calculations:

$$\text{Endotoxin (ng/ml)} = \frac{0.050}{A_{75} - A_{25}} \cdot \left(A_x + \frac{A_{75} - A_{25}}{2}\right) \text{ where}$$

A_{25} is the absorbance for the 0.025 ng/ml standard
A_{75} is the absorbance for the 0.075 ng/ml standard
A_x is the absorbance for the unknown sample.

To check for endotoxin contamination in the reagents, prepare a reagent blank A_3 (Tube 3), see below and calculate $A_2 - A_3$. If the test sample is coloured or opaque, use the readings A_3 and A_4 (Tubes 3 and 4) and calculate the absorbance: $A_1 - A_4 - (A_2 - A_3)$.

Table 3. Control blanks

Steps in procedure		Reagent blank Tube No 3	Sample blank Tube No 4
LAL (on ice)	µl	100	100
Sterile pyrogen-free water	µl	100	
Test sample	µl		100
Acetic acid 50%	µl	200	200
Substrate-buffer solution	µl	200	200
Mix		X	X

Samples other than water: Certain drugs and other substances may inhibit or enhance the enzymatic reaction. For every new product, the linear relationship between endotoxin concentration and absorbance should be proved by making a complete standard curve, using a pyrogen-free reference of the product. To choose a reference of the product, take some batches of good quality and follow the method Tube 1 - 4.

It is possible that some samples, eg blood products polynucleotides, solutions containing heavy metals or surfactants, or those of high ionic strength or osmolality will inhibit or enhance the enzymatic reaction. Such interferences may be decreased by diluting the sample with pyrogen-free water and then calculating with a dilution factor. If problems still exist, the method can be varied (within certain limits) in respect of incubation times. Patterns of inhibition or enhancement different from those of the traditional LAL gelation test may be found.

Blood and plasma fractions: Blood should be sampled on heparin (approx 25 IU/ml). Each batch of anticoagulant should be checked for endotoxin contamination. To avoid degradation of endotoxin, it has been recommended to keep the blood on ice (5). Platelet rich plasma (approx 500 000 platelets per µl, prepared by centrifuging at approx 200 g at 4°C for ten minutes, should be used because the platelets may bind more endotoxin than the other blood corpuscles (6). Inhibitors which may interfere with the activation are destroyed by a dilution 1 + 9 with sterile pyrogen-free water and heat treatment (75°C) for five minutes. Then the sample is left at room temperature for fifteen minutes before use. Shake vigorously before use. A standard curve may be prepared according to Table 4 by using a 0.250 ng/ml endotoxin solution (dilute 1 + 7 of the stock solution). The assay is performed according to Table 2 except that a considerable prolongation of the first incubation time is needed (eg from

Table 4. Preparation of standards in plasma

Endotoxin conc ng/ml of plasma	Plasma µl	Endotoxin 0.250 ng/ml µl	Water µl
0.000[x)]	200	0	1800
0.0313	200	25	1775
0.0625	200	50	1750
0.0938	200	75	1725
0.125	200	100	1700

x) to be used as a blank

nine to twentyfive minutes). If the plasma is bilirubic, hemolytic or opaque, use a plasma blank containing 100 µl of plasma dilution 1 + 9, 300 µl water and 200 µl acetic acid. A similar control procedure as for water may also be used.

RESULTS AND DISCUSSION

The method has mainly been used for the determination of endotoxin in water and is optimized for this system. By using a LAL to which no extra constituents have been added, it was possible to stepwise optimize the concentration of Ca^{2+}, Mg^{2+}, Na^{+}, K^{+}, and Tris buffer as well as pH. These data made it possible to make a LAL solvent which was possible to use in combination with all types of LAL tested (Table 5).

Table 5. Reaction parameters for commercial LAL

Batch No		Dilution volume	Incub time	Slope of stand curve	Amount of LAL
MA	RD418	1	8	0.8	50
"	RD526	1	9	0.9	50
"	L12978	10	10	0.5	50
MALL	01J	2	12	0.9	60
"		2	12	0.5	60
WOR	78K200	5	18	0.6	20
"	70M485	4	18	0.8	25
SIG	20F-0160	5	16	0.8	50

LAL from M A Bioproducts (MA), Mallinckrodt (MALL), Worthington (WOR), and Sigma (SIG) were tested. Dilution volume is given in ml and incubation time in min. The slope of standard curve is given in A units per 0.075 ng/ml. Amount of LAL compared to clotting test is given in %.

The parameters that have to be tested for each batch are: 1) the LAL concentration (usually about 50% of that recommended for the clotting assay) and 2) the incubation time that gives an A between 0.5 and 0.9 for 0.075 ng/ml endotoxin (usually 10-15 mins).The activity usually develops according to Fig. 1 which means that a rather narrow time range is expected.

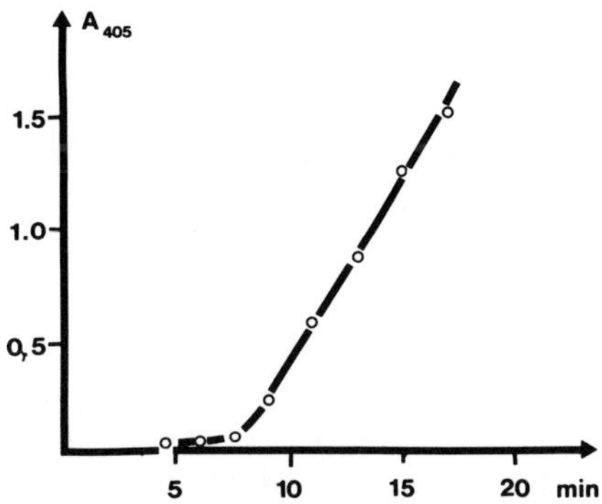

Fig. 1. Development of enzyme activity in a LAL + endotoxin incubate.

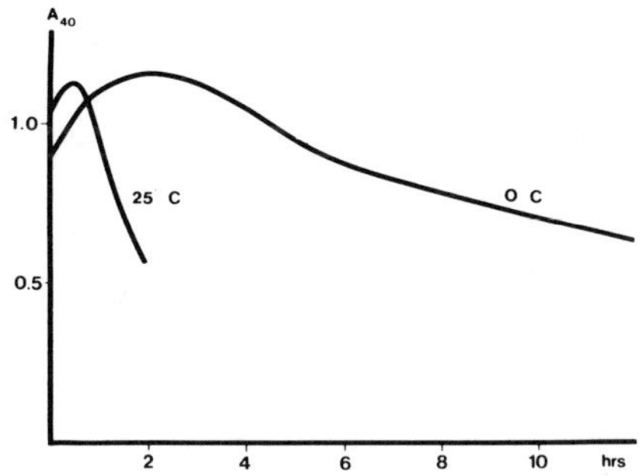

Fig. 2. Stability pattern of LAL. Activity obtained plotted against time of storage.

The stability pattern is shown in Fig. 2 but varies between batches. It was considered to be appropriate to keep the LAL on ice for 30 mins after reconstitution and then use it within one or sometimes a couple of hours. As the stability of the LAL changed considerably between batches, it was often considered important to use standards in each series of assays.

By changing the pH and NaCl concentration in the substrate reaction buffer, optimal conditions were obtained.

The amount of substrate used in the assay satisfies two conditions: 1) $S_o > 2 K_m$ (K_m = 0.14 mmol/l) and 2) the rate of substrate hydrolyzes decreases with about 5% or less during the assay.

The activation of enzyme evidently stops when the substrate + buffer solution is added. This is probably caused by dilution and pH change effects. This is also shown by the fact that comparatively large amounts of endotoxin, eg 0.2 ng/ml, (in the assay) can be added to the substrate without any effect on the results.

In order to avoid endotoxin contamination, the whole assay is carried out in an endotoxin free (heat treated) test tube and the reaction is finally stopped by adding acetic acid before the sample is taken out for measurement in a photometer.

In our experience, the cuvette blank (cuvette + water) A_o = 0.045-0.050, the reagent blank A_3 = 0.07-0.10, and the blank reading A_2 = 0.09-0.13 but at prolonged incubation time it may increase almost up to 0.20. The samples blank (A_4-A_3) must not exceed 0.4 and if it does, the sample should be diluted (use dilution factor in calculations).

When a number of pharmaceutical and parenteral products were tested, the standard curve differed from distilled water to different degrees, so it would be expected that the products to be tested should be used in the standardization procedure (see Materials and Methods).

When the reaction conditions are settled, a total standard curve should be made to check for linearity in the proposed range. For preparation of standards, see Table 6. A typical standard curve is shown in Fig. 3. The linear correlation coefficient should be >0.98 and the coefficient of variation of replicate samples below 5%.

Table 6. Preparation of standards for the standard curve

Endotoxin conc ng/ml	Stock sol 2 ng/ml µl	Stand sol 0.10 ng/ml µl	Stand sol 0.05 ng/ml µl	Stand sol 0.025 ng/ml µl	Pyrogen-free water or sample ref µl
0.100	100				1900
0.075		500	500		
0.050		1000			1000
0.025			500		500
0.0125				500	500

Note! All endotoxin dilutions must be thorougly shaken before use and kept at room temperature during the work

As can be seen in Fig. 4, the reaction is very sensitive to temperature. This can influence the day to day variation

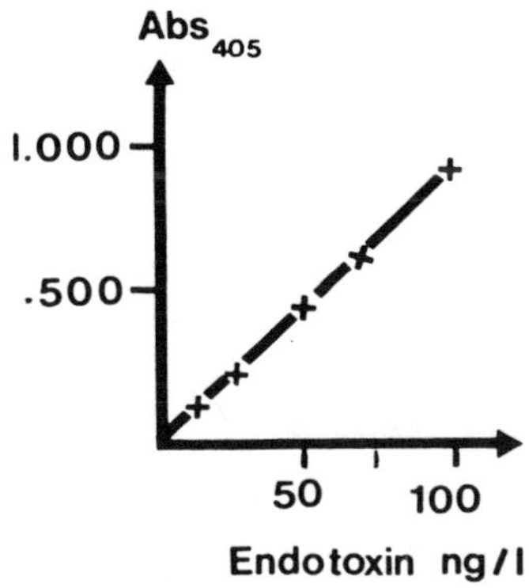

Fig. 3 Typical standard curve in water.

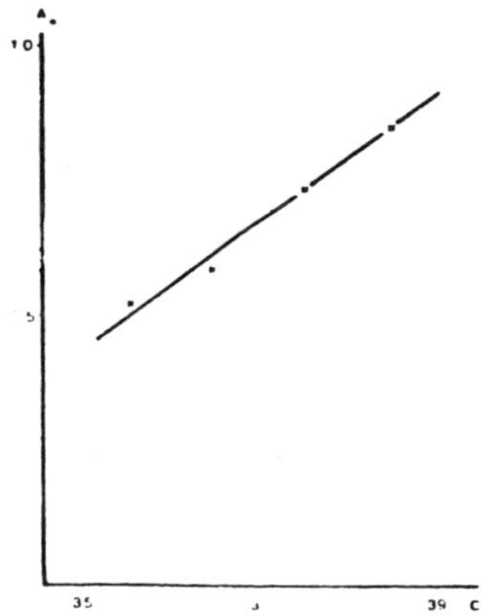

Fig. 4. Absorbance obtained at different incubation temperatures.

of a set of reagents.

When the endotoxin standards from two batches were tested, a vial to vial variability within each batch of less than 3% (C.V.) was found. The two batches differed, however, by 20%. That means, that the endotoxin standard should be titrated against some primary standard until a uniform production (C.V. <5%) is secured.

This chromogenic endotoxin assay is, besides from being quantitative, also independent of the conditions of the interaction on the coagulogen. This may possibly broaden the use.

To study how the blood corpuscles influence the assay, endotoxin (0.1 ng/ml) was added to whole blood (some hemolysis was observed) and platelet rich plasma. Platelet rich plasma was then obtained from the whole blood. Besides, the two plasmas were further centrifuged to obtain platelet poor plasma (approx 20,000 platelets per µl). Endotoxin was determined in the respective plasmas and our results did not show much difference (<20%) from a water solution to which the same amount of endotoxin was added. That means, that in our case only very little endotoxin was absorbed to the blood corpuscles. This is contradictory to others' results (6), so we still recommend the use of platelet rich plasma in the

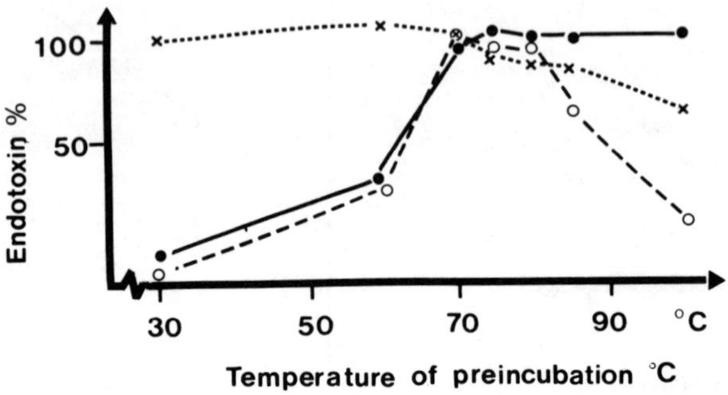

Fig. 5. Activity obtained after 5 min preincubation at different temperatures. Plasma was treated before endotoxin addition (-•-). Plasma + endotoxin was treated (--o--) and water + endotoxin was treated (···x···).

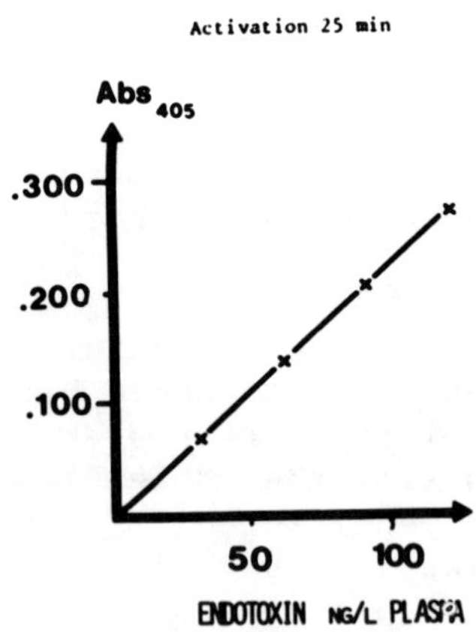

Fig. 6. Typical standard curve in plasma.

assay. Even if some endotoxin is lost at centrifugation, the amount in plasma may very well correlate with the clinical situation (5).

To make it possible to assay endotoxin in blood plasma or similar samples, it was necessary to get rid of protease inhibitors present. That was achieved by dilution with water followed by heating. As can be seen in Fig. 5, it was necessary to heat the plasma dilution to 70°C to get rid of the inhibitors. At that temperature, endotoxin is not destroyed but at temperatures above 80°C, significant amounts of activity is lost in water and still more in plasma. The heat treatment will also destroy most, probably all, protease activity that may have been present in pathological blood samples. The standard curve is shown in Fig. 6. 30 ng/l plasma corresponds to 2 ng/kg body weight which gives approx 0.6°C temperature rise in the rabbit test.

However, much work is still needed to develop and evaluate a clinical endotoxin assay.

REFERENCES

1. Levin J, Bang FB (1964). The role of endotoxin in the extracellular coagulation of Limulus blood. Bull Johns Hopkins Hosp 115:265.
2. Young NS, Levin J, Prendergast RA (1972). An invertebrate coagulation system activated by endotoxin: evidence for enzymatic mediation. J Clin Invest 51:1790.
3. Ohki M, Nakamura T, Morita T, Iwanaga S (1980). A new endotoxin sensitive factor associated with hemolymph coagulation system of horseshoe crab (Limulidae). FEBS Letters 120:217.
4. Iwanaga S (1978). Chromogenic substrates for horseshoe crab clotting enzyme. Its application for the assay of bacterial endotoxins. Haemostasis 7:183.
5. Thomas LLM, Sturk A, Kahlé LH, ten Cate JW (1981). Quantitative endotoxin determination in blood with a chromogenic substrate. Clinica Chimica Acta 116:63.
6. Scully MF, Newman YM, Clark SE, Kakkar VV (1980). Evaluation of a chromogenic method for endotoxin measurement. Thromb Res 20:263.

FRIBERGER SPEAKER
TITLE: A QUANTITATIVE ENDOTOXIN ASSAY UTILIZING LAL AND A CHROMOGENIC SUBSTRATE.

LEVIN: Could you tell us something about optimal ionic concentration?
FRIBERGER: We are using sodium chloride and magnesium chloride.
STOTT: When you heat the plasma to 70 degrees, do you allow it to cool back down before the addition of the LAL?
FRIBERGER: Yes, we let it cool down.
MORRISON: Could you tell me the relative sensitivity of the assays compared to a turbidimetric assay? Are the normal trypsin-like inhibitors effective in inhibiting your response?
FRIBERGER: I am afraid I cannot give you a good answer. We haven't checked out the trypsin inhibitors.
FIRCA: Your total incubation time was 15 to 20 minutes with the endotoxin. Have you tried to detect it after 1 or 2 hours?
FRIBERGER: No, definitely not.
NOVITSKY: Is there any reason?
FRIBERGER: By then we could have activiated most of the LAL. We were after the initial activation rate as some lysates have tremendous activity. It was not very convenient.
NOVITSKY: Have you tried to remove the coagulogen from any lysate?
FRIBERGER: We would very much like to have that done but we don't have enough experience with the LAL to do this.

Endotoxins and Their Detection With the
Limulus Amebocyte Lysate Test, pages 207–215

QUANTITATIVE TECHNIQUES FOR THE LAL TEST

John L. Sloyer,Jr.Ph.D., Laurel J.Karr,B.S.,
Thomas E. Stoneycypher,Jr.,B.S.
Dept. of Pediatrics, Univ. of Alabama, School of
Primary Medical Care
Huntsville, Alabama 35801

INTRODUCTION

Levin and Bang(1968) reported that an aqueous solution of lysed amebocytes from the horseshoe crab would form a gel matrix when reacted with as little as 50 picograms of Gram-negative bacterial endotoxin. False positive reactions did not occur with other bacterial toxins, normal plasma, or plasma constituents. The limulus amebocyte lysate (LAL), normally a clear solution, becomes a translucent gray in the presence of bacterial endotoxin. In addition to this appearance change it has been reported that the time for clot formation to occur is related to the concentration of endotoxin in the sample (Sloyer, et al 1980).

Spectrophotometric techniques have been described which quantitate endotoxin, however their use has not gained wide acceptance (Levin and Bang 1968, Levin, et al 1970). This is partially attributable to some difficulty encountered in constructing standard reference curves using endpoint optical density analysis. It was the purpose of this investigation to determine whether the rate of clot formation would be a more reliable measure of endotoxin concentration thereby eliminating the problem of endpoint analysis.

MATERIALS AND METHODS

All glassware and instruments were pyrogen-free. In addition pyrogen-free water was used in all dilutions.

LAL (Associates of Cape Cod, Inc., lot #52-58-233) was stored at 5 C. until reconstituted at which time it was dispensed into tubes in 0.1 ml aliquots and stored at -70 C. Storage time after reconstitution did not exceed two weeks. Endotoxin (lot #7) was stored at 5 C. A second endotoxin standard was prepared by ultrasonically disrupting a solution of E.coli (BBL lot A1NA). The sonicate was filtered (.45 um) and stored at 5 C.

Assays were performed by adding 0.1 ml of pyrogen-free water to a tube containing 0.1 ml of LAL. 0.2 ml of the sample to be tested or the standard was added to the tube. The mixture was gently shaken and introduced into a spectrophotometer equipped with a kinetic analyzer and temperature controller (Micromedic Systems, Inc.). The initial optical density 360 was recorded and the rate of OD change was monitored every 15 seconds. A kinetic ratio was determined which represented the ratio between maximum OD change and the time in minutes that the maximum delta O.D. occurred. This ratio was related to the endotoxin concentration of the standards and was used to construct reference curves for quantitation.

RESULTS

Fig. 1 displays the relationship of endotoxin concentration to reaction kinetics. Endotoxin standards ranging in concentration from 10 to 100,000 picograms per ml revealed proportionately maximum OD changes and time course as a function of concentration. The LAL used was listed as having a sensitivity of 60 picograms consequently no OD change was detected in the sample with only 10 picograms per ml. The relationship of the kinetic ratio to endotoxin concentration is shown in Fig. 2. Since a direct relationship existed, all subsequent data has been related to such a standard curve in order to evaluate factors affecting this type of kinetic assay. Unless otherwise stated the reaction was monitored at 37 C. The reproducibility of the assay at 37 C. is summarized in Table 1.

Table 1. Reproducibility of kinetic assay for endotoxin

Test Run	Endotoxin (pg/ml)		
	A	B	C
1	3467	277	13
2	2994	352	14
3	4839	364	17
4	3804	363	17
5	3583	274	
Mean	3737	330	15
S.D.	683	43	2
% variation	18%	13%	13%

As can be seen the reproducibility over the range of concentrations assayed varied from 13% to 18%. The reagents were dispensed with Eppendorf pipettes since the usual serological type were found to be completely unsatisfactory.

Storage of LAL in plastic tubes rather than glass was found to influence the concentration of endotoxin detectable. A comparison of the type of tube used for storage is shown in Table 2. LAL stored in plastic appeared to inhibit the reaction and LAL stored in this manner detected from between 26% to 91% of the endotoxin which was detected by LAL stored in glass tubes.

Table 2. Effect of type of storage container on sensitivity of LAL to endotoxin

Endotoxin Lot	Type of storage		
	Plastic	Glass	% p/g
ACC #4	122,307*	134,819	91%
	11,391	25,658	44%
	874	1,921	45%
	30	40	75%
ACC #7	29,009	80,196	36%
	6,023	9,613	63%
	751	2,148	35%
	15	58	26%

* picograms per ml

Fig. 1 LAL Clot Formation

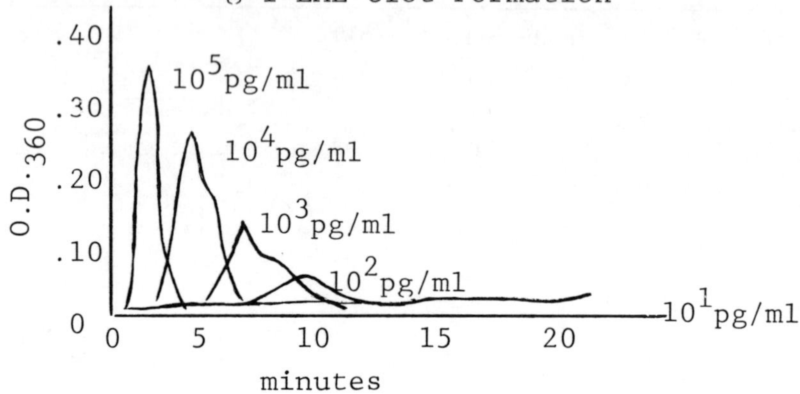

No reaction occurred at 25 C. but at 30 C. and 37 C., a reaction was observed at different rates as evidenced by the time it took for a maximum OD change to occur (Table 3).

Fig. 2 LPS Standard Plot

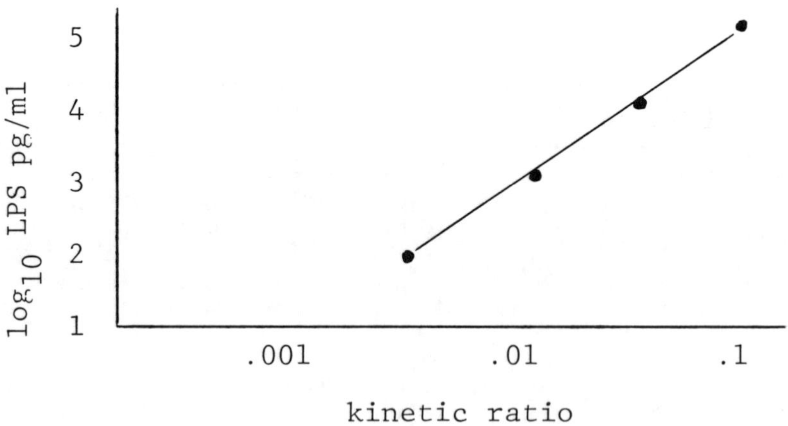

Table 3. Effect of temperature on reaction time to detect endotoxin

	Reaction time at:	
Concentration	30 C.	37 C.
270 pg/ml	32 min.	17 min.
2,300	22 min.	10 min.
12,000	16 min.	9 min.

Generally it took twice as long for a reaction to achieve maximum gel formation at 30 C. compared to 37 C. regardless of endotoxin concentration.

A measure of the kinetic reaction of a solution containing 2,300 pg/ml at 30 C. and 37 C. is shown in Fig. 3. At 30 C. the reaction took longer to initiate and when activated proceeded at a slower pace. The slope of each reaction reflected the rate of the reaction. The slopes for 30 C. and 37 C. as a function of endotoxin concentration are summarized in Table 4.

Table 4. Comparison of reaction rate at 30 C. and 37 C. at different endotoxin concentrations

	Slope of kinetic reaction	
Endotoxin	30 C.	37 C.
270 pg/ml	0.53	0.76
2,300	0.69	1.97
12,000	1.30	2.69

Fig.3 LAL Kinetics-2300 pg/ml

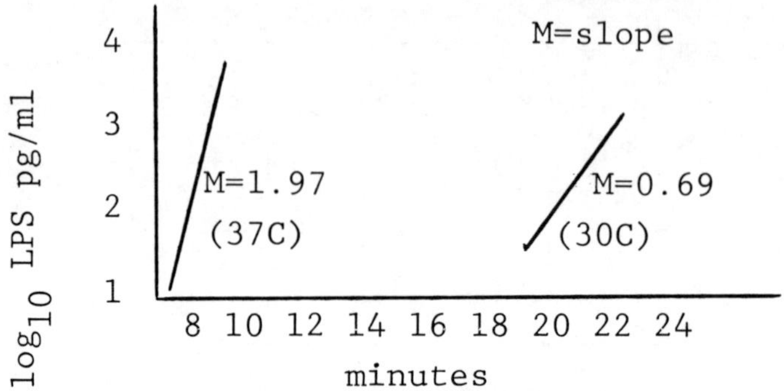

It is clear that concentrations as low as 270 pg/ml can be equally detected at 30 C. or 37 C. From Fig. 4, it can be seen that sensitivity of the test is not compromised by either temperature and that a reaction need not go to maximum OD to determine endotoxin concentration.

Fig.4 LAL Kinetics-Reaction slope

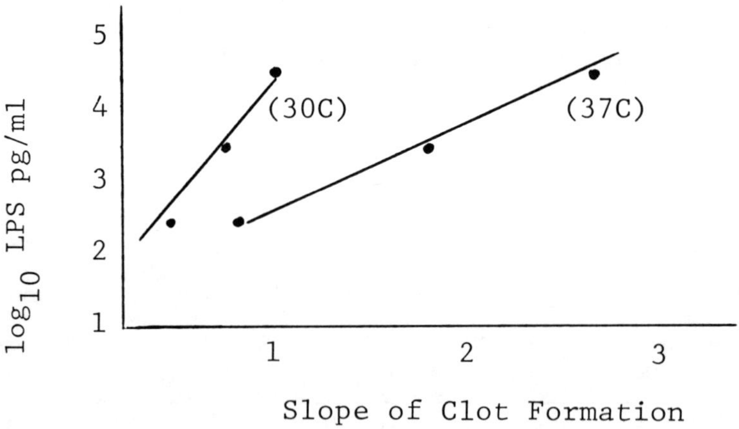

Although we have little difficulty in quantitating endotoxin in plasma-free solutions at 37 C., we have found that at 30 C. plasma-bound endotoxin is more readily detectable as shown in Table 5.

Table 5. Effect of incubation temperature on detection of plasma-bound endotoxin

Plasma dilution	% added endotoxin:	
	30 C.	37 C.
1:10	21.8% (4)	15.8% (7)
1:100	91.1% (10)	74.8% (8)

If plasma is diluted 100-fold and assayed at 30 C., there is approximately a 22% increase in the amount of endotoxin detected compared to that obtained at 37 C.

DISCUSSION

This investigation has provided a method for quantitating endotoxin using spectrophotometry. Although this is not the first report to use such a method it is the first one to relate endotoxin concentration to a change in OD rather than an endpoint OD. The advantage of using such a method is that it is independent of the initial color of the solution being tested. This makes it somewhat easier to adapt to clinical samples such as urine, and plasma. No problem appears to exist for using such samples as CSF or colorless pharmaceuticals. Since a considerable difference in sensitivity was associated with type of storage container we recommend that plastics not be used. On the other hand, while a reduction in reaction temperature from 37 C. to 30 C. resulted in a slowing of the reaction, there did not appear to be a reduction in the sensitivity of the LAL to endotoxin detection. The reproducibility of the assay at concentrations lower than 300 pg/ml was approximately 13% which appears to be acceptable. This level of reproducibility was achievable with Eppendorf pipettes and could be improved using automated reagent dispensers. Ordinary serological pipettes were found to be unsatisfactory.

The observation that the slopes of reactions were proportional to endotoxin concentration suggests that a quantitative estimate could be obtained without having to wait for the reaction to go to completion. This type of kinetics lends itself to automation more easily than the kinetic ratio described in this report. However the use of such a method and the effect it may have on reproducibility remain to be determined.

Lastly we have provided a method whereby endotoxin can be detected in plasma although not at a 100% level. Perhaps this will induce others to further explore the technique for detecting plasma-bound endotoxin.

REFERENCES

Levin J, Bang FB (1968) Clottable protein in Limulus: its localization and kinetics of its coagulation by endotoxin. Thromb diath hemorrh 19:186.

Levin J, Tomasulo PA, Oser RS (1970) Detection of endotoxin in human blood and demonstration of an inhibitor. J Lab Clin Med 75:903.

Sloyer JL, Karr LJ, Stalons D (1980) Rapid kinetic assay for bacterial endotoxin. Abstract, ASM.

SLOYER SPEAKER
TITLE: QUANTITATIVE TECHNIQUES FOR THE LAL TEST.

REICHLER: How quickly can the rate of the reaction or the slope be determined?
SLOYER: We assume that once the reaction begins, it will take three points to determine the slope. However, since the reaction is complete within five minutes, there does not seem to be much advantage in determining the slope from other than the initial three points.
REICHLER: Sometimes you could actually see more than one slope in three of those reactions. Which one do you work with?
SLOYER: Once the reaction starts, we use the first three points. Therefore, we use the first slope.
SPENCER: Do you know whether your slope remains uniform if you are using different lots of endotoxin and different lots of LAL reacting with a different product?
SLOYER : No I do not, since we have examined only two lots of LAL from Associates of Cape Cod.

Endotoxins and Their Detection With the Limulus Amebocyte Lysate Test, pages 217–230

LAL-REACTIVE MATERIAL ASSOCIATED WITH HEMODIALYSIS MEMBRANES

Loretta A. Carson and Norman J. Petersen

Centers for Disease Control
4402 North Seventh Street
Phoenix, Arizona 85014

Over the past 10 years, studies conducted by our laboratories of pyrogenic reactions in patients undergoing maintenance hemodialysis have provided a mass of epidemiologic evidence that the etiologic pyrogen is endotoxin (LPS) produced by gram-negative water bacteria (Carson, et al 1979; Favero, Petersen, et al 1974; Favero, Carson et al 1974). Empirical data based on the ability to terminate outbreaks of pyrogenic reactions by reducing or eliminating gram-negative bacterial or endotoxin contamination in fluids associated with dialysis procedures suggested that under certain conditions, endotoxin in the dialysis fluids could pass through the dialysis membrane in the absence of apparent leaks (Hindman, et al 1975; Petersen, Boyer, et al 1978; Petersen, Carson, et al 1980). However, overt pyrogenic reactions are only one aspect of the possible complications attending maintenance hemodialysis; others include headache, myalgia, nausea and vomiting, post-dialysis fatigue, leukopenia and leukocytosis, and respiratory-distress syndrome (Bok, et al 1980; Dumler, Levin 1979; Hindman, et al 1975; Holdsworth, et al 1978). Over the past year and a half, further studies conducted in our laboratories and by other investigators have suggested that another source of LPS, namely that derived from the dialyzers themselves, may be associated with at least some of these accessory symptoms.

In typical laboratory experimental setups, which initially were used largely in experiments attempting to demonstrate the passage of endotoxin across dialysis membranes, closed-loop circuits were aseptically created using a blood pump and sterile, pyrogen-free blood lines. In the dialysate loop, generally either reverse osmosis product water containing high levels of naturally occurring gram-negative water bacteria, or a dialysis solution prepared with nonpyrogenic dialysis concentrate (Dialytrate, Broemmel Pharmaceuticals) to which was added purified bacterial LPS (E. coli 055:B5, Difco) was recirculated through the dialysate compartment at a rate of 500 ml/min. Sterile pyrogen-free distilled water (Sterile Water for Irrigation, USP, Cutter Medical) or saline (0.9% Sodium Chloride Irrigation, USP, Cutter Medical) was recirculated through the blood compartment at a rate of 200 ml/min, after an initial 500-ml volume of water or saline was discharged to waste. Samples from the blood and dialysate loops were collected at various time intervals and assayed for endotoxin by the Limulus amoebocyte lysate (LAL) gel-clot method (Sullivan, et al 1976), using lysate (Pyrotell, Associates of Cape Cod (ACC)) with lot sensitivities ranging from .08-.125 ng/ml when tested against E. coli EC-2 reference endotoxin standard (Bureau of Biologics, Food and Drug Administration (FDA)).

During some of the early tests it was noted that levels of endotoxin in the closed-loop dialysate systems decreased with time. For example, a volume of dialysate containing 100 ng/ml of purified LPS (E. coli 055:B5) was circulated through the dialysate compartment of a hollow fiber (HF) dialyzer. After 4.5 hrs, the endotoxin levels in the circulated dialysate had decreased to 0.08 ng/ml. This suggested that LPS: (1) was inactivated, (2) was deposited within the dialysate compartment itself, or (3) was dialyzed into the blood loop. It was considered doubtful that the endotoxin was inactivated since LPS is highly resistant to a variety of physical and chemical treatments (Tsuji, Harrison, 1979). Dialysis into the blood compartment was a potentially exciting explanation for the loss of LPS on the dialysate side. However, while field and laboratory studies strongly suggest that endotoxin can pass an intact membrane, in this particular experiment consistently only heavy turbidity was observed in the LAL test with blood-side samples, with a degree of gelation insufficient to meet the criterion for positive test results. In subsequent tests using other dialyzers, firm gelation in fluids from the blood compartments was demonstrated in only 2 of 9 test situations. Thus, the more likely explanation for the massive loss of LPS appeared to be deposition in the dialysate compartment itself, with the LPS being adsorbed or somehow bound to the surfaces in the compartments. It is possible also that some alteration in the structure of the LPS might have occurred during recirculation through the dialysate compartment which reduced its ability to initiate true gel formation in the LAL assays.

Subsequently, a new lot of lysate from ACC was obtained which was capable of detecting lower levels of LPS than the lysate used previously, i.e., having a stated titer of .008 ng/ml, and a measured titer of .01 ng/ml against the EC-2 reference standard LPS. At the time, field studies were being conducted to assist a local hospital experiencing problems of bacterial and LPS contamination in the water-treatment system of the hospital's acute-care hemodialysis unit. Although no pyrogenic reactions were observed in patients undergoing hemodialysis procedures, a sample of dialysate showed an LPS level of 1.8 ng/ml and, unexpectedly, a sample of venous blood from a coil dialyzer (Travenol Ultra-flo[R] II) showed a level of 0.06 ng/ml of LAL-reactive material. (Throughout this paper the terms LPS or endotoxin will be used interchangeably with the term LAL-reactive material, on the presumption that the LAL test is specific for LPS (Tomasulo 1979), at least at the concentrations being reported here, and that the LAL test results are not simply false positives.) Laboratory tests were conducted using a new dialyzer from the same lot as the one used by this patient. A solution of sterile, pyrogen-free 0.9% saline was circulated through the blood compartment at 200 ml/min for 105 min, and a total of 144 ng of LPS was extracted from the dialyzer, for a leaching rate of 1.4 ng/min. In other tests with this dialyzer the passage of purified LPS from dialysate into the blood compartment could not be demonstrated.

Subsequent laboratory tests using various HF and parallel plate dialyzers (PPD) produced by different manufacturers have demonstrated that new hemodialyzers release LPS to fluids passed through the blood compartments. In most of the dynamic leaching tests conducted, as is illustrated by results presented in

Figure 1, endotoxin appears to be released at one rate during the first 15-min period of recirculation and at a lower rate during subsequent 15-min periods. Moreover, static leaching tests using various dialyzers show that significant masses of endotoxin are potentially available

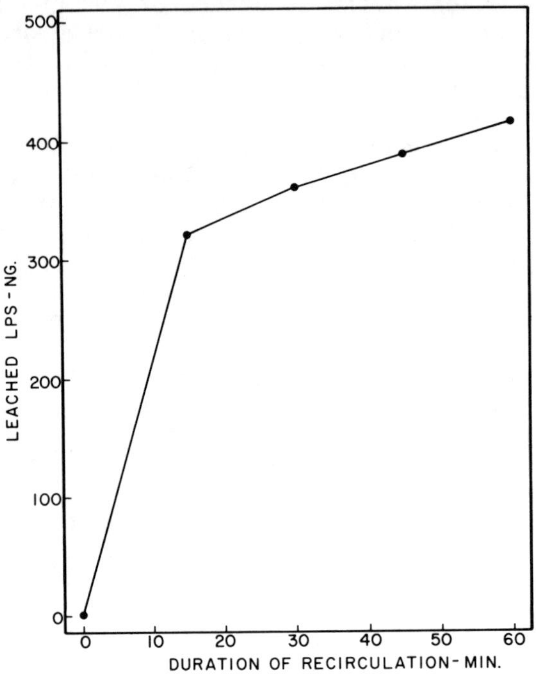

Fig. 1. Leaching rate of LAL-reactive material from a hollow fiber dialyzer during 60 min of saline recirculation.

for release. For example, tests on the blood compartments of Cobe HF[R] 130 and PPD 1.3 dialyzers (Table 1) showed mean values of approximately 6000 and 8200 ng/dialyzer. A value of 9000 ng was obtained with a Cordis-Dow C-DAK[R] 1.8, although interestingly, both static and dynamic tests of Cordis-Dow C-DAK[R] 3500 or 4000 dialyzers have shown no detectable endotoxin (<0.01 ng/ml). The hollow fiber used in these types of dialyzer is manufactured using a

Table 1: Mean Total Mass of LAL-Reactive Material Leached from Blood and Dialysate Compartments of New Dialyzers During a 16-hr Static Soak

	Total Leached LPS - ng/dialyzer	
Compartment	HF Dialyzer (n = 2)	PP Dialyzer (n = 2)
Blood	6000	8160
Dialysate	108	144

different process with respect to the involvement of process water than that used for most other dialyzer membranes. Additionally, published reports as well as the advertising for this dialyzer suggest that the commonly observed "wash out" syndrome as well as neutropenia is observed less frequently in patients using this type of dialyzer than in patients using other dialyzers (Dumler, Levin 1979; Shin, et al 1980).

Table 2 presents summary data of the results of LAL assays of 71 samples of saline used to prime first-use hemodialyzers under clinical conditions. Significant

Table 2: Levels of LAL-Reactive Material in Priming Saline Recirculated Through New Hemodialyzers Prior to Dialysis

Dialyzer Type	n	Mean(ng/ml)	Range(ng/ml)
Cobe PPD 1.6	17	0.01	<0.01- 0.02
Travenol CF^R 12.11	22	33	0.03-160
Travenol CF^R 1500	14	29	1.9 -160
Travenol CF^R 2300	2	8.8	7.5 - 10
Cordis-Dow C-DAK^R 1.8	6	7.5	3.0 - 30
Cordis-Dow C-DAK^R 2.5	2	1.8	0.5 - 3.0
Cordis-Dow C-DAK^R 3500	3	< 0.01	----
Cordis-Dow C-DAK^R 4000	5	< 0.01	----

differences in endotoxin levels existed both between individual dialyzers of the same type, as evidenced by the range of values, and between different types of dialyzers. Table 3 shows the distribution of 23 of the 71 samples, to illustrate two points: (1) Differences between levels among the same types of dialyzers cannot be explained by differences in the duration of recirculation, and (2) reused dialyzers release significantly lower levels of LPS than new dialyzers of the same type.

Hemodialysis procedures include rinsing the blood compartment with a single pass of 500 ml of saline to waste followed by closing the blood line circuit, recirculating approximately 250 ml of saline for 30 min at 200 ml/min, and using anywhere from 10 ml to 250 ml of this

Table 3: Levels of LAL-Reactive Material in Priming Saline Recirculated Through Hemodialyzers Prior to Dialysis

Dialyzer Type	Sample Number	No. of Times Used	Minutes of Recirculation	LPS ng/ml
Cobe	1	0	67	0.02
PPD 1.6	2	0	14	0.02
Travenol	3	0	30	160
CF^R 12.11	4	1	28	0.45
	5	1	30	0.03
	6	0	30	160
	7	0	20	60
	8	0	95	30
	9	0	53	30
CF^R 1500	10	0	49	30
	11	0	44	30
	12	0	30	30
	13	0	36	15
	14	0	66	5
	15	0	24	5
	16	0	30	3.8
	17	1	45	0.24
	18	2	35	0.03
CF^R 2300	19	0	33	10
	20	0	15	7.5
Cordis-Dow				
C-DAK^R 2.5	21	0	23	0.45
C-DAK^R 4000	22	0	89	< 0.01
	23	0	34	< 0.01

priming saline in the initiation of dialysis. Because endotoxin is characteristically released from a new dialyzer at a rapid rate during the early period of fluid flow, the patient may receive a bolus of priming fluid containing levels of endotoxin at the start of dialysis which frequently exceed the 1 ng/kg of body weight dose considered to be the threshold pyrogenic dose for man (Tsuji, Steindler, et al 1977). In one instance where the saline contained 160 ng/ml, the priming volume delivered resulted in the patient receiving a bolus of 40,000 ng of LPS. It was proposed that the priming procedure be modified so that the saline recirculating through the blood compartment during preparation of the patient be discharged to waste and replaced with 500 ml of fresh saline just before initiation of dialysis. This procedure was tested at a local dialysis center and showed that the mean bolus dose of endotoxin delivered to the patient was reduced 95% (Table 4).

Table 4: Effect of Replacing Recirculated Saline Used to Prime New Hemodialyzers with Fresh Saline Prior to Dialysis

	Recirculated Saline		Fresh Saline		
Test No.	Min of Recirculation	Endotoxin ng/ml	Min of Recirculation	Endotoxin ng/ml	Percent Reduction
1	30	160	0	0.12	99.9
2	20	60	0	0.12	99.8
3	53	30	0	0.48	98.4
4	30	30	0	0.48	98.4
5	95	30	0	1.9	93.7
6	44	30	0	1.9	93.7
7	30	160	0	15	90.6
8	30	3.8	0	0.9	76.3

Tests conducted by other investigators have confirmed that the LAL-reactive material as obtained from the hemodialyzers is nonpyrogenic in the USP rabbit pyrogen test. This would suggest either that the pyrogenicity of the LPS being measured is markedly lower than that of the FDA reference endotoxin standard, or that the LPS recovered from dialyzers may differ significantly from the biologically active LPS associated with patient pyrogenic reactions (Tomasulo 1979; Wachtel, Tsuji 1977).

During the initial investigation of LPS associated with Cordis-Dow dialyzers, levels ranging from .32-1.28 ng/ml as determined with ACC lysate (ACCL) could not be detected using lysate reagent (Pyrogent) obtained from Mallinckrodt, Inc. (ML). A series of tests on fluids from various dialyzers (Table 5) showed that with ML no gel

Table 5: Comparative Reactivity of Commercial Sources of *Limulus* Lysate Reagent Against Fluids Extracted from Hemodialyzers

Dialyzer	Source of LAL-Reactive Material	LAL Reactivity (ng/ml)	
		ACCL[a]	ML[b]
Cordis-Dow C-DAK[R] 1.8	64-hr soak in saline	60	ND[c]
	16-hr soak in saline	3.7	ND
	15-min prime in saline	.64	ND
	15-min prime in saline	1.28	ND
	15-min prime in CDW[d]	.32	ND
Travenol CF[R] 2300	60-min soak in CDW	60	ND
Erika HPF-100	15-min soak in saline	0.6	ND
Gambro GF-120	15-min soak in saline	0.08	ND

[a] ACCL = Associates of Cape Cod lysate
[b] ML = Mallinckrodt lysate
[c] ND = Not detectable
[d] CDW = Distilled water for irrigation, Cutter Medical

formation was observed in the presence of what appeared to be levels as high as 60 ng/ml of LPS when measured with ACCL. Three different lots of ML reagent were used, with sensitivities ranging from .004 to .016 as determined with EC-2 and ACC no. 13 reference standards, so that negative results did not appear to be a simple matter of lysate sensitivity. These results were reminiscent of the earlier studies with ACCL where markedly increased turbidity was observed but no true gel formation of blood-side samples was obtained. Moreover, LAL-reactive material was now being detected with the newer ACCL at levels that should have been detectable with the old ACCL, which suggested that differences in manufacturing procedures can markedly affect lysate responses to endotoxin components.

It was suggested that there might be inhibition due to an endotoxin-binding or masking agent in the saline rinse fluid, and fluids were accordingly tested with Pyrosperse, a metallo-modified polyanionic dispersant (Mallinckrodt, Inc.) which is designed to reverse binding interference with endotoxin in blood and blood products. Treatment of either distilled-water or saline rinses of the blood compartments of hemodialyzers with Pyrosperse reagent still did not result in positive gelation with ML.

Other tests conducted in our laboratories indicated no inhibition by the product in samples of C-DAK^R 1.8 or CF^R 2300 fluids spiked with *E. coli* 055:5B or EC-2 when tested either with ACCL or ML. However, some inhibition of ML reactivity was obtained in test systems using dialyzer fluids spiked with the natural gram-negative bacterial flora obtained from the storage tank of a laboratory water-softening unit. In other experiments, samples of filtered tank water containing 10 ng/ml of natural environmental LPS and of *E. coli* 055:B5 solutions containing 80 ng/ml were primed for 30 min in the blood compartment of a C-DAK^R 1.8 dialyzer. LAL assays showed that ML detected only 24%-36% of the LPS as compared to levels obtained with ACCL. The nature of the inhibition observed with ML lysate reagent has still not been fully resolved.

A possible explanation for the nonpyrogenic nature of the LAL-reactive material in new dialyzers is that endotoxin molecules are degraded by alkylation during ethylene oxide (EtO) sterilization of the dialyzer. Reductions ranging from 75%-91% in the pyrogenicity of *E. coli* endotoxin following 6.5 hrs exposure to EtO at 50% relative humidity have been demonstrated using the USP rabbit pyrogen test (Tsuji, Harrison 1979). It was further suggested that such LPS degradations may affect the specificity of LAL reagents. A series of tests were conducted to examine the effect of EtO sterilization on the LAL reactivity of dialyzer-associated components. C-DAK[R] 1.8 dialyzers were leached until no LAL-reactive material could be detected using ACCL. The dialyzers were soaked overnight with a solution containing 80 ng/ml of _E. coli_ 055:B5 to allow attachment of LPS to the dialyzer membrane. One dialyzer was treated with EtO, and air-rinse fluids and 15-min prime fluids from both dialyzers were tested by LAL assay. No significant differences were observed in levels of LPS recovered from EtO and non-EtO treated dialyzers. Samples of _E. coli_ 055:B5 treated with EtO at the same time lost 50% of their reactivity in the LAL assay. Crude phenol-extracted LPS preparations of _Acinetobacter calcoaceticus_ and _Pseudomonas aeruginosa_ lost 94% and 74% of their reactivity, respectively.

Overall, our investigations to date have demonstrated the presence of apparently nonpyrogenic LAL-reactive material in most new dialyzers. Although its origin is not known, we have speculated that it may be present in the raw materials used in membranes or, more likely, in the water used in the production of membranes. It is of interest to note that LAL assays have shown a marked decrease in the levels of endotoxin leached from dialyzers in recent months when compared with levels reported previously. For instance, the mean level of endotoxin measured in the priming saline from 12 Travenol CF[R] 12.11 dialyzers used between 1/23/81 and 2/24/81 was 28 ng/ml compared to a mean level of only 0.31 ng/ml in saline from nine similar dialyzers used between 4/13/81 and 6/2/81. This 99% reduction in levels from cuprophan dialyzers was matched by a 98% reduction (6.5 ng/ml vs. 0.13 ng/ml) in samples collected from Cordis-Dow C-DAK[R] 1.8 and 2.5 dialyzers made of regenerated cellulose.

These findings have been confirmed in controlled dynamic leaching experiments conducted in the laboratory on cuprophan dialyzers from three different companies and on Cordis-Dow regenerated cellulose dialyzers. Figure 2 compares typical leaching curves for "high" and "low" cuprophan dialyzers. Because this change in the levels of endotoxin appeared to take place rather dramatically at about the beginning of 1981, the newer "low" units are

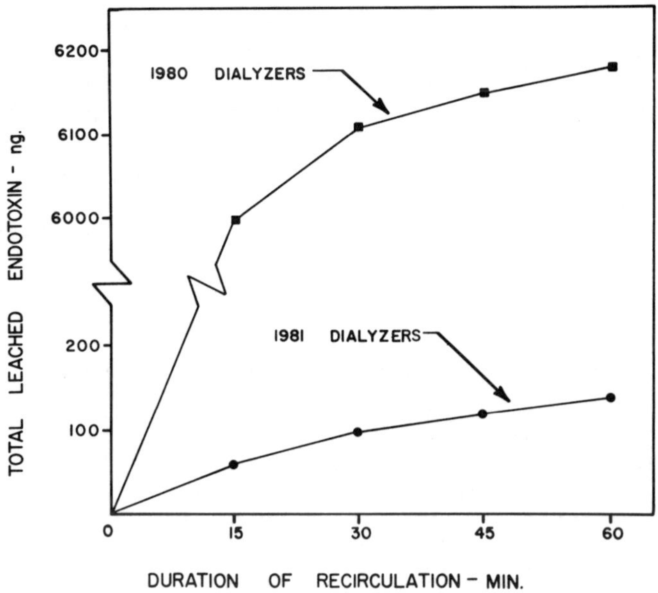

Fig. 2. Comparison of typical leaching rate curves of LAL-reactive material from hollow-fiber cuprophan dialyzers manufactured in 1980 and 1981.

referred to as 1981 dialyzers and the older "high" units as 1980 dialyzers. As can be seen from the shapes of the two curves, the reduction in LAL-reactive material was primarily the result of eliminating that source of material leached at a high rate during the first few minutes of use. We have further speculated that the coincidental reduction in levels of LAL-reactive materials in dialyzers manufactured by different firms may have resulted from a change in manufacturing procedures prompted by the first reports of our findings during the fall of 1980. While information regarding this subject has been exchanged with several manufacturers, questions concerning the source of the LAL-reactive material, why it is not pyrogenic, or how the levels have been reduced have not been resolved.

If it can be confirmed that the nonpyrogenic LAL-reactive material present in most new dialyzers is the cause of some of the presently unexplained clinical manifestations in patients using these devices, steps should be taken to ensure that levels of LAL-reactive material are reduced during the production and assembly of the membrane. Levels of LAL-reactive material in the final product could be monitored with LAL assays of fluids collected by either static or dynamic leaching tests, using lysate capable of detecting degraded or nonpyrogenic endotoxin.

REFERENCES

Bok DV, Pascual L, Herberger C, Sawyer R, Levin NW (1980). Effect of multiple use dialyzers on intradialytic symptoms. Proc Clin Dial Transplant Forum (in press).

Carson LA, Petersen NJ, Favero MS (1979). Use of the *Limulus* amoebocyte lysate assay system for detection of bacterial endotoxin in fluids associated with hemodialysis procedures. In Cohen E (ed): "Biomedical Applications of the Horseshoe Crab (*Limulidae*)", New York: Alan R Liss, Inc, p 453.

Dumler F, Levin N (1979). Hemodialysis induced leukopenia and hypoxemia - unrelated events. Arch Intern Med 139:1103.

Favero MS, Carson LA, Bond WW, Petersen NJ (1974). Factors that influence microbial contamination of fluids associated with hemodialysis machines. Appl Microbiol 28:822.

Favero MS, Petersen NJ, Boyer KM, Carson LA, Bond WW (1974). Microbial contamination of renal dialysis systems and associated health risks. Trans Amer Soc Artif Intern Organs XX:175.

Hindman SH, Favero MS, Carson LA, Petersen NJ, Schonberger LB, Solano JT (1975). Pyrogenic reactions during haemodialysis caused by extramural endotoxin. Lancet 2:732.

Holdsworth SR, Fitzgerald MG, Hosking CS, Atkins RC (1978). The effect of maintenance dialysis on lymphocyte function. Clin Exp Immunol 33:95.

Petersen NJ, Boyer KM, Carson LA, Favero MS (1978). Pyrogenic reactions from inadequate disinfection of a dialysis fluid distribution system. Dial Transplant 7:52.

Petersen NJ, Carson LA, Favero MS, Marshall JH, Aguero SM (1980). Removal of bacteria and bacterial endotoxin from dialysis fluids by sorbents. Amer Soc Artif Intern Organs 3:6.

Shin J, Matsuo M, Shinko S, Fujita Y, Inoue S, Sakai R, Nishioka M (1980). A study on hemodialysis leukopenia using various dialyzers. J Dial 4:51.

Sullivan JD, Valois FW, Watson SW (1976). Endotoxin: The *Limulus* amoebocyte lysate system. In Bernheimer AW (ed): "Mechanisms in Bacterial Toxicology", New York: John Wiley α Sons, Inc, p 218.

Tomasulo PA (1979). Correlation of *Limulus* amoebocyte lysate assay with accepted endotoxin assays. In Cohen E (ed): "Biomedical Applications of the Horseshoe Crab (*Limulidae*)", New York: Alan R Liss, Inc, p 293.

Tsuji K, Harrison SJ (1979). *Limulus* amoebocyte lysate - a means to monitor inactivation of lipopolysaccharide. In Cohen E (ed): "Biomedical Applications of the Horseshoe Crab (*Limulidae*)", New York: Alan R Liss, Inc, p 367.

Tsuji K, Steindler KA, Harrison SJ (1977). *Limulus* amoebocyte lysate assay for detection and quantitation of endotoxin in a small volume parenteral product. Appl Environ Microbiol 40:533.

Wachtel RE, Tsuji K (1977). Comparison of *Limulus* amoebocyte lysates and correlation with the United States Pharmaceutical pyrogen test. Appl Environ Microbiol 33:1265.

CARSON SPEAKER
TITLE: LAL-REACTIVE MATERIAL ASSOCIATED WITH HEMODIALYSIS MEMBRANES.

WATSON: Your paper raises many questions. Are we dealing with a true endotoxin, an inactivated endotoxin or some other compound which is giving a false positive reaction? Dr. Liu from BoB analyzed your material in a mass spectrometer. He has evidence to strongly suggest that your material does contain an endotoxin. We have found other cases where a solution will give a negative rabbit test and a strong LAL test. Possibly the endotoxin is conjugated with a protein which masks the active sites of the LPS and thus it does not appear to be pyrogenic. Possibly when this material is reacted with LAL, the inhibitor becomes disassociated from the endotoxin or in some way the active sites of the LPS become exposed. The question is should we permit endotoxins to be present in parenterals even though they do not appear to be pyrogenic in the rabbit. We are going to face the question many times; that is, do you trust the rabbit fever test or the LAL test.
NEELY: The questions which were raised are very important but the data you have may suggest something of even more importance, that is, clinicians using dialysis equipment should rinse it off first so the problem doesn't exist. Is this material extractable with water?
CARSON - Yes and we have made recommendations to the association of the dialysis people suggesting that the prime be discarded. If this is done, we get a 95% reduction in level so they are moving in this direction.
LEVIN: In regards to the pyrogenicity problem, there are at least two papers in the literature which report pyrogenic reactions associated with dialysis in which endotoxin was detected both in the dialyzer and the patient. This may be an example of the patient telling you what you want to know, i.e., there is an endotoxin passing through the dialysis membrane.
CARSON: We have not found one case where there was a pyrogenic reaction where a new dialyzer was involved. Almost invariably it is a used dialyzer, and in these cases the disinfectant fluids were the source of the pyrogen.

Endotoxins and Their Detection With the Limulus Amebocyte Lysate Test, pages 231–246

A CORPORATE APPROACH TO IN-PROCESS AND END-PRODUCT TESTING WITH THE LAL ASSAY FOR ENDOTOXIN

Frederick C. Pearson III, Marlys E. Weary, and Roger Dabbah

TRAVENOL LABORATORIES, INC.
6301 Lincoln Avenue
Morton Grove, IL 60053

INTRODUCTION

A decade ago, Travenol's management decided to evaluate the *Limulus* amebocyte lysate (LAL) test as a potential replacement for the USP rabbit pyrogen test. The following summarizes this activity and presents the current status of the in-house LAL program.

Manufacture of *Limulus* Amebocyte Lysate

During 1972, Travenol established a *Limulus* bleeding installation in Beaufort, SC. To obtain hemolymph, crabs caught by local fishermen are bled by inserting a needle into the cardiac sinus. Three hundred milliliters of hemolymph are collected in a unit containing 300 ml of 0.025M EDTA in 3% (w/v) NaCl to prevent amebocyte coagulation. The hemolymph is immediately centrifuged and the supernatant is decanted. The amebocytes are washed in 3% NaCl, centrifuged again, and lysed with nonpyrogenic distilled water (water: cells 5:1).

The amebocyte lysate is extracted and stored frozen until tested for potency and pooled. Lysate raw material that passes potency testing is shipped to Round Lake, IL (Hyland Diagnostics, a Division of Travenol Laboratories), for bulking, filling, lyophilization, and distribution to domestic and international LAL testing facilities (Table 1).

Table 1. Travenol LAL Testing Laboratories

Domestic	Cleveland, MI	Hyland, Glendale, CA
	Hays, KS	Carolina, PR
	Morton Grove, IL	Jayuya, PR
	Mountain Home, AR	Aibonito, PR
	North Cove, NC	Maricao, PR
International	Australia	Israel
	Belgium	Italy
	Brazil	Mexico
	Canada	Norway
	Columbia	South Africa
	England	Spain
	France	Singapore
	Ireland	

LAL Test System (Lowry Protein-Colorimetric Method)

The Travenol Optical Density-Lowry Protein method is based on increasing concentrations of endotoxin precipitating proportionally more lysate protein. The amount of lysate-specific protein (coagulogen) that is precipitated can be quantified by a simple Lowry protein determination (Nandan and Brown, 1977).

Equal volumes (0.1 ml) of test sample and lysate are mixed in 10 x 75-mm, pyrogen-free test tubes, incubated at 37°C for one h, and then centrifuged for ten min at 1375 x g. The supernatant is removed by vacuum aspiration. The amount of lysate-specific precipitated protein is determined using the Oyama and Eagle (1956) modification of the Lowry protein assay (Lowry et al, 1951). Using a flow-through spectrophotometer, the optical density (OD) is recorded at 660 nm and the results related to a standard reference curve.

Positive and negative controls must be run with each test to rule out false positives (contamination) or false negatives (inhibition of the clotting reaction by the test sample). Negative controls consist of solutions, identical in chemical composition to the test samples, that have previously been tested using LAL and shown to be nonpyrogenic. Solutions can be used as negative controls if the OD of the precipitable protein is less than 0.2.

Positive controls consist of three concentrations (200, 50, and 12 pg/ml) of an *E. coli* endotoxin standard (055:B5, Difco Laboratories, Detroit, MI), prepared in the same solution as the negative control, that are used to generate an OD standard curve. Test samples with OD's at or above 50 pg/ml are considered to have failed.

To qualify a product for LAL testing, inhibition and/or enhancement of the LAL reaction by the product must be ruled out. The negative product control must have an OD of less than 0.2, which rules out enhancement of the reaction (false positive). There must be a differential of 0.2 OD units between the negative control and 50-pg/ml control points, as well as between the 50-pg/ml and 200-pg/ml control points, obviating LAL inhibition (false negative). When a product tests within these parameters, it is qualified.

An endotoxin-water curve, an endotoxin-spiked product curve, and a human serum albumin curve are used to ensure adequate performance of the LAL reagent and endotoxin, lack of enhancement or inhibition, and performance of the Lowry protein reagents, respectively.

The Travenol LAL procedure was introduced in production plants early in 1974, rapidly becoming the most important quality control tool for in-process pyrogen testing of raw materials, plant water, and parenteral solutions. For the first time, a method was available for monitoring nonpyrogenic levels of endotoxin that remained undetected when the rabbit test was used.

STATUS OF LAL TESTING

Testing Medical Devices

During the past decade, several device manufacturers began using LAL for testing medical devices and administration sets (Weary and Baker, 1977; Ferry, 1979). Further impetus was given to LAL test development by a joint announcement of the Bureau of Biologics (BOB) and the Bureau of Medical Devices (BMD) in the Federal Register of Nov 4, 1977, indicating that the LAL test was approved for use as a final pyrogen test of biological products and medical devices provided that manufacturers: (1) submit adequate data establishing the equivalence of their LAL test procedure and the USP rabbit pyrogen test; and (2) obtain approval for product license amendments.

While the former request seems a reasonable one, it is difficult for a single manufacturer to establish LAL/rabbit test equivalency considering the inherent variability of the rabbit assay and rabbit colony sensitivity. To obviate this problem, the Health Industry Manufacturers Association (HIMA) undertook a collaborative study to evaluate a dose of *E. coli* 055:B5 standard endotoxin that would be equivalent to the pyrogenic level in the rabbit test and, thus, be used as a final release limit for medical devices (HIMA, 1979). Using 12-rabbit pyrogen-test colonies from regulatory, reference, and pharmaceutical laboratories, the study concluded with 95% confidence that a 50% pass/fail result would occur at some level of *E. coli* 055:B5 in excess of 0.1 ng/ml (1.0 ng/kg). Thus, the study recommended that a reasonable release limit for medical devices would be 0.1 ng of endotoxin/ml, which establishes a built-in safety factor over the rabbit test.

Following the HIMA study, the "Bureau of Medical Devices Draft Guidelines for Adoption of the Limulus Amebocyte Lysate Test for Pyrogenicity of Devices" (1980) was issued. This document requires the manufacturer to demonstrate the following: (1) the sensitivity of the LAL test and the ability of testing personnel to consistently reproduce LAL end points, (2) the lack of enhancement or inhibition of the LAL reaction by product, and (3) comparable results using the LAL and rabbit tests.

Typical data from five Travenol manufacturing facilities demonstrating the equivalence of the LAL and rabbit tests or the former's superiority can be seen in Table 2. In most cases, the LAL assay failed materials at the 50-pg/ml concentration, but the rabbit test consistently passed them. Only at the Morton Grove and Alliston facilities were rabbit colonies sensitive enough to fail endotoxin-spiked saline at 100 pg/ml. These data also clearly indicate that the Travenol OD procedure is two to four times more sensitive when saline and water solutions, the most commonly used rinse solutions for pyrogen testing of medical devices, are tested.

Table 2. Comparative Pyrogen Test Results with *E. coli* 055:B5 in Water (Cumulative Rabbit Temperature Increases Expressed in Celsius Degrees)

	Sample No. 1 200 pg/ml			Sample No. 2 100 pg/ml			Sample No. 3 50 pg/ml		
	8-Rabbit Test	LAL Tests Plain	LAL Tests Isotonic	8-Rabbit Test	LAL Tests Plain	LAL Tests Isotonic	8-Rabbit Test	LAL Tests Plain	LAL Tests Isotonic
Alliston, Canada	+6.4 (Fail)	Fail	Fail	+4.0 (Fail)	Fail	Fail	+0.1 (Pass)	Fail	Fail
Cleveland, MI	+4.7 (Fail)	Fail	Fail	+1.8 (Pass)	Fail	Fail	+3.1 (Pass)	Pass	Pass
Hays, KS	+2.6 (Pass)	Fail	Fail	+1.4 (Pass)	Fail	Fail	+3.1 (Pass)	Pass	Fail
Morton Grove, IL	+6.0 (Fail)	Fail	Fail	+4.3 (Fail)	Fail	Fail	+2.6 (Pass)	Fail	Pass
Mountain Home, AR	+4.1 (Fail)	Fail	Fail	+3.1 (Pass)	Fail	Fail	+4.1* (Fail)	Pass	Fail
Mountain Home, AR	+2.9 (Pass)	Fail	Fail	+2.9 (Pass)	Fail	Fail	+1.0 (Pass)	Pass	Pass
North Cove, NC	+5.1 (Fail)	Fail	Fail	+2.6 (Pass)	Fail	Fail	+2.0 (Pass)	Fail	Pass
Carolina, PR	+0.9 (Pass)	Fail	Fail	+1.6 (Pass)	Fail	Fail	+1.2 (Pass)	Fail	--

* 2 rabbits each had increases of more than 1.0°C.

Travenol has conducted end-product release testing on medical devices since Apr 24, 1978, when permission was received from the BMD. The status of the corporation's end- product testing program for general product lines, including medical devices, is summarized in Table 3.

Final-Release Testing of Large Volume Parenterals with LAL

On Jan 18, 1980, the Bureau of Drugs (BOD) issued "Draft Guidelines for the Validation of Limulus Amebocyte Lysate Test as an End-Product Pyrogen Test for Human and Veterinary Injectable Drugs Other than Licensed Biologicals." Responses from industry had to be made to the Hearing Clerk by Mar 18, 1980. Some of industry's requests were incorporated in a subsequent draft guideline published in 1981.

Equivalency of LAL and USP Rabbit Pyrogen Tests

To establish the equivalence of the LAL endotoxin assay and the rabbit test, or LAL's superiority, a number of parallel tests have been conducted using both assays

Table 3. Regulatory Status of Travenol LAL Testing Program in the United States

Task	FDA Agency*	Notification Date
LAL License	BOB	Jun 24, 1978
Release of Medical Devices	BMD	Apr 24, 1978
Release of Non-NDA LVP's	BOD	Nov 1, 1978
Release Testing of Blood Collecting Systems	BOB	Nov 8, 1978
Release of Blood Fractions - 25% Albumin	BOB	Pending

* BOB indicates Bureau of Biologics; BMD, Bureau of Medical Devices; BOD, Bureau of Drugs.

with specified levels of endotoxin. A large body of data, collected over several years, allows direct comparison of the two test systems at or near the pyrogenic threshold level. These data support the argument that the LAL assay is equivalent to, or more sensitive than, the rabbit test in detecting endotoxin. The data in Table 4 were obtained from blinded trials conducted concurrently at several Travenol facilities. This comparative study provides information on the relative performance of the 3-rabbit test and the LAL OD assay. In all cases, the LAL test was equivalent or superior to the rabbit test. In addition, as can be seen in Table 5, 5% dextrose and 0.9% sodium chloride were selected for equivalency testing. Samples were tested in plants that routinely manufacture these solutions, and at the Travenol Corporate R&D facility. The unknowns sent to each of the sites consisted of an E. coli 055:B5 endotoxin reference standard that gave concentrations of 195, 95, and 49 pg/ml when diluted in the products.

Table 4. Comparison of Results from Multisite Blinded Pyrogen Tests

Endotoxin (pg/ml)	Rabbit Test		LAL Test	
	Pass	Fail	Pass	Fail
195	0	4	0	4
98	2	2	0	4
50	4	0	2	2

Table 5. Comparative Pyrogen Tests of Travenol Products Spiked with Various Concentrations of E. coli 055:B5 Endotoxin

Product	Endotoxin (pg/ml)	Rabbit Test Pass	Rabbit Test Fail	LAL Test Pass	LAL Test Fail
Normal Saline	195	0	4	0	4
	95	2	2	0	4
	49	4	0	2	2
5% Dextrose	195	0	4	0	4
	95	0	4	1	3
	49	4	0	1	3

The Travenol LAL test designates 50 pg of reference standard endotoxin per ml as the screening pass/fail limit. At Travenol, this concentration in 0.9% sodium chloride and 5% dextrose passes the rabbit test (see Table 5). Both the LAL and rabbit tests failed all samples containing 195 pg of endotoxin per ml. However, at the lower concentrations, the LAL assay was more sensitive and failed a greater number of samples.

In early 1978, a study was initiated to determine equivalence between the Travenol LAL OD test and the rabbit test. Because they represent a cross section of Travenol solutions, the following products were used: 5% dextrose injection, 10% dextrose and 0.9% sodium chloride, 0.45% sodium chloride, 1.5% glycine, lactated Ringer's injection, 0.9% sodium chloride, and sterile water for injection. Comparative rabbit and LAL tests were performed under normal daily test conditions. All products tested were Travenol lots that had previously passed LAL testing. _E. coli_ 055:B5 endotoxin was added to 20 samples of each product, and the samples were submitted for rabbit and LAL testing. A different Travenol product was tested each week for a total of eight weeks, so that comparative rabbit and LAL tests were run for five days on each product (Table 6).

Based on comparative data, which demonstrate that the LAL assay is as sensitive as, or more sensitive than, the rabbit assay for LVP's, Travenol made a submission to the BOD requesting permission to release LVP solutions using

Table 6. Comparative Pyrogen Tests of Travenol LVP's Spiked with E. coli 055:B5

Product	Endotoxin (pg/ml)	Rabbit Test Pass	Rabbit Test Fail	LAL Test Pass	LAL Test Fail
5% Dextrose	35	10	0	10	0
(injection)	75	9	1	0	10
0.9% Sodium	35	10	0	9	1
Chloride	75	7	3	4	1
0.45% Sodium	35	10	0	10	0
Chloride	75	9	1	4	6
Lactated Ringer's	35	10	0	10	0
(injection)	75	4	6	4	6
Sterile Water for	35	10	0	9	1
Injection	75	8	2	1	9
10% Dextrose & 0.9%	35	10	0	9	1
Sodium Chloride	75	8	2	7	3
1.5% Glycine	35	9	1	9	1
	75	6	4	6	4

LAL (summer 1978). In the absence of their approval, the BOD was informed that Travenol would begin releasing "grandfather" LVP solutions by Nov 1, 1978, and Travenol has been doing so in the US since that time. The corporation still awaits an official approval to perform final-release pyrogen testing by LAL. However, approval elsewhere has been forthcoming; on Mar 2, 1979, Canada approved final release of LVP's by the Travenol LAL procedure, and on Feb 25, 1981, South Africa and Zimbabwe.

Over the past several years, a great deal of confidence has been generated in the LAL test as a final release test for pyrogens. While several hundred thousand LAL tests have been conducted by Travenol, and several hundred have required retest in rabbits, in no case has there been an LAL pass and a rabbit failure (Table 7).

Final-Container Testing of Biologicals

Although LAL has been generally accepted for pyrogen testing of many LVP's and most medical devices, a number of reports in the scientific literature remain divided on the applicability of the LAL assay for the final release

Table 7. Summary of 1980 Rabbit and LAL Tests in Travenol Domestic and International Plants, 1980

	Domestic	International
Number of rabbit tests	19,761	18,423
Number of 5- or 8-rabbit repeat tests	97	282
Number of pyrogenic 8-rabbit tests	8	6
Number of LAL tests	140,316	73,036
Number of 3-tube LAL repeat tests	1,205	1,203
Number of LAL repeat test failures	230	137
Number of 8- or 12-rabbit test failures resulting from LAL test failures	1	14

of plasma fractions. Several investigators have reported on the use of LAL as a means of detecting endotoxin in plasma fractions (Hochstein and Seligman, 1979; Cooper et al., 1972; and Pearson et al., 1981).

The joint announcement by the BOB and BMD gave added impetus to the LAL testing program at Hyland Therapeutics, a division of Travenol Laboratories, where the quantitative LAL methodology had been introduced during the summer of 1976 to evaluate a pyrogen problem with 25% albumin. Goals were set to validate the LAL system and establish equivalence with the rabbit assay for testing 25% human serum albumin (HSA), 5% HSA, human purified protein fraction (HPPF), and antihemophilic factor (AHF).

Two lots of HSA (25%) were spiked with 0, 10, 50, and 100 pg of *E. coli* 055:B5 endotoxin per ml, and LAL tests were conducted on both lots on each of 30 consecutive work days. The LAL OD method clearly differentiated between levels of the commercial endotoxin (Table 8). Analysis of the data indicated that the test-to-test and day-to-day variations were minor.

To establish the relative sensitivity of the tests, a total of 957 parallel rabbit and LAL tests were run on 251 lots of 25% HSA. Results of these studies were used

Table 8. Results of Parallel LAL OD Tests Performed on Two Lots of 25% HSA Spiked with Endotoxin

	Endotoxin (pg/ml)			
	0	10	50	100
Lot B				
Mean	0.275	0.383	0.670	0.911
SD	± 0.037	± 0.041	± 0.087	± 0.129
CV	0.13	0.11	0.13	0.14
Lot C				
Mean	0.198	0.320	0.585	0.818
SD	± 0.041	± 0.046	± 0.100	± 0.130
CV	0.21	0.14	0.17	0.16

to establish a pg/ml endotoxin cutoff point based on a dose-response relationship between endotoxin and rabbit fever response. When a 50-pg/ml LAL limit was applied, 902 samples passed both tests. However, 26 samples that passed the rabbit test failed the LAL assay, indicating its greater sensitivity. The rabbit test failed 26 samples that also failed the LAL test, and in 3 cases, the LAL test passed samples that failed the rabbit test. The latter may be due to the variability of the rabbit test. When the LAL cutoff value was increased from 50 to 60 pg of endotoxin per ml, an additional 8 lots of albumin passed that had also passed the rabbit test. When 70 pg/ml was used as a cutoff, an additional 7 lots passed (Table 9). Analyses of these data indicated that 60 pg/ml is an acceptable, yet conservative, limit for 25% HSA in its final container. The agreement between the rabbit and LAL tests at this cutoff value was 97.81%.

To evaluate the relative sensitivity of the rabbit and LAL tests, the percentage of disagreement between the two tests was computed. The test more likely to yield a failure was judged to be the more sensitive. Using this criterion, at a 60-pg/ml LAL cutoff, 14.3% of the product that failed the rabbit test passed the LAL test, but 85.7% of the product failing the LAL test passed the rabbit test. The LAL OD procedure for detection of endotoxin in 25% HSA is, consequently, more sensitive than the currently accepted rabbit test (Table 9).

Table 9. Results of Parallel Tests Conducted on 25% HSA with the LAL and Rabbit Tests

	Rabbit - Pass	Rabbit - Fail
LAL Cutoff = 50 pg/ml		
LAL - Pass	902	3
LAL - Fail	26	26
LAL Cutoff = 60 pg/ml		
LAL - Pass	910	3
LAL - Fail	18	26
LAL Cutoff = 70 pg/ml		
LAL - Pass	917	3
LAL - Fail	11	26

Based on these studies and others, Travenol's Hyland Therapeutics Division made a submission to BOB on Nov 8, 1978, requesting final release of 25% HSA using LAL. Although a number of questions were subsequently answered, the submission remains unapproved by BOB. Nonetheless, Hyland remains optimistic that the request will be approved when the Bureaus of Medical Devices, Drugs, Biologics, and Veterinary Drugs release the LAL guidelines for end-product testing.

In-Process Testing with LAL

One of the most valuable uses of the LAL assay is for in-process pyrogen testing of raw materials and water for production. Such a rapid, sensitive, and relatively inexpensive assay permits the monitoring of subpyrogenic levels of endotoxin, and provides for the detection of rising pyrogen levels before critical contamination occurs.

Normal serum albumin (NSA, 25%) is an outstanding example of a product that contained relatively high levels of endotoxin (nonpyrogenic in rabbit). From the mid-1970's to 1978, Hyland Therapeutics adapted the Travenol LAL OD procedure to human blood fractions, and BOB began monitoring blood fractions released using LAL. Data are given in Table 10 (Hyland Therapeutics is represented by the letter E). As can be seen, Hyland's 25% NSA contained an average of 4.0 ng/ml endotoxin in 1975, but contained only 0.05 ng/ml just three years after adoption of the LAL assay. The LAL test has

Table 10. Average Endotoxin Content of 25% NSA (Human) in ng/ml*

Manufacturer	Year 1975	1976	1977	1978
A	1.0(66)	4.0(68)	3.0(35)	1.0(87)
B			0.6(43)	0.5(10)
C	4.0(67)	3.0(42)	2.0(61)	2.0(120)
D	0.1(28)		0.6(69)	0.5(70)
E	4.0(148)	3.0(252)	0.8(44)	0.05(161)
F			0.8(6)	1.4(10)
G			0.6(7)	0.7(19)
H			2.0(5)	2.0(19)
I			1.0(28)	
J			0.6(30)	
K	8.0(4)		1.0(19)	
L			0.4(18)	0.4(25)

* ( ) total lots tested. Source: HD Hochstein et al.

proven to be a critical tool in troubleshooting pyrogen problems in other pharmaceutical products as well.

CONCLUSION

Over 500,000 LAL tests have been conducted by Travenol during the past five years. Fewer than 500 of these tests failed product at the 50-pg/ml endotoxin alert limit, which would necessitate retest by the rabbit assay. During this time, there have been only 3 instances (Hyland) in which the LAL test passed and the rabbit test failed a test substance; however, the LAL test consistently fails test materials at the 50-pg/ml alert limit, and the rabbit test consistently passes them. Thus, LAL has proven to be a specific, accurate, and sensitive replacement for the rabbit test for use with medical devices, LVP's, and plasma fractions. While final release testing of LVP's is permitted only in Canada, South Africa, and Zimbabwe, and testing of biologicals has yet to be sanctioned, it is with confidence and enthusiasm that Travenol anticipates worldwide regulatory acceptance for final release of appropriately qualified health-care products by LAL testing.

REFERENCES

Cooper JF, Hochstein HD, Seligman EB, Jr. (1972). The *Limulus* test for endotoxin (pyrogen) in radiopharmaceuticals and biologicals. Bull Parent Drug Assoc 26:153.

Draft guidelines for the validation of the *Limulus* amebocyte lysate test for human and veterinary injectable drugs other than licensed biologicals. Federal Register, Jan. 18, 1980, p 3668.

Ferry EW (1979). Stagnation to anticipation - an over view of LAL use in the pharmaceutical industry. In Cohen E (ed): "Biomedical Applications of the Horseshoe Crab (Limulidae)," New York: Alan R. Liss, p 339.

HIMA Collaborative Study for the Pyrogenicity Evaluation of a Reference Endotoxin by the USP Rabbit Test. HIMA Document Series, Vol. 1, No. 7, Washington, DC: Health Industry Manufacturers Association, 1979.

Hochstein HD, Seligman EB, Jr. (1979). *Limulus* lysate testing of biologic requirements for the rabbit pyrogen test. In Cohen E (ed): "Biomedical Applications of the Horseshoe Crab (Limulidae)," New York: Alan R. Liss, p 501.

Licensing of Limulus lysate: Use as an alternative for the rabbit pyrogen test. Federal Register, Nov. 4, 1977, p 57749.

Lowry OH, Rosenbrough NJ, Farr AL, Randall RJ (1951). Protein measurement with the Folin phenol reagent. J Biol Chem 193:265.

Mascoli CC, Weary ME (1979). *Limulus* amebocyte lysate (LAL) test for detecting pyrogens in parenteral injectable products and medical devices: Advantages to manufacturers and regulatory officials. J Parent Drug Assoc 33:81.

Nandan R, Brown DR (1977). An improved *in vitro* pyrogen test: To detect picograms of endotoxin contamination in intravenous fluids using *Limulus* amebocyte lysate. J Lab Clin Med 89:910.

Pearson FC, Bachich J, Srigley W, Magalong E, Greff E, Abroson F, Garcia D (1981). Applications of the *Limulus* amebocyte lysate assay to testing of human albumin. Am Soc Indust Micro (in press).

Oyama VI, Eagle H (1956). Measurement of cell growth in tissue culture with a phenol reagent (Folin Ciocalteau). Proc Soc Exp Biol Med 91:305.

Weary ME, Baker B (1977). Utilization of the *Limulus* amebocyte lysate test for pyrogen testing large volume parenterals, administration sets, and medical devices. Bull Parent Drug Assoc 31:128.

Weary M, Pearson FC (1980). Pyrogen testing with *Limulus* amebocyte lysate. Med Dev Diagn Indust 2:34.

PEARSON SPEAKER
TITLE: THE RELATIVE PYROGENICITY OF ENVIRONMENTAL ENDOTOXIN AS MEASURED BY THE LIMULUS AMEBOCYTE LYSATE ASSAY AND THE USP RABBIT PYROGEN TEST.

ENZINGER: You presented data showing that a product containing 1,000 ng/kg endotoxin passed the USP test in one case but failed in another case. Did you by chance isolate those organisms to find out what species they were?
PEARSON: The preliminary work we did on bacterial isolation and identification was limited.
ENZINGER:- If you look at environmental isolates in manufacturing plants you would rarely find E. coli, Salmonella or Shigella instead you would find Pseudomonas and Chromobacter, or similar non-fermenters. These species probably wouldn't elicit a greater pyrogenic response in the rabbit than E. coli. I think this knowledge must be factored in when we are talking about safety factors in the amount of allowable endotoxin in parenteral solutions.
PEARSON: Our data on environmental endotoxins is very preliminary. I see nothing wrong with using purified endotoxins if we recognize we have created a significant safety factor when we use them.
MORRISON: Are environmental endotoxins less potent because they were derived from different species than the purified endotoxins employed? If these environmental endotoxins could be purified, would their potency increase and be equal to the present purified endotoxin standards?
PEARSON: Our data indicate Pseudomonas endotoxin, prior to purification, is potent in LAL but relatively impotent in rabbits, however after purification this endotoxin becomes very potent in rabbits.
MORRISON: Let me repeat the question. Are environmental endotoxins less toxic or are they on a per weight basis equally toxic but they are diluted out by the fact that you have non-toxic contaminants present?
PEARSON: The pyrogenicity of Pseudomonas sp. is far less than you would expect based on its potency with LAL. We know if you extract the endotoxin from this species, the pyrogenicity increases greatly. This suggests the more purity the greater toxicity.

TSUJI: Are you trying to say that you obtained higher reactivity in rabbits than with LAL when you used the purified LPS than the impure LPS? Our experience indicates that LAL activity and rabbit activity are closely related regardless of the purity of endotoxin.
PEARSON: Are you talking about what you would consider relatively purified lipopolysaccharides or are you talking about environmental endotoxins?
TSUJI: Whole cell environmental versus purified LPS.
PEARSON: With *E.coli*?
TSUJI: Both *E. coli* and *Pseudomonas*. When you talk about environmental endotoxins, you must also consider sources other than water, for example, dust particles and human associated microorganisms. You have only considered environmental endotoxins from water.
PEARSON: No, I am not saying there is more reactivity in the rabbit than with LAL when endotoxins are purified. Our experience is just the opposite. With environmental endotoxin there is more LAL reactivity than would be predicted when the sample is tested by rabbit. We speculate that the gap between rabbit and LAL is closed as endotoxins become purified. The data presented represents water from a great variety of sources from raw materials in process to end-product test.
ENZINGER: Possibly there is no real difference between environmental and purified endotoxin. In working with environmental endotoxins you are encountering different genera of bacteria than with purified endotoxins. Maybe the differences observed stem from differences in activity of endotoxins from various genera. Endotoxins from *Pseudomonas* may be more pyrogenic than those obtained from *E. coli*. It is not clear whether *E. coli* cells would be less pyrogenic than purified endotoxin derived from this organism.
PEARSON: We know little about organisms responsible for the presence of endotoxins in our products.
MASCOLI: The waters you tested were reused water supplies of that community and therefore they contained endotoxins from many different species of bacteria.
MORRISON: I would like to emphasize what I consider a very relative and important point that you made and that is the safety factor introduced by the fact that purified endotoxin is more reactive with LAL than environmental endotoxin.

Endotoxins and Their Detection With the Limulus Amebocyte Lysate Test, pages 247–260

DETECTION OF *LIMULUS* AMEBOCYTE LYSATE REACTIVE MATERIAL IN CAPILLARY *FLOW* HEMODIALYZERS

Frederick C. Pearson, Marlys Weary, Joyce Bohon

Travenol Laboratories, Inc.
6301 Lincoln Avenue
Morton Grove, IL 60053

INTRODUCTION

Microgram amounts of *Limulus* amebocyte lysate reactive material (LAL-RM), a *material that* does not produce a reaction in the USP rabbit pyrogen test, have been found in recirculated saline flushes of capillary flow dialyzers. This article deals with a series of preliminary studies intended to characterize the LAL-RM, identify its source, and determine whether it is, in fact, pyrogenic.

PRELIMINARY STUDIES

Recognition of Background "Endotoxin" in Hemodialyzers

Traditionally, saline washes from the blood compartments of capillary flow dialyzers have had LAL endotoxin equivalent levels above the 50-pg/ml alert limit used by Travenol Laboratories. These dialyzers were released using the USP rabbit pyrogen test. Among the thousands of dialyzer lots produced by Travenol since 1970, no USP rabbit pyrogen failures have been seen. In Sep 1978, a study was designed to establish the relationship between the Travenol LAL test (using a pass/fail concentration of 100 pg/ml referenced to *E. coli* 055:B5) and the USP rabbit test. When 100 lots *of dialyzers* were evaluated with this limit, saline rinse solutions consistently passed both tests.

The results of this study were submitted to the Bureau of Medical Devices, and Travenol was granted permission to release this product using LAL with a 100-pg/ml limit.

Center For Disease Control (CDC) Studies

In the Jan-Mar 1980 Quarterly Report issued by CDC, data were presented showing as much as 6,000 ng of "endotoxin" leaching from the blood compartments of new dialyzers during a 16-hour static soak. If this "LAL-reactive material" was received as a bolus injection by a 70-kg recipient, the individual would receive a dose of 85.7 ng/kg, 10 times that expected to induce a pronounced fever if the substance were endotoxin, However, because the LAL-RM shows little or no ability to induce fever, it cannot be identified as an endotoxin until further characterization.

Norman Peterson of CDC also shared data on Travenol dialyzers in which recirculation time ranged from 15 to 66 min, and LAL-RM concentrations ranged between 5.0 and 30 ng/ml on first-use dialyzers. However, reused hemodialyzers exhibited dramatically less LAL-RM, suggesting that significant leaching took place during original hemodialysis. These data can be seen in Table 1.

Table 1. Concentration of LAL-RM in Travenol CF® Dialyzers

CF Model No.	Status*	Recirculation Time (minutes)	LAL-RM Conc. (ng/ml)
2300	N	15	7.5
1500	N	66	5.0
1500	N	49	30.0
2300	N	33	10.0
1500	N	36	15.0
1500	R	24	5.0
1500	R	45	0.24
1211	R	30	0.03
1211	R	28	0.45

* N indicates new unit; R indicates unit reused once. Source: N. Peterson and L. Carson, CDC (personal communication).

Confirmation of CDC Findings

To reproduce the CDC studies, Travenol initiated a dialyzer recirculation experiment using two unused capillary flow dialyzers--CF® 1500 and CF® 1211.

Both dialyzers were primed with dialysate for 10 min at flow rates of 500 ml/min. Next, 1 liter of saline in a Viaflex® plastic bag was attached to the blood compartment of each unit. Following the instructions on the package insert, most of the saline was rinsed through each unit, and this discarded solution was measured and collected in depyrogenated containers for evaluation with LAL. About 685 ml of priming solution was retrieved from the CF 1500 dialyzer, and about 775 ml from the CF 1211. The remaining priming solution was allowed to recirculate within the blood chamber of each unit at a flow rate of 200 ml/min. At specified intervals, 10-ml samples were withdrawn using sterile, depyrogenated equipment, and endotoxin levels were quantitated using the LAL test with a gel clot end point. Results for the CF 1500 dialyzer are shown in Table 2. The CF 1211 had 1.6 ng/ml of LAL-RM in the discarded prime and 6.4 ng/ml in all ensuing samples tested.

Table 2. Concentration of LAL-RM in Saline Flushed Through a CF® 1500 Dialyzer

Sample	Recirculation Time	Approximate Amount of Sample (ml)	LAL-RM Conc. (ng/ml)
Sample 1 (discarded saline prime solution)	None	685	1.6
Sample 2	1 hour	10	6.4
Sample 3	2 hours	10	25.6
Sample 4	3 hours	10	25.6
Sample 5	Static soak overnight	10	25.6
Remainder	Static soak overnight	275	25.6
		1000	

By calculating the amount of LAL-RM found in the discarded prime, each of the samples, and the saline remaining in each unit at the end of each study, it appears that the CF 1500 contained at least 9.0 micrograms of LAL-RM, and the CF 1211 approximately 1.5 micrograms. Comparable levels of LAL-RM were found in other capillary flow dialyzers evaluated. Three Asahi dialyzers contained 32 ng of LAL-RM per ml, while a fourth contained 64 ng/ml. Two Erica dialyzers were found to contain 40 and 80 ng/ml.

Nonpyrogenicity of LAL-RM

In an effort to further establish a relationship between LAL-RM and rabbit pyrogenicity, extractions were made by recirculating saline through the blood compartments of various dialyzers for 3 hours. The extracts were then drained from each dialyzer and assayed using the LAL and USP rabbit tests. As shown in Table 3, at concentrations of up to 200 ng/ml (2000 ng/kg), LAL-RM is not pyrogenic in the USP rabbit test. Three-rabbit fever test results ranged between 0.0° and 0.4°C in this study, and a marginal rabbit pyrogen test failure would require a total three-rabbit response of 1.4°C. Based on the Health Industry Manufacturers Association (HIMA) study on rabbit pyrogenicity, one would expect 100% rabbit fever failures at a dose of 10 ng/kg, if the material were pyrogenic.

Table 3. Relationship of Pyrogenicity and LAL-RM from CF® Dialyzers

Sample	LAL-RM Conc. (ng/ml)	USP 3-Rabbit Test Increases (°C)
1500-5M1786	100	-0.2, 0.0, +0.2 = +0.2
1500-5M1786	200	-0.2, -0.2, 0.0 = 0.0
1511-5M1787	200	+0.1, +0.1, +0.1 = +0.3
1511-5M1787	38	0.0, +0.1, +0.1 = +0.2
1200-5M1782	38	0.0, +0.1, 0.0 = +0.1
1200-5M1782	160	+0.1, -0.2, -0.3 = +0.1
1211-5M1784	80	0.0, +0.1, 0.0 = +0.1
1211-5M1784	40	-0.2, +0.1, +0.1 = +0.1
2300-5M1788	160	0.0, 0.0, -0.2 = 0.0
2300-5M1788	160	-0.2, +0.3, +0.1 = +0.4

Source of LAL-RM

From shelf samples of the previous 17-months' production, 29 lots of Enka cuprophane hollow fibers were evaluated using the Travenol LAL assay. Approximately 3 g of fibers from each lot were soaked for 20 min in 50 ml of normal saline--a ratio of fiber to saline approximating that occurring in a normal rinse procedure for the hollow-fiber dialyzer, finished-product pyrogen test. Approximate quantitation of LAL-RM in each rinse was achieved by serially diluting each sample twofold with normal saline. The OD value of the dilution closest to that of the 50-pg/ml endotoxin control was multiplied by the dilution factor to produce the endotoxin equivalent value (ng/ml). All but two of the 29 lot samples failed the LAL test at a cut-off point of 0.4 ng/ml. The 2 samples that passed initially exhibited higher OD values as they were diluted, indicating that an inhibitor may have been present in the undiluted sample. The level of LAL-RM in the 29 samples ranged from 0.4 to 12.8 ng/ml.

The study indicates two significant facts: (1) LAL-reactive responses are associated with the cuprophane fiber material; and (2) the level of activity has been reasonably consistent in the fiber lots for the 17 months sampled.

Additional studies indicate that LAL-RM is not only associated with cuprophane fibers but also predominantly associated with the blood-path compartment. When LAL-RM was measured on the dialysate side of three CF 1500 dialyzers, LAL results were consistently less than 0.5 ng/ml. Although the blood-path discard of these dialyzers measured between 0.4 and 7.0 ng/ml, LAL-RM ranged between 20 and 80 ng/ml in saline recirculated through the blood path for two hours. The results of this study are given in Table 4. These data strongly suggest that LAL-RM is associated with the blood-path side of cuprophane fibers.

Interaction of LAL-RM with LAL

The LAL-RM is also atypical with respect to reactivity with various commercial LAL preparations. Mallinckrodt LAL consistently fails to detect this material, whereas Travenol, Associates of Cape Cod, and Microbiological Associates lysates consistently detect LAL-RM. Travenol LAL

Table 4. LAL-RM on Either Side of Three Capillary Flow Dialyzer Membranes

Origin of Extract	Total (ng)	LAL-RM (ng/ml)	Volume (ml)
Blood Path	4,000	20	200
Blood Path Discard	630	7	900
Dialysate Side	180	0.1	1800
Blood Path	15,200	80	190
Blood Path Discard	360	0.4	900
Dialysate Side	800	0.4	2000
Blood Path	3200	20	160
Blood Path Discard	360	0.4	900
Dialysate Side	320	0.2	1900

appears to consistently indicate the highest concentration of LAL-RM, followed by that of Associates of Cape Cod, and Microbiological Associates (Table 5). LAL-RM is also detected by lysate prepared from the Japanese horseshoe crab, *Tachypleus*. Work conducted at CDC; Martin Luther King, Jr. Hospital; Associates of Cape Cod; and Mallinckrodt also shows that Mallinckrodt LAL does not detect LAL-RM.

Table 5. Comparison of LAL Clot End Points with Saline Rinses from Travenol CF® 1500 Dialyzers

	LAL End Point (ng/ml)			
Rinse	Travenol	Associates of Cape Cod	Microbiological Associates	Mallinc-krodt
1	200	100	5	0
2	200	50	10	0
3	500	100	10	0

However, Mallinckrodt LAL can detect LAL-RM if subjected to chloroform extraction. It should be noted that, while the LAL produced by Associates of Cape Cod is chloroform-extracted, the Travenol preparation is not. Consequently, the extraction procedure cannot be assumed to account for differences. A personal communication from CDC indicated that, when LAL-RM is phenol-extracted, the substance is detectable with Mallinckrodt LAL. Because phenol is commonly used in the extraction of endotoxin from gram-negative bacteria, it is possible that phenol unmasks an active site required for interaction with Mallinckrodt LAL. However, this remains speculation.

Using the Travenol procedure or Mallinckrodt lysate with a gel clot end point, E. coli 055:B5 endotoxin spiked in the dialyzer solution can be recovered on a quantitative basis. However, the use of Pyrosperse, a preparation sold by Mallinckrodt to overcome LAL test inhibition, failed to produce a positive LAL result with unspiked dialyzer wash solution when Mallinckrodt LAL was used.

Effect of Cuprophane Processing on Endotoxin

Acidic and basic conditions are known to decrease or eliminate the biological activity of endotoxin. Because both acid and base are used in the production of cuprophane fibers for hemodialyzers, the activity of E. coli 055:B5 endotoxin subjected to a simulated cuprophane process was evaluated. Endotoxin was prepared in water, and the solution adjusted to pH 13 using 10N NaOH. Endotoxin was held at this pH for 3.25 hours at room temperature, and then the pH was adjusted to less than 1.0 using 0.1N H_2SO_4. (Alkaline hydrolysis is used to prepare cotton linters for the spinning solution, and once fibers are produced, they are subjected to a number of baths including a 2-minute acid bath.)

The results of this study showed that E. coli 055:B5 endotoxin maintained its LAL reactivity throughout the process but exhibited approximately a tenfold decrease in rabbit pyrogenicity. A dose of 10 ng of treated E. coli 055:B5 per kg failed to elicit a USP rabbit pyrogen failure. Based on the HIMA study cited earlier, a strong rabbit pyrogen test failure should have been observed.

Even at a concentration of 100 ng/kg, the sum of the three-rabbit fever responses was only 1.7°C. These data (Table 6) suggest that the LAL-RM could represent a degraded form of endotoxin that induces LAL reactivity, but has lost its ability to provoke a rabbit fever response at the doses tested.

Table 6. Impact of Cuprophane Production Process on the Pyrogenicity and LAL Reactivity of *E. coli* 055:B5 Lipopolysaccharide

LPS Concentration (ng/ml)	Rabbit Dose (ng/kg)	3-Rabbit Result (°C)
100	1000	2.4
10	100	1.7
1	10	0.2

Animal Safety Studies

Because endotoxin can elicit numerous biological reactions in addition to LAL gelation, a study was undertaken to evaluate some *in vivo* responses to LAL-RM that are known to be altered by *iv* injection of endotoxin. Doses of concentrated LAL-RM (1000 ng/kg) were administered to rats and compared to the responses to saline (negative control) and *E. coli* 055:B5 (positive control). Various hematological, coagulation, and blood chemistry tests and complement activation were monitored over time. Although the WBC and RBC counts and coagulation were altered in the endotoxin positive control group, no differences were noted in those rats receiving LAL-RM, except for RBC's, which decreased. These results are summarized in Table 7.

Table 7. Biological Activity of Concentrated LAL-RM From Capillary Flow Dialyzers

Test	Toxicity/Activity
Mouse Safety Test	None
Mitogenicity	None
Neutropenia/Leucocytosis	None
Blood Chemistry Alterations	None
Coagulation Alteration	None
Complement Activation	None

The decrease in RBC's may not be a significant finding because cells may have undergone hemolysis, and endotoxin is not known to induce hemolysis of RBC's. This study continues to support the notion that LAL-RM is not toxic at concentrated doses in animals. It should be repeated when significantly more concentrated LAL-RM is available.

One-half milliliter of concentrated saline extract from dialyzers containing 200 ng of LAL-RM per ml was injected iv into each of five mice (5000 ng/kg for each mouse). None of the animals exhibited abnormal reactions initially or at 4, 24, and 48 hours after injection, clearly demonstrating that LAL-RM is nontoxic at high doses.

DISCUSSION

LAL-RM has been found in saline rinses of hemodialyzers. Various commercial preparations of LAL were found to react with this material. The exception was Mallinckrodt LAL. Concentrated LAL-RM is nonpyrogenic in the USP rabbit pyrogen test at doses of up to 2,000 ng/kg, nontoxic in the USP mouse toxicity assay at 5,000 ng/kg, and does not alter hematology, liver-function enzymes, coagulation, or complement activity at 1,000 ng/kg in rats.

Based on the results of the HIMA study on rabbit pyrogenicity, one would expect that 100% of the rabbit colonies in the U.S. would exhibit a pyrogenic response at a dose of 10 ng of standard *E. coli* 055:B5 endotoxin per kg. Travenol's Morton Grove rabbit colony has been qualified to detect as little as 1.0 ng of the same endotoxin standard per kg. Thus, LAL-RM, while highly reactive with LAL, exhibits no pyrogenicity at levels up to 2000 times the dose of standard endotoxin required to elicit a pyrogenic response. While this lack of pyrogenicity is not typical of endotoxin, doses of between 2.0 and 100.0 ng of "environmental endotoxin" per kg pass the USP rabbit pyrogen assay in the same colony qualified at 1.0 ng/kg (Pearson et al., 1982). Possibly, LAL-RM in capillary flow dialyzers is an extreme case of "environmental endotoxin." Also, over 200 ng of *Pseudomonas aeruginosa* (Piromen®) endotoxin per kg is required to elicit a pyrogenic response in rabbits (Pearson et al., 1982). The purified lipopolysaccharide of *Rhodopseudomonas* requires microgram doses to cause pyrogenicity in rabbits (Galanos et al., 1977).

Our studies with E. coli treated with NaOH for several hours are consistent with the finding of Kraft (1973), who demonstrated that endotoxin treated with base lost disproportionately less LAL activity. Detoxification of endotoxin by NaOH is also supported by the studies of Neter (1956) and Nowotny (1969). Rylander (personal communication) of the University of Gothenburg, Sweden, has conducted studies which demonstrate that free lipid A is relatively reactive in the LAL assay but unreactive in the rabbit. This could explain why LAL-RM is not pyrogenic in the rabbit but is reactive in LAL. Watson (personal communication) of the Woods Hole Oceanographic Institution has evaluated LAL-RM. Preliminary evidence using gas chromatography and mass spectrometry indicates that LAL-RM is an endotoxin.

There are other data which suggest that LAL-RM in dialyzers is not an endotoxin:

1. LAL-RM is not inhibited by polymyxin B when activity is evaluated using LAL.
2. Saline rinses of LAL-RM from hemodialyzers are significantly inhibited by the rinse solution because OD increases with dilution. When E. coli 055:B5 endotoxin is spiked in these rinses, its detection with LAL is not inhibited.
3. LAL-RM does not stimulate the production of endogenous pyrogen at 30 ng/ml, whereas 5 ng of E. coli endotoxin per ml in the same system does (personal communication, Charles Dinarello, Tufts University School of Medicine).
4. LAL-RM (100 ng/ml) does not stimulate macrophage killing of SV_{40}-transformed fibroblasts in culture. Significant killing was seen at 0.1 ng/ml with Salmonella Re595 lipopolysaccharide (personal communication, David Morrison, Emory University).
5. Morita et al. (1981) have demonstrated the presence of a (1 ⟶ 3)-B-D-glucan-mediated coagulation pathway found in LAL, which is not activated by endotoxin. Because the hemodialysis membrane fiber is cellulose-derived, the alternative coagulation pathway could be activated by a similar glycan.

Based on the data presented in this paper, LAL-RM could be (1) a "degraded endotoxin" or an "environmental endotoxin," which is highly reactive with LAL but nonpyrogenic, or (2) a substance of unknown origin, perhaps a glycan, which induces LAL gelation. Extensive studies are underway to resolve these possibilities.

REFERENCES

Galanos C, Roppel J, Weckesser J, Rietschel ET, Mayer H (1977). Biological activities of lipopolysaccarides and lipid A from _Rhodospirillaceae_. Infect Immun 16:407.

Kraft AR, Jones RB, Homer LD, Merecki BE, Herman CM (1973). Detection of detoxified endotoxin by the _Limulus_ lysate assay. Surg For 24:68.

Morita T, Tanaka S, Nakamura T, Iwanaga S (1981). A new (1 ⟶ 3)-B-D-glucan-mediated coagulation pathway found in Limulus amebocytes. FEBS 129:318.

Neter E, Westphal O, Luderitz O, Gorzynski EA, Eichenberger E (1956). Studies of enterobacterial lipopolysaccharides. Effects of heat and chemicals on erythrocyte-modifying, antigenic, toxic and pyrogenic properties. J Immunol 76: 377.

Nowotny, A. (1969). Molecular aspects of endotoxic reactions. Bacteriological Reviews 33:72.

Pearson FC, Weary M, and Bohon J (1982). The relative pyrogenicity of environmental endotoxin as measured by the amebocyte lysate assay and the USP rabbit pyrogen test. In: "International Conference on Endotoxin Standards and LAL Use with Parenteral Drugs," New York: Alan R. Liss (in press).

PEARSON SPEAKER
TITLE: DETECTION OF LIMULUS AMEBOCYTE REACTIVE MATERIAL IN CAPILLARY FLOW HEMODIALYZERS.

NOVITSKY: One other piece of chemical information is available which suggests this material is lipopolysaccharide. Three samples were tested. There were two concentrated and lyophilized dialysate samples, one contained 200 ng/ml, the other 100 ng/ml of LAL reactive material (concentration of LPS as determined with Associates of Cape Cod, Inc., LAL). The standard contained 100 ng/ml 055:B5 (gravimetric and LAL analysis). These were sent to Dr. Liu, BoB, as unknowns. He tested the samples with GC/MS and did a complete analysis of all sugars. However, he was primarily concerned with ethanolamine as a marker for LPS. The 200 ng/ml dialysate sample showed approximately twice the amount of ethanolamine as the 100 ng/ml dialysate sample, and the 055:B5 standard sample was roughly equivalent in ethanolamine content at the 100 ng/ml dialysate sample. Additionally, 2 dialysate samples contained a lot of what Dr. Liu referred to as protein residual. That is, he found amino acids in the two dialysate samples. This was in addition to what looked like the sugars normally associated with LPS. He felt very strongly from the GC profiles that these sugar products originated from lipopolysaccharide.

NEELY: The comment that ethanolamine is present and that a clot is formed suggests the presence of thromboplastin or a thromboplastin-like material. Has this been looked into?

PEARSON:- No we have not, that is an excellent suggestion and we will do so.

RYLANDER - Did I hear you correctly, that the raw material for this was cotton lint?

PEARSON- Yes

RYLANDER. Cotton lint is a material which is heavily contaminated with gram-negative bacteria. My suggestion is that when you treat this lint in your manufacturing procedure, you split up your lipopolysaccharide material and you obtain Lipid A free in solution. Free Lipid A will then form micellae in which form it may be biologically inactive but may still be reactive with LAL although we haven't done these studies. This could explain discrepancies with the negative cell culture tests and the positive limulus test that you have just shown.

PEARSON: That would also explain the lack of pyrogenicity.

RYLANDER: Yes, definitely.

LEVIN: Before you get into trouble trying to define endotoxin by its activity you should recall an old paper by Nowotny. He partially detoxified endotoxin before the LAL days. He showed that in the four assays he conducted there was no correlation whatsoever. There is no reason to believe that a partially degraded endotoxin will give you relative equivalent data in different assays.

MILLS: To sum up what you have said so far where does your data leave us as far as a substitute pyrogen assay? You have shown a material at extremely high concentrations that won't elicit a rabbit febrile response but in certain lysates you get gelation. This presents a severe problem of correlation.

PEARSON: Do you think there is anything in Mallinckrodt's manufacturing process that could be responsible for the difference observed between the LAL's reactivity to the dialysate? Could you possibly be removing an LAL component in your processing?

MILLS: No. I don't know how you manufacture your LAL but I do know the difference between our (Mallinckrodt's) and Associates of Cape Cod's LAL is that ACC uses a chloroform extraction step and we do not. Chloroform extraction does remove alot of material. Our current process does not remove any material. There is no extraction in our process.

PEARSON: Our LAL is an unadulterated LAL differing only in that we use EDTA in the collecting solution rather than NEM.

MORRISON: Would you think it's fair to demand a one-to-one correlation of bacterial product activity between LAL and rabbit test? Because if you don't, you really don't have a problem.

PEARSON: I agree. I think that you really can't make a one-to-one comparison. One example to emphasize this point is environmental endotoxin. Actually what you have is a large safety factor. The problem that we are still left with is the fact that all LAL products do not react similarly to the dialysate.

MORRISON: You're right. This is a different problem and a much more serious problem than correlating the LAL test with the rabbit pyrogen test.

MILLS: I disagree. I think we have to have correlation between the LAL and rabbit test to have a useful test.

MASCOLI: I must agree with both Drs. Morrison and Levin that the LAL test is doing exactly what it's intended to do, that is, alerting us that something is amiss. The serious problem here is not the lack of correlation with the rabbit test but the fact that one LAL does not react. If this is not an endotoxin, then at least we have erred on the side of the patient.

MILLS: Shouldn't the patients have responded to such a large amount of endotoxin from these dialyzers. And since they didn't respond, we are in fact not erring on the side of the patient as Dr. Mascoli has just said.

CARSON: I said the patients did not respond with a pyrogenic response to the dialysate; I can't say if they didn't respond in some other manner.

MILLS: I still have a problem with the usefulness of the LAL test if it cannot be correlated with the rabbit test.

Endotoxins and Their Detection With the Limulus Amebocyte Lysate Test, pages 261-267

LAL ASSAY FOR PLASMAPROTEINS

H. Ronneberger

Research Laboratories of

Behringwerke AG D 355 Marburg (W.-Germany)

Human plasmaproteins are being used more and more freqently for prophylaxis and therapy in patients. Besides this there are some special preparations produced from animal blood, such as anti-human-lymphocyte-globulin, which is used therapeutically after organ transplantations. The most important drugs obtained from human blood are albumin, gammaglobulin preparations for intramuscular and above all for intravenous application , factors of the coagulation system, such as factors VIII and IX, or prothrombin, or factors of the complement system, such as the C1-inactivator. A new plasmaprotein, antithrombin III, has recently been introduced in therapy in order to successfully inhibit intravascular coagulation during liver failure caused by poisoning through mushrooms or chemical compounds such as carbon tetrachloride. The clotting factor XIII, which is used in the therapy of alterations of wound healing is produced from human placenta. The classic assay for examining plasmaproteins injected into humans for possible pyrogenicity is the pyrogen test in rabbits which has been carried out for many years. This in vivo biological assay is very suitable to prevent endangering patients by biological preparations contaminated with lipopolysaccharides of gram negative bacteria. This is particularly the case when this assay is performed under standardized conditions. For this reason we exclusively use specified pathogen-free rabbits fed on a standard diet and housed in air conditioned animal rooms (Ronneberger 1977). For years we have not noted any severe adverse reactions due to endotoxins in plasmaproteins. This means that the existing assay for pyrogenicity is safe and reliable. Generally, the dose of plasmaproteins tested in the rabbit is determined by the dose therapeutically used in humans calculated per kg body weight. In most cases it lies between 1 and 10 ml/kg. In the Limulus assay only 0.1 ml are tested. Due to the higher sensitivity of this in vitro assay, the detection of toxicologically not relevant amounts of endotoxin can be achieved.

Therefore the search for alternative methods does not seem to be very urgent and is only of interest if considerable benefits can be derived. Such benefits would be faster and safer test results or a decrease of the high costs of the pyrogen test in rabbits or a decrease in the number of experimental animals. This is of special importance with plasmaproteins derived from human blood, because these preparations are antigens for rabbits. That means that the animals can be used only once for pyrogen tests. After repeated use, anaphylactic reactions may influence the fever reaction in a positive or negative way.

When using the Limulus assay for the pyrogenicity examination of plasmaproteins, a number of questions are of interest. How much is the amount of LPS in the most important preparations? The problem of threshold values has to be discussed in that concern. Are there fractions which inhibit or accelerate gel formation in the Limulus test? How could these inhibitions be overcome? Which assay would be best for testing this group of biological drugs? Should macro-, micro- or turbidimetric methods be used? An assortment of the most important plasmaprotein preparations produced by the Behringwerke was examined as to their endotoxin content by means of the Limulus test. As a rule, a complete lack of LPS can not be expected in all these biological drugs. However, the amount of endotoxin must be so low that the danger of adverse reactions can be excluded. The aim should be not to exceed the sensitivity of the assay in rabbits, taking into account the doses given to humans or rabbits, because here the greatest experiences exist (Greisman, Hornick 1969). Referring to a standard endotoxin with identical strength of the FDA reference endotoxin EC2 with 5 EU/ng we found that the sensitivity of our rabbit strain is between 1 and 10 ng/kg LPS. With the Limulus test only a few pg of the same LPS can be detected.

Applying the classical method of the Limulus assay, we mixed 0.1 ml of the undiluted preparations with 0.1 ml LAL and incubated in a water-bath at 37°C for 1 hour. On finding a positive reaction, the quantitative estimation of LPS was assayed by serial twofold dilutions of the sample. The endotoxin concentration was calculated by multiplying the lysate sensitivity with the reciprocal of the last sample dilution yielding gel formation. In all cases, the level of sensitivity of the lysate was confirmed by a same serial dilution of the standard endotoxin.

All samples pyrogenic in rabbits were also positive in the Limulus test. On the other hand, 29 samples, negative in the rabbits led to gel formation due to the higher sensitivity of this assay as compared to the animal test. One sample lot of antithrombin III, containing pyrogens, proved to be an exception. In this case, endotoxin was not detected because the gel formation did not take place.

Examining those lots of human albumin which were pyrogen-free in rabbits we found that most of them contained about 80 pg/ml of endotoxin. Here we regard the LPS content of 320 pg/ml as a limit. 5S-immunoglobulins contained traces of LPS between 20 and 80 pg/ml. Examining pyrogen free 7S immunoglobulins for intravenous use endotoxin amounts between 40 and 320 pg/ml were found. With these preparations the limit of 320 pg/ml LPS also applies.

When establishing endotoxin threshold values, the dose employed in the rabbit test should be the criterion. If in this assay the sensitivity is higher than 5 EU/kg or 1 ng/kg EC2, higher doses should not be allowed. As another limit the human dose/kg could be discussed, in cases that show fundamental differences to the rabbit dose. In the investigation of albumin and i.v.-gammaglobulins rabbits are injected with 3 ml/kg body weight and therefore a limit of 350 pg/ml EC2 seems to be appropriate. Investigations by Steere and co-workers (Steere, Rifaat, Seligman 1978) showed that albumins with concentrations of 4 ng/ml or more induced fever in human patients. Therefore, a safe limit must be distinctly below this amount. The same rule should apply to the other plasmaproteins even though our experience here is not wide enough to allow us to make a final decision.

The examination of inhibition of the Limulus reaction was performed by adding LPS to the sample and comparing it with LPS-spiked isotonic saline. Twofold serial dilutions of both endotoxin spiked preparations were examined in the

Limulus test. The plasmaprotein sample showed no inhibitory effects if in both dilution series the identical endpoint of gel formation was detected. If, however, the endpoint in the sample dilution series was lower than in that of the saline, then inhibition of the Limulus reaction was assumed. Albumin, gammaglobulins for i.m. and i.v. application, plasmaprotein solutions, fibrinogen, antithrombin III, prothrombin, factors VIII and IX, anti-human-lymphocyte-globulin and anti-thymocyte-globulin were examined for inhibitory effects. All these samples were pyrogen-free in rabbits. All together 78 lots of these 13 preparations were assayed. The assay was performed with the product and with the product plus 10 ng/ml standard LPS-spike. The sensitivity of the Limulus lysate to this endotoxin lay between 7 and 13 pg/ml. We prefer lysates with the highest possible sensitivity but this choice could be argued about.

During the last years Hochstein and co-workers (Hochstein, Seligmann, Marquina, Rivera 1979) investigated more than 1500 lots of human albumin from different manufacturers. In many lots they found inhibition of the Limulus assay. Chloroform treatment did not manage to get rid of this defect completely.

Some lots however did not show inhibitory effect. Thus albumin can sometimes mask endotoxin, whereby fundamental differences between the products of different manufacturers may be found.

We investigated 8 albumin lots of our production for inhibition of the test. In 5 of them we could not demonstrate inhibition, one showed a slight alteration of one twofold dilution and 2 such alterations for 2 twofold dilutions. 4 of 12 examined lots of plasma-protein-solution showed a masking of the added LPS of 1 to 2 dilution steps. Inhibition was also seen with immunoglobulins caused by specific binding of endotoxin by antibodies. 22 lots with 12 lots i.v.-gammaglobulin preparations were examined, 5 of them showed an inhibition of gel formation of one to 2 twofold dilution steps. This inhibitory effect was overcome by dilution of the samples. It was also possible to overcome it by denaturation through heating the samples but it was not possible to overcome it with dispersing agents. No inhibitions or acceleration of gel formation was observed by factors VIII, IX, XIII and by prothrombin concentrate. Examining two pyrogen-free lots of fibrinogen-concentrate after addition of LPS, we found that no gel formation took place. Except for the fact that this drug contains 5 % glucose, no reason was found for a negative influence on the Limulus reaction. Perhaps the protein together with the glucose had altered the distribution of the added endotoxin. Anti-human-lymphocyte and thymocyte-globulins derived from animal plasma did not cause any inhibition of the gel formation.

Antithrombin III is the most important inhibitor of the coagulation system, and therefore, an inhibition of gel formation can be expected in the Limulus assay. Indeed this can be demonstrated, after adding endotoxin to antithrombin III it can no longer be found. Even a spike of 1 mg LPS did not lead to gel formation. After a denaturation of the protein by heating the added LPS was found and in comparison to the placebo control only a slight residual inhibition was noted. In contrast to this result, native antithrombin III that was not pyrogen-free gave an increase of body temperature in the rabbit test.

Because of these unpredictable alterations of gel formation, the Limulus test is not suitable, at least at present, as release assay for final products of plasmaproteins. The rabbit test cannot as yet be substituted by the in vitro assay. Nevertheless important applications for this method are in-process-controls and examination of equipment and materials used during production (Cooper, Neely 1980). To overcome inhibition various methods are familiar. Apart from chloroform treatment which offers insufficient results, samples can be treated with special dispersing agents (Pyrosperse® Mallinckrodt) or shortly boiled. This third method is only possible up to a protein concentration of 5% . With 20% albumin, a solid mass will be produced that cannot be assayed in the Limulus test. All these procedures have not been sufficiently examined to date. Therefore, those methods can only be employed for samples from in-process-controls and not used for the release of final products.

It was reported (Sullivan, Watson 1975) that heparin inhibits the Limulus reaction unless calcium and sodium chloride are added to the lysate. With a commercially available reagent (Pyrogent®) we found inhibition of gel formation with 50 U/ml or higher amounts of heparin.

Some clotting factor preparations, e.g. PPSB, may contain 10-15 U heparin /ml as stabilizing agent. This amount of anticoagulant does not inhibit a positive endotoxin reaction of LPS-spiked clotting factors, but this could depend on the type of lysate used in the test. For this reason it is of vital importance in the investigation of this type of plasmaproteins to perform an inhibition test before calculating the content of endotoxin in a given sample or when changing the type of lysate.

The source of contamination with endotoxins during certain production steps can quickly be found. Testing these kind of samples some peculiarties must be taken into consideration because the in process samples may contain constituents not found in end products and these could perhaps have a negative influence on the assay. Thus it is possible that specimens of intermediate steps of alcohol fractionation may contain traces of residual ethanol which according to our experience, do not have an adverse effect on gel formation if they do not exceed 2 %. A LPS-spike can be refound in isotonic saline and in albumin.

Intermediate samples can also contain the residual anticoagulant EDTA leading to alteration of the Limulus reaction by binding of calcium ions. Addition of higher amounts than 0.2 mg/ml EDTA to LPS-spiked water and albumin led to prevention of gel formation. Usually the content of anticoagulant in these in-process samples is far lower, so that an alteration of the reaction is not to be expected.

These 2 examples demonstrate that a thorough investigation must bear such components in mind and that for this reason an examination of the inhibitory effect is sometimes necessary. Recapitulating one can say that to date the Limulus assay for plasmaprotein preparations is a suitable in-process test saving time, money and laboratory animals. Because of its high sensitivity, this assay can increase the quality of this group of drugs. Further research and accumulation of experience are however necessary to solve the problems of threshold values, inhibition, and the introduction of new techniques.

REFERENCES

Cooper JF, Neely ME (1980). Validation of the LAL test for end-product evaluation. Pharmaceutical Technology 4,6:72.

Greisman SE, Hornick RB (1969). Comparative pyrogenic reaction of rabbit and man to bacterial endotoxin. Proc Soc Exp Biol Med 131:1154

Hochstein HD, Seligmann EB, Marquina RE, Rivera E (1979). Limulus amebocyte lysate testing of normal serum albumin (human) released in the United States since 1975. Develop Biol. Standard 44:35.

Ronneberger HJ (1977). Comparison of the pyrogen tests in rabbits and with Limulus lysate. Develop biol Standard 34:27.

Steere AC, Rifaat MK, Seligmann EB et al. (1978). Pyrogen reactions associated with the infusion of normal serum albumin (human). Transfusion 18:102

Sullivan JDJr, Watson SW (1975). Inhibitory effect of heparin on the limulus test for endotoxin. Journ Clin Microbiol 2:151.

Endotoxins and Their Detection With the Limulus Amebocyte Lysate Test, pages 269-280

Use of the Limulus Amebocyte Lysate Endotoxin Assay in Plasma Fractionation.

A.M. Cundell, Ph.D., S. Spiegelman, M.S., M.H. Stryker, Ph.D.
Blood Derivatives Program
New York Blood Center, Inc.
Melville, N.Y. 11747

The experience of the Blood Derivatives Program with the Limulus Amebocyte Lysate (LAL) Assay to monitor Endotoxin contamination of raw materials, equipment, in-process materials, and final products is outlined. The test was found suitable for use in an aggressive program to prevent the manufacture of pyrogenic lots of Normal Serum Albumin (Human) from pooled human plasma. The program consisted of monitoring equipment and the various stages of the cold ethanol fractionation-diafiltration process for the presence of Endotoxin. Data will be presented on the correlation between the LAL and Rabbit Pyrogen test for 80 lots of Albumin produced during 1980.

The feasibility of the LAL test for use with Antihemophilic Factor, Immune Serum Globulin, and Interferon is discussed.

Introduction

The New York Blood Center Plasma Laboratories located in Melville, Long Island were completed in the fall of 1979. The Labs routinely process in excess of 6,000 liters of plasma each week. The human blood plasma is separated from whole blood collected from voluntary blood donations in the Greater New York/New Jersey Metropolitan region or is plasma recovered from expired whole blood supplied from American Red Cross Regional Centers. The plasma derivatives produced are Albumin (5 and 25%), Immune Serum Globulin and Coagulation Factor VIII (Antihemophilic Factor). The Quality Control Department is also responsible for the release of

Interferon manufactured at the Pilot Plant of the New York Blood Center in New York City.

In this publication we will report our experience with the Limulus Amebocyte Lysate (LAL) assay for the detection of bacterial endotoxin in raw materials, components, in-process samples, and final container material. Questions that we will attempt to answer are: 1) Can the LAL test be used to monitor in-process steps in the manufacture of plasma derivatives, and 2) Is the LAL test predictive of the pyrogenicity of the final products?

MATERIALS AND METHODS

Albumin Production

The plasma fractionation method employed at The New York Blood Center to produce albumin is a modified cold ethanol fractionation procedure with a diafiltration step to remove all traces of alcohol and to allow the adjustment of salt content to a physiological range. After the albumin solution has been adjusted to meet specifications for protein content, pH and salt content it is sterile filtered and aseptically filled into sterile, depyrogenated glass vials in a clean room. The vials of albumin are stoppered, sealed and terminally pasteurized to inactivate any hepatitis virus that may be present ($60 \pm 0.5^{\circ}C$ for 10 hours).

Antihemophilic Factor Production

Antihemophilic Factor (Coagulation Factor VIII) derived from human plasma is effective in the prevention of severe bleeding episodes attributable to Factor VIII deficiency. Factor VIII is concentrated from Fresh Frozen Plasma by thawing at $1^{\circ}C$ and harvesting the insoluble residue, cryoprecipitate. In this form the activity is associated with cold insoluble fibrinogen, fibronectin, globulins, and adventitous plasma proteins.

The AHF is purified by cold extraction of the cryoprecipitate with a suitable buffer to remove the adventitous plasma proteins, and then warm extraction to solubilize the AHF. Other procoagulants are removed by adsorption to finely dispersed alumina followed by centrifugation. Additional purification is obtained by a low temperature precipitation of most of the fibronectin in the AHF solution. The product is concentrated by ultrafiltration, sterile filtered, aseptically filled into 100 ml vials and lyophilized. AHF is reconstituted with sterile water for injection by the patient immediately prior to intravenous infusion.

Immune Serum Globulin

Immune Serum Globulin is a sterile solution containing gamma globulin antibodies derived from human plasma by cold ethanol fractionation. It is suitable for intramuscular injection when immune prophylaxis or passive immune intervention is prescribed.

Interferon

Human Leukocyte Interferon is a sterile aqueous solution used as an injectable in clinical trials for the treatment of viral diseases and cancer. It is prepared by incubating a culture of human leukocytes in the presence of Sendai virus which stimulates the production of Interferon by the leukocytes. The Interferon is purified from a cell-free extract of the medium, sterile filtered, and aseptically filled into 2 ml containers.

Limulus Amebocyte Lysate Assay

Reconstituted endotoxin standard (E.C. 3 & 7) is diluted ten-fold and 0.8 ml of this dilution is added to 9.2 ml of sterile pyrogen-free distilled water (0.8 ng/ml). Twofold dilutions from 0.4 to 0.0125 ng/ml endotoxin are prepared to standardize the Lysate (Associates of Cape Cod). An appropriate series of twofold dilutions of the test sample is made. To 0.1 ml of the diluted samples within a 10 x 75mm flint glass test tube is added 0.1 ml of lysate. Similar tubes of the reference standard and negative controls are dispensed. The tubes are incubated for one hour in a 37°C tube heating unit. After the incubation period, the tubes are removed individually and inverted through 180°. If the gel does not break when the tube is inverted, the tube is scored as positive. The product of the sensitivity of the lysate and the highest dilution where clot formation occurs is the endotoxin level in the test sample.

A number of sample preparation schemes have been developed to test raw materials and in-process material for endotoxin content. In general, solutions of chemical raw materials are prepared 0.3N or isoosmolar in the pH range 6.0 to

8.0. The pH is adjusted using sterile, pyrogen-free 0.3N HCl or 0.3N NaOH solution.

Alcohol or fractionation samples containing alcohol must be evaporated to remove the alcohol for the assay to be successful. For example, 100 ml of 95% alcohol is evaporated to dryness in a rotary evaporator and the residue resuspended in 5 ml of pyrogen-free water. Filter rinses containing alcohol are diluted with WFI, evaporated to 5 ml and assays are run on 2- to 8-fold dilutions (Table 1).

Other samples, such as Water for Injection (WFI), WFI rinses of cleaned-in-place (CIP) tanks, ultrafiltration unit filtrate, permeate and rinses, filter aid rinses, filter rinses, and steam condensate of sterile holding tanks are diluted to 8-fold. In-process and final container samples are diluted to 64-fold or until an end point is reached (See Table 2).

Rabbit Pyrogen Test

Five rabbits are selected from sham tested rabbits (weight $\geq$1.5 Kg.) that have been quartered in-house for at least 7 days. They are weighed, placed in restraining cages, the rectal probes are inserted, and the right ear over the vein shaved. Three of the rabbits, whose control temperatures are within 1°C of one another and do not exceed 39.8^{o}C, are selected. The predetermined volume of test sample is injected into the vein with a fresh disposable syringe for each rabbit. The rectal temperatures are recorded after 1, 2 and 3 hours and the individual maximum temperature rises and their sum are calculated. The test is negative for pyrogens if no rabbit shows a rise in temperature $\geq 0.6^{o}$C and the sum of the individual maximum temperature rises does not exceed 1.4^{o}C.

The dosage of each product to be injected into the rabbit is found in Table 3.

Results

Albumin: During the first calendar year of operation of the Plasma Processing Laboratory 55 lots of 5% albumin (3 pyrogenic lots) and 25 lots of 25% albumin were produced for submission to the Bureau of Biologics. The mean endotoxin levels of the non-pyrogenic lots were 0.12 $\pm$ 0.19 and 0.39 $\pm$ 0.42 ng/ml respectively. These results were achieved by an aggressive program of endotoxin monitoring and control of chemical raw materials, equipment rinses, water for injection, and final products with the LAL assay. Incoming chemical raw materials and components (stoppers, bottles and seals) routinely are found to be pyrogen-free. Fractionation tanks are subject to a clean-in-place (CIP) procedure and rinsed with hot water for injection (WFI). These tank rinses are tested for the presence of endotoxin and released if pyrogen-free ($\leq$0.05 ng/ml). The water for injection is maintained at 80°C in the bulk storage tank and the recirculating loop within the fractionation area. Samples are drawn from each heat exchanger and tested for microbial count and endotoxin on a daily basis. The ultrafiltration unit (Romicon 20 cartridge unit) is cleaned in place with caustic, sanitized with dilute hypochlorite solution and stored charged with 0.35% phosphoric acid when not in use. Romicon rinses, filtrate and permeate are routinely monitored for endotoxin and microbial counts. The unit is broken down and the components cleaned when the tests suggest a microbial buildup. Condensate from steam-sterilized bulk tanks is tested and released prior to sterile filtration of albumin into the tank. Filter aid and filter rinses are also tested. The 5% albumin bulk solution immediately prior to sterile filtration must contain 0.4 or less ng/ml of endotoxin (25% albumin $\leq$1.6 ng/ml Endotoxin) before the Authorization to Fill the final containers is issued. If test samples are found to contain endotoxin the raw material is not released to Production, equipment is recleaned or resterilized, or the albumin solution is depyrogenated.

A review of the data of the endotoxin levels in the albumin, pre- and post-pasteurization supports the view that for moderate-low concentrations of endotoxin (2 ng/ml) pasteurization will reduce the endotoxin level (25% albumin, bulk 0.80, final container 0.40 ng/ml, significant at the 1% level, and 5% albumin, bulk 0.37, final container 0.13 ng/ml, significant at the 1% level). Decisions to proceed with the fill of an albumin lot may require a Pilot pasteurization of material from the lot.

Final container albumin is tested for Pyrogen using the Rabbit Pyrogen Test. With the 5% lots (N=55) tested with endotoxin levels less than 1 ng/ml, all of the lots passed the Rabbit Pyrogen Test. However, no relationship between endotoxin level and the sum of the individual maximum temperature rises was apparent. Also, no correlation was apparent with the 25% lots (N=25) although the endotoxin level was over three times higher than the 5% lots due to concentration by ultrafiltration.

Antihemophilic Factor

Data for 10 lots of AHF is presented in Table 4. Two of the 10 lots failed the Rabbit Pyrogen Test (Dose 12 u/kg). Microbial analyses of in-process samples from the AHF purification reveal a consistent low level of microbial contamination of the in-process material prior to sterile filtration. Reports in the literature suggest that 0.1 ng/ml of endotoxin corresponds to 10^3 bacterial cell wall equivalents per ml (Jorgenson, et al., 1973). Attempts are being made to depyrogenate in-process material prior to sterile filtration. It would be advantageous to be able to monitor the endotoxin level during the AHF purification and actively intervene to depyrogenate the material. Preliminary experiments with surface charged depyrogenation membrane filters (Zetapor, AMF Cuno) have been unsuccessful. Passage through the filters did not reduce the Factor VIII activity and failed to remove endotoxin.

Immune Serum Globulin

Data for 6 lots of Immune Serum Globulin are presented in Table 5. All six lots passed the Rabbit Pyrogen Test (Dose 1 ml/kg. of rabbit weight). Again the correlation between the endotoxin level and the pyrogenicity was weak.

Interferon

Data for 10 lots of Leukocyte Interferon is presented in Table 6. Nine of the 10 lots failed the Rabbit Pyrogen Test (Dose 0.5 ml/kg). The correlation of the LAL endotoxin and the Rabbit Pyrogen Tests was not good. The source of the pyrogen is under investigation. Possible problem areas are the bacterial contamination of cell cultures and chromatography columns and the presence of leukocyte pyrogens in the product.

Discussion

Our experience with the LAL-endotoxin assay has shown that it is an inexpensive, rapid, sensitive and reproducible test particularly suitable for the monitoring of endotoxin during protein fractionation to prevent the manufacture of pyrogenic lots. The assay is invaluable in monitoring the cleaning of equipment and checking bulk solutions before the authorization to fill albumin lots. The fractionation tanks and sterile bulk tanks used in large scale production can be shown to be endotoxin-free using LAL assays. The success of the ultrafiltration system, a major innovation in the Plasma Processing Laboratory, is dependent on our ability to monitor microbial growth within the unit. The large surface area and complexity of the storage tank, transfer valves, pumps and ultrafiltration hollow fiber cartridges present many places for microbial growth and the concommitant release of endotoxin.

Endotoxin determination of final container albumin is not reported in the release protocols that are sent to the Bureau of Biologics. As a result of the endotoxin monitoring the majority of our 5% albumin lots submitted to the Bureau during 1980 and the first 6 months of 1981 contain 0.1 ng of endotoxin per ml or less (77 of 90 lots, 86%). All of the lots found to contain 1 ng per ml or less passed the Rabbit Pyrogen Test. It is our contention that lots could be released on the basis of the result of the LAL endotoxin assay, demonstration of compliance with Good Manufacturing Practices, the 100% Visual Inspection after a 14 day quarantine, and the Final Container sterility test. A cutoff point for endotoxin level, e.g. 0.5 ng/ml, could be established and only lots exceeding that value would be required to be put through the Rabbit Pyrogen Test. Further refinements of the LAL-endotoxin assay appear possible. The use of chromogenic substrates may be useful for endotoxin determinations in plasma and cryoprecipitate. A preliminary evaluation of the microtechnique for endotoxin testing suggests that it would give comparable results to the standard assay with less subjectivity and saving of lysate (Gardi and Arpagaus, 1980). The ability of the dye solution to be drawn up the capillary tube only when no gelling of lysate has occurred is more obvious than the degree of clotting of the lysate in a 75 x 100mm tube. In excess of 15,000 LAL-endotoxin assays are run at the Plasma Processing Laboratory each year. The microtechnique has the potential of reducing the amount of reagents used in the tests to a tenth of the current usage. This would represent a significant cost savings.

To date, our experience with the LAL-endotoxin assay monitoring in-process samples and final products of Antihemophilic Factor, Immune Serum Globulin and Interferon are limited. These products may contain constituents that inhibit or mimic test responses. The nature of manufacturing processes, e.g. lack of protein fractionation at high alcohol concentration, absence of a pasteurization step, the use of gel filtration chromatography, the incompatability of the products with a 14 day quarantine and inspection, all increase the possibility of pyrogens in the products.

References

Gardi, A. and G.R. Arpagaus, 1980. Improved Microtechnique for Endotoxin Assay by the Limulus Amebocyte Lysate Test. Anal. Biochem. 109: 382-385.

Jorgenson, J.H., H.F. Carvajal, B.E. Chipps and R.F. Smith, 1973. Rapid detection of gram-negative bacteriuria by use of Limulus endotoxin assay. Appl. Microbiol. 26: 38-42.

Table 1. Processing Materials and Components Samples - NSA (H)

Sample	Microbial Count	LAL Assay
a) Processing Material		
Water for Injection (WFI)	Daily	Daily
Cold Tapwater	Weekly	Weekly
Hot Tapwater	Weekly	——
Clean Steam Condensate	Weekly	Weekly
Pharm-Air	Weekly	——
b) Incoming Components		
Vials	Every 5 lots	Every 5 lots
Vial Stoppers	Every 5 lots	Every 5 lots
Chemical Raw Materials	As Received	As Received
Alcohol	As Received	As Received

Table 2. In-Process Control Samples - NSA(H)

	Microbial Count	LAL Assay (Stat)
Pooled Plasma	X	
IV-4 Filtrate		X
V-Rework	X	X
Cuno Filter Wash		X
Filter Aid Wash		X
Romicon Rinse	X	X
Initial Concentrate	X	X
Permeate	X	
Diafiltered Concentrate	X	X
Final Concentrate	X	X
Cuno Filter Wash		X
Cuno Filtered Albumin	X	X
Reworked Albumin Pool	X	X
Sodium Adjusted Albumin	X	X
Bulk Solution	Sterility Test	X Q.A.
Pasteurized Final Container	Sterility Test	X Q.A.

Table 3. Rabbit Pyrogen Test Dosages

Product	Dosage (ml/kg)
25% NSA(H)	3.0
5% NSA(H)	10.0
AHF	10 u/kg
ISG	1.0
Interferon	0.5

Table 4. Correlation between LAL Endotoxin Assay and Rabbit Pyrogen Test for 10 Experimental lots of AHF.

Lot #	Endotoxin Level (ng/ml)	Rabbit Pyrogen Test (10 u/kg) T1	T2	T3	$\Sigma\Delta T^{o}C$
L1-81	0.8	0.2	0.4	0.2	0.8
L2-81	1.6	0.1	0.0	0.2	0.3
L3-81	2 - 20	1.2	0.9	0.5	2.6 (fails)
L4-81	0.8	0.2	0.0	0.5	0.7
L5-81	0.2	0.5	0.5	0.2	1.2
L6-81	6.4	0.5	0.5	0.2	1.2
L7-81	0.2	0.0	0.0	0.0	0.0
L8-81	6.4	0.3	0.2	0.1	1.5 (fails)
L9-81	1.6	0.8	0.0	0.0	0.8 (fails)
L1-80	3.2	0.0	0.6	1.2	1.8 (fails)

Table 5. Correlation between LAL Endotoxin Assay and Rabbit Pyrogen Test for Immune Serum Globulin (N=6, Dose 1 ml/kg of rabbit).

Lot #	Endotoxin Level (ng/ml)	Rabbit Pyrogen Test T1	T2	T3	$\Sigma\Delta T^{o}C$
2B16A	7.68	0.0	0.1	0.1	0.2
2B16B	3.84	0.2	0.0	0.1	0.3
1B11A	0.48	0.0	0.3	0.3	0.6
1B11C	0.24	0.2	0.5	0.0	0.7
5B12A	7.68	0.3	0.0	0.3	0.6
5B12B	7.68	0.1	0.0	0.1	0.2

Table 6. Correlation between LAL Assay and Rabbit Pyrogen Test for Leukocyte Interferon (N=10, Dose 0.5 ml/kg).

Lot #	Endotoxin Level (ng/ml)	Rabbit Pyrogen Test T1	T2	T3	$\Sigma\Delta T^{o}C$
109	0.8	0.4	0.2	0.2	0.8
110	6.4	1.0	1.2	0.5	2.7
111	0.4	0.1	1.0	0.9	2.0
112	6.4	1.3	0.7	0.7	2.7
113	1.6	1.1	0.4	0.9	2.4
114	1.6	1.9	1.3	1.5	4.7
115	1.6	0.2	1.3	0.0	1.5
116	0.8	0.9	0.7	0.9	2.5
117	6.4	1.5	1.3	0.6	3.4
118	6.4	1.2	1.4	2.4	5.0

Endotoxins and Their Detection With the Limulus Amebocyte Lysate Test, pages 281-285

LEUCOCYTE ACTIVATION OF THE LIMULUS AMEBOCYTE LYSATE

Peter C.Fink[1], Rolf G.Wolkersdorfer[2]

Medizinische Hochschule Hannover, [1]Institute for Clinical Chemistry, [2]Division of Immunology and Transfusion Medicine, Karl-Wiechert-Allee 9, D-3000 Hannover, Germany

INTRODUCTION

The question of specific activation of Limulus amebocyte lysate (LAL) by endotoxin has been challenged (Elin et al. 1978; Fink et al. 1981 a,b; Koster et al. 1978; Stumacher et al. 1973; Wildfeuer et al. 1974). A correlation between LAL positivity and elevated leucocyte counts in synovial fluids (Elin et al. 1978), in peripheral blood of neonates, parturients (Fink et al. 1981 a,b), and of patients with liver cirrhosis (Prytz et al. 1976) has been documented.

Extending those observations the purpose of the present investigation was to examine the relationship between substance(s) derived from purified leucocyte subpopulations and LAL positive reactions.

Additionally, it was of interest to study the influence of colony stimulating activity (CSA), that has been demonstrated to induce the maturation of granulocytic and monocytic precursors from committed stem cells (Monner et al. 1981; Cline & Golde, 1977) on LAL reactivity.

MATERIALS AND METHODS

Isolation of leucocyte subpopulations

Granulocyte and monocyte/lymphocyte populations from peripheral blood of healthy adults were separated according to Olofsson et al. 1980 with modifications by discontinuous density gradients of polyvinylpyrollidone coated silicia gel (Percoll). To obtain a pure lymphocyte preparation, monocytes were depleted from the monocyte/lymphocyte mixture by plastic adherence and iron fillings (Fink et al. 1977). Thereafter, the cells were resuspended in 0.15 mol/l NaCl, serially diluted, and the filtered (0.22 µm) supernatants (0.5 ml) tested for LAL reactivity. Morphological analysis of isolated cells was performed by cytocentrifuge smears stained with Türk's solution to determine the percentage of granulocytes, monocytes and lymphocytes. Viability of cells was checked by staining with trypan blue. More than 98% consistently excluded the dye.

Preparation of human placental conditioned medium (HPCM)

Briefly, human placenta obtained by Caesarean section, was cut into 5 g pieces and cultivated at 37°C in a fully humidified atmosphere with 5% CO_2 in RPMI 1640 containing penicillin, streptomycin and amphotericin. After 7 days incubation cellular elements were removed by centrifugation and filtration (0.22 µm). Subsequently, HPCM was dialyzed against 0.01 mol/l sodium phosphate buffer pH 6.8 at 4°C and absorbed by calcium phosphate gel (BDH Chemicals Ltd., England, Prod.No. 4438). The supernatant was collected and concentrated by lyophilization. In pyrogen-free water resuspended HPCM was filtered (0.22 µm) and stored at - 196°C until used. HPCM containing CSA was confirmed by bone marrow cultivation (manuscript in preparation).

Limulus amebocyte lysate test

Lyophilized LAL (ACC Lot No.52-76-245) and endotoxin standard (Salmonella abortus equi, Novo-Pyrexal,0.1 µg/ml) were used and the technique performed as described elsewhere (Fink et al. 1981 b). LAL reactivity of filtered cell supernatants as well as $HPCM_{CSA}$ was examined under identical conditions.

RESULTS AND DISCUSSION

All filtered supernatants from pure granulocyte (100%) and lymphocyte/monocyte (90% / 10%) preparations activated LAL in a dose dependent manner as shown in Fig.1

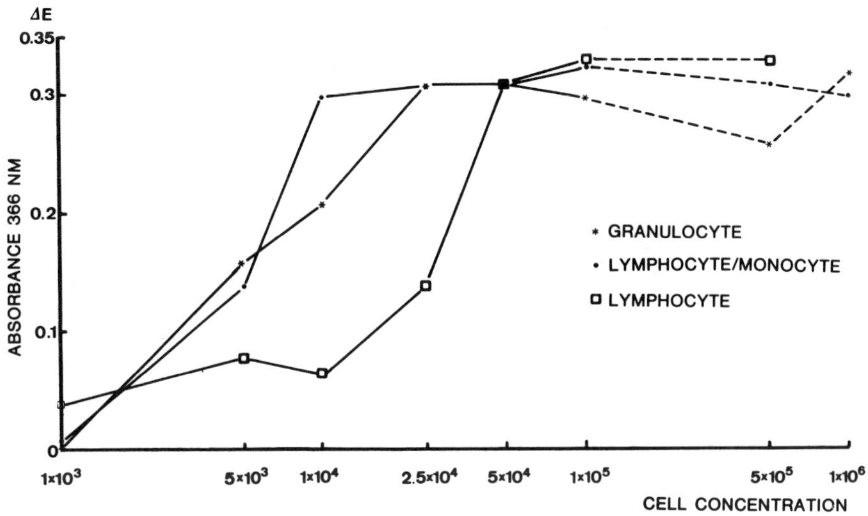

Fig.1. Activation of LAL by leucocyte subpopulations (cells/µl).

In contrast, filtered supernatants from granulocyte/ monocyte depleted lymphocyte (100%) preparations revealed less LAL activating potency than those found from granulocyte or monocyte/lymphocyte preparations for the corresponding range of cell concentrations (5×10^3/µl to 2.5×10^4/µl).

Fig.2. depicts the influence of prepared $HPCM_{CSA}$ on LAL. Compared to the LAL activation induced by Novo-Pyrexal (10 - 200 pg/ml) $HPCM_{CSA}$ concentrations from 3.97 ng/ml upward were necessary to obtain a similar absorbance profile.

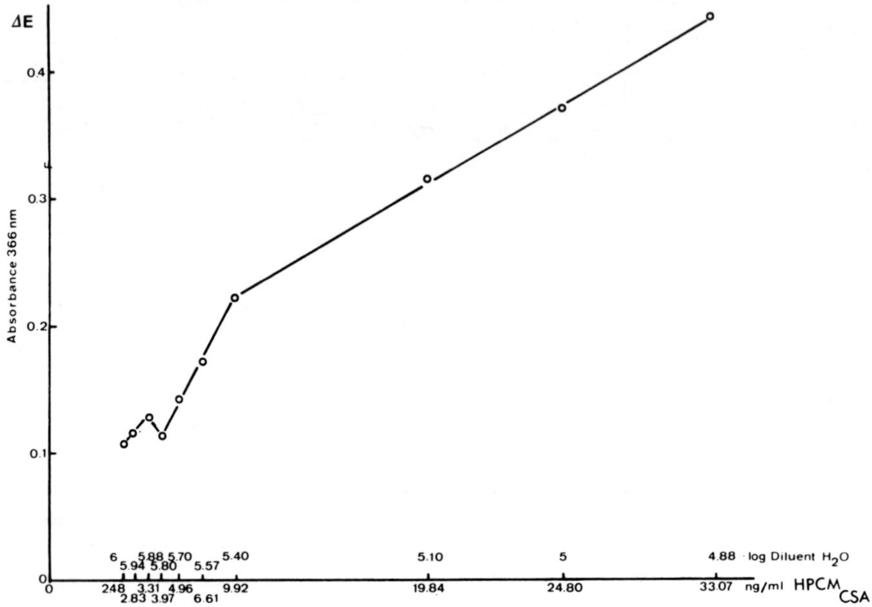

Fig.2. Activation of LAL by $HPCM_{CSA}$

The presented data suggest that leucocyte derived substance(s) beside endotoxin activate LAL.
In this context, LAL positivity induced by $HPCM_{CSA}$ requires further clarification and seems to be related to myeloproliferative activities shown in stem cell culture experiments.

REFERENCES

Cline MJ, Golde DW (1977). Mobilization of hematopoietic stem cells (CFU-C) into the peripheral blood of man by endotoxin. Exp Hemat 5:186.
Elin RJ, Knowles R, Barth WF, Wolff SM (1978). Lack of specificity of the Limulus lysate test in the diagnosis of pyrogenic arthritis. J Inf Dis 137:507.
Fink PC, Schedel I, Peter HH, Deicher H (1977). Inhibition of spontaneous and antibody dependent

cellular cytotoxicity by sera and isolated anti-globulin preparations from rheumatoid arthritis patients. Scand J Immun 6:173.

Fink PC, Kozak J (1981a). Determination of free lipid A and endotoxin. In Willoughby DA, Giroud JP (eds): "Inflammation:mechanisms and treatment", MPM Press Ltd International Medical Publishers, Lancaster, England.

Fink PC, Lehr L, Urbaschek RM, Kozak J (1981b). Limulus amebocyte lysate test for endotoxemia:Investigations with a femtogram sensitive spectrophotometric assay. Klin Wochenschr 59:213.

Koster F, Levin J, Walker L, Tung KSK, Gilman RH, Rahaman MM, Majid MA, Islam S, Williams RC (1978). Hemolytic-uremic syndrome after shigellosis. Relation to endo-toxemia and circulating immune complexes. N Engl J Med 298:927.

Monner DA, Gmeiner J, Mühlradt PF (1981). Evidence from a carbohydrate incorporation assay for direct activa-tion of bone marrow myelopoietic precursor cells by bacterial cell wall constituents. Infect Immun 31:957.

Olofsson T, Gärtner I, Olsson I (1980). Separation of human bone marrow cells in density gradients of poly-vinylpyrollidone coated silicia gel (Percoll). Scand J Haematol 24:254.

Prytz H, Holst-Christensen J, Körner B, Liehr H (1976). Portal venous and systemic endotoxaemia in patients without liver disease and systemic endotoxaemia in patients with cirrhosis. Scand J Gastroenterol 11:957.

Stumacher RJ, Kovnat MJ, McCabe WR (1973). Limitations of the usefulness of the Limulus assay for endotoxin. N Engl J Med 288:1261.

Wildfeuer A, Heymer B, Schleifer KH, Haferkamp O (1974). Investigations on the specificity of the Limulus test for the detection of endotoxin. Appl Microbiol 28:867.

ACKNOWLEDGMENT

This work has been supported in part by grant Fi 281/3-3 from the Deutsche Forschungsgemeinschaft.

Endotoxins and Their Detection With the Limulus Amebocyte Lysate Test, pages 287-299

APPLICATION OF THE *LIMULUS* AMEBOCYTE LYSATE ASSAY IN OUTBREAKS OF PYROGENIC REACTIONS ASSOCIATED WITH PARENTERAL FLUIDS AND MEDICAL DEVICES

Anita K. Highsmith, Roger L. Anderson, and James R. Allen, Center for Infectious Diseases, Centers for Disease Control, Public Health Service, U. S. Department of Health and Human Services, Atlanta, Georgia 30333

INTRODUCTION

Fever is a common symptom among hospitalized patients. When a high, spiking fever occurs in an acutely ill patient or one receiving multiple types of invasive procedures, the differential diagnosis of the cause of the fever can be lengthy, and identifying the source can be difficult. A major concern is that the fever may be caused by infusion of a product contaminated either with viable bacteria or with endotoxin. We have used the *Limulus* amebocyte lysate (LAL) test in our laboratory as one epidemiologic tool to assist in solving the problems and to extend our knowledge about the frequency of bacterial and/or endotoxin contamination of fluids and medications administered to patients or medical equipment and devices used on patients. In this paper we will explore general information about contamination of fluids and devices and provide further information about some of the investigations in which we have been involved. The utility of the LAL assay for endotoxin testing will be discussed.

BACKGROUND

The true incidence and causes of pyrogenic reactions associated with parenteral fluids and medical devices is not known. Three nationwide epidemics involving parenteral fluids occurred in this country between 1971 and 1981 (4,16,23), but reported localized outbreaks have been relatively infrequent. A summary of published reports by Maki (13,14,15) suggests that the incidence of intravenous (IV)-associated septicemia is less than 1 in 1000 and that the rate of contamination of IV fluids in various hospitals ranges from 0.9% to as high as 9%. A few studies reported 20% to 30% of IV fluids sampled during an epidemic to be intrinsically contaminated (15).

In most cases, epidemiologic investigations have revealed bacterial contamination of IV fluids and devices to be due to inadequate procedures, or techniques, or to contamination of in-use equipment. Hospital outbreaks of acute pyrogenic reactions have been reported in which microbial contamination could not be documented. The patients had symptoms indistinguishable from sepsis. Typically, before the onset of illness, the affected patients had had a procedure such as hemodialysis, or cardiac catheterization, or had received a common parenteral product.

One of the major roles of the Hospital Infections Program at the Centers for Disease Control (CDC) is to assist in the investigation of disease outbreaks. During a 2-year period 6 outbreaks of pyrogenic reactions were investigated; three involved infusion therapy and three were associated with medical devices and/or infusion (Table 1). Each of the outbreaks is described and the selective use of the LAL test in detecting endotoxin or endotoxin-like substances from suspect specimens is discussed.

DESCRIPTION OF THE EPIDEMIOLOGIC INVESTIGATIONS

Outbreak 1. Unknown pathogen, LAL-positive. An outbreak of pyrogenic reactions associated with intravenous administration of a commercial preparation of cephalothin (as an admixture) occurred at a large referral hospital. Because of the possibility that a commercial product which was distributed nationwide might be contaminated, the CDC was invited to join the Food and Drug Administration in an investigation.

In this outbreak, the LAL assay proved to be a reliable and reproducible laboratory marker, making it possible to examine rapidly all IV products prepared for

TABLE 1. APPLICATION OF THE LAL TEST TO SELECTED OUTBREAKS OF PYROGENIC REACTIONS INVESTIGATED BY HOSPITAL INFECTIONS PROGRAM, CDC

Year	Epidemic Mechanisms	Epidemic Pathogen(s)	LAL Assay
1977	Intravenous infusion of cephalosporin antibiotic product	Unknown*	+
1977	Intravenous lipid emulsion	*K. pneumoniae* *S. epidermidis*	Not done
1979	Intravenous infusion	Culture negative	-
1977	Contaminated transducer heads used with disposable chamber-dome	Polymicrobial	Not done
1979	Lumbar myelography	Culture negative	-
1979	Ventricular shunt extension tubing	*P. aeruginosa*	+

*Nonviable bacillary cell forms were detected by electron microscopy.

60 patients. Twenty-nine bags of epidemiologically implicated IV fluids were found to be LAL positive and rabbit pyrogen positive. Sixteen had been administered to patients who had reactions and another 13 would have almost certainly caused reactions if administered.

Available *Limulus*-positive infusion fluids from the outbreak, sterile distilled water, cephalothin, and normal saline were sterile.

Gram stains of LAL-positive IV fluid were examined before and after centrifugation, all Gram stains appeared negative for bacterial forms. However, when aliquots were concentrated by membrane filtration, using a 22-um Nuclepore* filter (Nuclepore Corp., Pleasanton, CA), gram-negative rods were visible. Flagella staining of these materials was negative.

Electron microscopic studies were conducted on LAL-positive and -negative IV fluids, commercial

*Use of trade names is for identification only and does not constitute endorsement by the Public Health Service, U.S. Department of Health and Human Services.

endotoxin, and saline in various combinations. Bacterial forms were present only in the LAL-positive IV admixture containing 2 g cephalothin that was associated with a patient reaction.

Gas-liquid chromatography determinations did not reveal any differences between the LAL-positive and -negative samples.

Outbreak 2. Known pathogen, LAL not done. Thirteen cases of bacteremia developed in 33 infants receiving a commercial intravenous lipid emulsion while hospitalized in a neonatal intensive care unit. Our participation in this investigation began after simultaneous cultures of blood and the lipid emulsion infusion were positive for the same microorganism. Subsequent microbiologic culturing of unused and pharmacy-prepared fluid was negative. LAL assay was not used in this study for the following reasons: 1) it appeared epidemiologically that the entry of microorganisms into the administration system most likely occurred during infusion, and 2) the reliability of LAL testing of lipid emulsion fluid was unknown at the time of the outbreak.

Outbreak 3. Culture negative, LAL negative. Three hospitalized patients had severe pyrogenic reactions suggesting adverse reaction to intravenous therapy during and immediately after surgery. The infusion fluid administered to each patient was immediately examined by LAL and culture. LAL-negative test results were available within 3 hours after receipt of in-use and unopened suspect infusion fluids. Cultures were negative, and subsequent evaluation of clinical histories, the epidemiologic data, and laboratory results confirmed the initial results from the LAL tests that intravenous therapy was not the cause of the postoperative fevers.

Outbreak 4. Known pathogens, LAL not done. Thirty-seven hospitalized patients with primary nosocomial bacteremia caused by several species of *Pseudomonas* (5), *Enterobacter* (2), *Acinetobacter* (2), and *Citrobacter diversus* were identified in intensive care units (in 1 hospital) over a 9-month period. A case-control study demonstrated exposure of case patients to blood pressure-monitoring to be statistically significant ($p=.006$). In addition, a portion of the transducer system was found to be heavily contaminated, and cultures of many of the transducers contained 1 or more of the bacterial strains involved in the outbreak. Organisms implicated in the outbreak were also isolated from enviromental samples taken in the utility room where intravascular pressure monitoring systems were assembled. Samples from the infusion fluid and medical devices were not subjected to LAL testing, since the infusion fluid was culture-negative and not directly implicated in the outbreak and the environmental samples and pressure monitoring system were known to have polymicrobial contamination.

Outbreak 5. Culture negative, LAL negative. A severe leptomeningeal reaction developed in 3 patients after lumbar myelography. Because of the possibility of intrinsic contamination, the radio-contrast media, instruments, and equipment used during myelography were examined microbiologically and by LAL. All materials were found to be endotoxin-negative and culture-negative for bacteria. The patient complications may have been the result of hypersensitivity-like reactions to the iodinated contrast dyes used in the myelogram procedures or due to irritation from detergent contaminants residue on syringes and needles.

Outbreak 6. Known pathogen, LAL-positive. Five patients had *P. aeruginosa* meningitis or abdominal abscesses after ventricular shunt neurosurgery. The ventricular shunt extention tubing was prepared from nonsterile commercial tubing, rinsed with tap water in the neurosurgical operating room and autoclaved before being implanted. Hospital personnel knew that on a previous occasion the autoclave used for this sterilization had malfunctioned. *P. aeruginosa*, serotype 011, was isolated from the patients, the tap water, and the water faucet aerator. It was postulated that *P. aeruginosa*, 011, from the tap water contaminated the tubing before it was autoclaved, that autoclaving did not sterilize the tubing and that this contaminated implant was the source of infection. The tap water (culture positive for *P. aeruginosa*) was LAL-positive; however, the shunt tubing was not available for testing.

METHODS FOR COLLECTION, MICROBIOLOGIC SAMPLING, AND LAL TESTING OF IN-USE INFUSION FLUIDS AND MEDICAL DEVICES

The six outbreaks described represent the variety of situations that may occur in a hospital and demonstrate the use of the LAL in the rapid detection of contamination. However, although most pyrogenic reactions are due to Gram-negative bacteria, reactions can be the results of intrinsic or extrinsic contamination of Gram-positive bacteria, fungi, or possibly, chemicals. A laboratory should not rely solely on any one test when examining a suspect fluid or device. A single test may be more rapid than another, but a full range of tests is required of each sample. The potential sources for contamination of infusion fluids (2,12,13) are shown in Figure 1. Methods for processing in-use infusion and medical devices for microbiologic culturing and LAL testing are given in Tables 2 and 3.

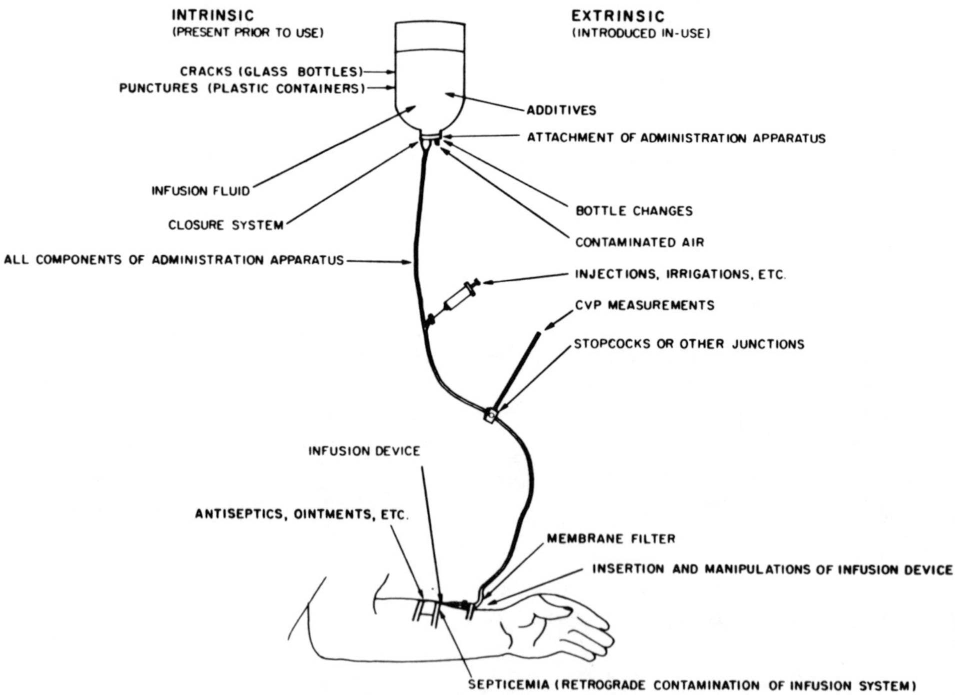

Figure 1 POTENTIAL MECHANISMS FOR CONTAMINATION OF IV INFUSION SYSTEMS

TABLE 2. METHOD FOR LAL TESTING OF IN-USE INFUSION FLUID IN THE HOSPITAL

1. Record all information on the label of each fluid, including special instructions and all lot numbers.

2. Cap the administration set with a sterile ensheathed needle and transport the entire delivery set promptly to the laboratory in a clean plastic bag.

3. Sampling for endotoxin substance and microbial growth should be conducted in a monitored laminar flow hood or a minimally occupied laboratory.

4. Withdraw 10 to 20 ml of fluid from the administration set using a depyrogenated syringe and needle.
   a. Place 3-5 ml of fluid into a screw-cap depyrogenated tube and test for endotoxin as described (24,25).
   b. Culture 0.1 ml quantitively on the surface of a fresh trypticase soy agar plate containing 5% sheep's blood or 1.0 ml in a pour plate.
   c. Inject remaining fluid into a blood-culture bottle.
   d. Gram stain a smear of the original fluid and of the centrifuged aliquot.

5. Add an equal volume of double-concentrated brain heart infusion broth enriched with 0.5% beef extract to remaining fluid in the container--or if the container is empty, add 50 ml, mix, and incubate the container.

6. Attempt to obtain and culture microbiologically and test by LAL remaining samples of any other fluids the patient may have received in the preceding 24 to 48 hours.

7. Incubate all containers aerobically at $36 \pm 1^{\circ}$ C for at least 5 to 7 days before discarding. If a blood product that may have produced illness is being sampled, specimens should also be incubated at 17°C to 20°C and anaerobically at $36 \pm 1^{\circ}$C.

8. All isolates should be fully identified, through species, and tested for susceptibility to a panel of antimicrobials. All isolates, especially of gram-negative bacilli, should be retained until the scope of the problem has been fully elucidated.

9. If clinical and epidemiologic data suggest manufacturer-related contamination of a commercial product (intrinsic contamination), the local and Federal public health authorities (the Food and Drug Administration and the CDC) should be immediately informed. Remaining unused products should probably not be sampled by the hospital but retained for evaluation by these agencies.

TABLE 3. LAL TESTING OF MEDICAL DEVICES

1. Determination of Lysate Sensitivity
    a. Use *E. coli* 055:B5 endotoxin or equivalent.
    b. Demonstrate test capable of detecting pyrogenic levels equivalent at least to 0.1 ng/ml and $\leq$0.02 ng/ml for medical devices having cerebrospinal fluid contact.
2. Preparation of Sample
    a. Record all label information including lot number for each test or suspect sample.
    b. Depending on device, select appropriate method for preparing an eluate or extract.
        1) By flushing or immersing in nonpyrogenic rinse solution (20-40 ml of nonpyrogenic saline or water depending on the size of the device).
           Hold rinse solution in fluid pathway 1 hour at room temperature
        2) By disassembling or cutting the device into pieces prior to extraction by immersion. Minimum extraction time should be 15m at 37°C or 1h at room temperature.
        3) Maintain extract under conditions appropriate for endotoxin stability until it is tested.
3. LAL Test
    Perform test as described in *U. S. Pharmacopoeia XX*.

DISCUSSION

In summary, we have shown the LAL to be an important test in rapidly detecting contamination of suspect infusion fluids and medical devices. Pyrogens of importance in large volume parenterals and medical devices are frequently endotoxin in nature (17).

The LAL is currently the most sensitive assay for bacterial endotoxin. Some investigators, however, have questioned the specificity of the test. After the outbreak involving serum albumin (23), Steere examined by LAL and rabbit-pyrogen test 662 vials of 25% normal serum albumin all of which have been administered to patients (22): samples were obtained from 13 lots produced by 5 manufacturers. None of the patients were considered to have albumin-induced pyrogen reaction but 443 of the 662 vials (65%) gave a positive LAL. We have also observed in a laboratory study that regular insulin from an unusued vial was LAL-positive (opened and tested by LAL after being held for 18h at 4 °C and 25 °C). The question remains one of the correct interpretation of LAL-positive tests when a parenteral is involved. However, Pearson reports that false-positive LAL reactions are usually due to substances other than endotoxin and can be ruled out by using appropriate controls (19,20). Mascoli and Wearys (17,26) have compared the LAL and rabbit-pyrogen test and reported that some pyrogen detected by LAL were rabbit-test-negative. Some rabbit tests that initially failed to detect pyrogens were later found on retest to be positive; these assays always were detected and confirmed by LAL.

In the absence of a suspected problem, it should not be necessary for hospital laboratories to use the LAL test for infusion fluids and medical devices before use. Manufacturers of these products are first responsible for pyrogen-free materials. Earlier in these proceedings, Ross, Bureau of Medical Devices, reported that more than 30 firms have been approved to use the LAL test for end-product release of medical devices, including those for short-term patient exposures such as water used in respiratory care.

When infusion fluids or devices are implicated in an outbreak, it is important to maintain an aliquot of the fluid or fluid rinse in a depyrogenated tube until all laboratory results are complete. If the sample is LAL-positive, the amount of endotoxin per ml should be calculated from the endpoint. Care must be taken to eliminate minute quantities of endotoxin from all equipment used to prepare samples for testing.

In a recent study (1) we compared 3 laboratory methods, to detect microorganisms in artificially contaminated infusion fluid (dextrose 5% in quarter strength saline): the standard pour plate (SPP), adenosine triphosphate (ATP), and LAL. The SPP was more sensitive in detecting microorganisms but required 24 h for the results. Although the LAL and ATP both rapidly detected contamination, the LAL was more sensitive than the ATP method under the test procedures used.

With the acceptance of the LAL as an official test, more hospitals and laboratories are using this test to rapidly detect contamination of equipment and supplies used in hemodialysis (3,5,6), in cardiac (10,11), and hepatic catheterization (Trapana, Y., personal communication) and in respiratory therapy (6,21). In addition, its usefulness in detecting contamination of infusion fluids has been documented, and it is being used for examination of specimens from patients for rapid detection of infections such as bacteriuria (18) and meningitis (8).

Although the number of reported epidemics remains relatively low, these numbers do not reflect the incidence of endemic pyrogenic reactions occurring in hospitalized patients. These numbers may change with increased surveillance of nosocomial infections and major changes in the population of hospitalized patients, i.e., age, underlying diseases, use of antimicrobials, types of surgical procedures, infusion therapy, hemodialysis, and types of invasive vascular devices.

REFERENCES

1. Anderson RL, Highsmith AK, Holland BW (1981). Comparison of standard pour plate, adenosine triphosphate, and Limulus Amebocyte Lysate for the detection of microbial contamination in intravenous fluids. In: "Abstracts of the Annual Meeting of the American Society for Microbiology," Dallas: 1981 American Society for Microbiology, p 308.
2. Buxton AE, Highsmith AK, Garner JS, West CM, Stamm WE, Dixon RE, McGowan, Jr, JE (1979). Contamination of intravenous infusion fluid: effects of changing administration sets. Ann Intern Med 90:764.

3. Carson LA, Peterson NJ, Favero MS (1979). Use of the *Limulus* amoebocyte lysate assay system for detection of bacterial endotoxin in fluids associated with hemodialysis procedures. In Cohen E (ed): "Biomedical Applications of the Horseshoe Crab (*Limulidae*)," New York: Alan R. Liss Inc., p 453.
4. Center for Disease Control (1973). Septicemias associated with contaminated intravenous fluids. Morbidity and Mortality Weekly Rep 22 (13):115.
5. Favero MS, Carson LA, Bond WW, Peterson NJ (1971). *Pseudomonas aeruginosa*: growth in distilled water from hospitals. Science 173:836.
6. Favero MS, Peterson NJ, Carson LA, Bond WW, Hindman SH (1975). Gram-negative water bacteria in hemodialysis systems. Health Lab Sci 12:321.
7. Highsmith AK, Anderson RL, West CM, Dixon RE (1979). Characterization of an endotoxin-like substance recovered from an intravenous antibiotic solution involving pyrogenic reactions. In Cohen E (ed): "Biomedical Applications of the Horseshoe Crab (*Limulidae*)," New York: Alan R. Liss, p 465.
8. Jorgensen JH, Lee JC (1978). Rapid diagnosis of gram-negative bacterial meningitis by the *Limulus* endotoxin assay. J Clin Microbiol 7:12.
9. Jorgensen, JH, Smith RF (1973). Rapid detection of contaminated intravenous fluids using the *Limulus* in vitro endotoxin assay. Appl Microbiol 26:521.
10. Kundsin RB, Walter CW (1980). Detection of endotoxin on sterile catheters used for cardiac catheterization. J Clin Microbiol 11:209.
11. Lee RV, Drabinsky M, Wolfson S, Cohen LS (1973). Pyrogen reactions from cardiac catheterization. Chest 63:757.
12. Mackel DC, Maki DG, Anderson RL, Rhame FS, Bennett JV (1975). Nationwide epidemic of septicemia caused by contaminated intravenous products: mechanisms of intrinsic contamination. J Clin Microbiol 2:486.
13. Maki DG (1977). Growth properties for microorganisms in infusion and methods of detection. In Phillips I, Meers PD, D'Arcy PF (eds): "Microbiological Hazards of Intravenous Therapy," Lancaster, England: MTP Press, p 14.

14. Maki DG (1980). The prevention and management of device-related infection in infusion therapy. J Med 11:239.
15. Maki DG (1981). Epidemic nosocomial bacteremias. In Wenzel RP (ed): "CRC Handbook of Hospital Acquired Infections," Boca Raton: CRC Press, Inc., p 371.
16. Maki DG, Rhame FS, Mackel DC, Bennett JV (1976). Nationwide epidemic of septicemia caused by contaminated intravenous products. I. Epidemiologic and clinical features. Am J Med 60:471.
17. Mascoli CC, Weary ME (1979). *Limulus* amebocyte (LAL) test for detecting pyrogens in parenteral injectable products and medical devices: advantages to manufacturers and regulatory officials. J Parenteral Drug Assoc 33:81.
18. Nachum R, Shanbrom E (1981). Rapid detection of gram-negative bacteriuria by *Limulus* amoebocyte lysate assay. J Clin Microbiol 13:158.
19. Pearson FC, Weary M (1980). The *Limulus* amebocyte lysate test for endotoxin. Bio Science 30:461.
20. Pearson FC and Weary M (1980). The significance of *Limulus* amebocyte lysate test specificity on the pyrogen evaluation of parenteral drugs. J Parenteral Drug Assoc 34:103.
21. Reinhardt DJ, Nabors W, Kennedy C, Malecka-Griggs B (1981). *Limulus* amebocyte and direct sampling methods for surveillance of operating nebulizers. Appl Environ Microbiol (in press).
22. Steere AC, Rifaat MK, Seligmann EB, Hochstein HD, Friedland G, Dasse P, Wustrack KO, Axnick KJ, Barker LF (1978). Pyrogen reactions associated with the infusion of normal serum albumin (human). Transfusion 18:102.
23. Steere AC, Tenney JH, Mackel DC, Snyder MJ, Polakavetz S, Dunne ME, Dixon RE (1977). *Pseudomonas* species bacteremia caused by contaminated normal human serum albumin. J Infect Dis 135:729
24. Sullivan, Jr., JD, Valois FW, Watson SW (1976). Endotoxins: the *Limulus* amebocyte lysate system. In Bernheimer AW (ed): "Mechanisms in Bacterial Toxinology," New York: John Wiley and Sons, p 217.

25. United States Pharmacopoeia (1980). 20th ed, Easton: Mack Printing Co, p 902.

26. Weary, ME, Donohue G, Pearson FC, Story K (1980). Relative potencies of four reference endotoxin standards as measured by the *Limulus* amebocyte lysate and USP Rabbit Pyrogen Tests. Appl Environ Microbiol 40:1148.

Endotoxins and Their Detection With the Limulus Amebocyte Lysate Test, pages 301–317

APPLICATION OF THE LAL TEST AND THE CHROMOGENIC SUBSTRATE TEST TO THE DETECTION OF ENDOTOXIN IN HUMAN BLOOD PRODUCTS

Reiko Homma, Kazuo Kuratsuka and Kiyoto Akama

Department of General Biologics Control
The National Institute of Health
Tokyo, 141, Japan

INTRODUCTION

The Department of General Biologics Control in the National Institute of Health is charged with controlling the safety of various biological products. Over the past few years, both the kinds of human blood products and the number of lots have increased beyond a level easily tested by the rabbit pyrogen test, the test specified in the Minimum Requirements for Biological Products.

The main pyrogen in these products is endotoxin. Since the *Limulus* Amebocyte Lysate (LAL) test has been developed as an alternate to the rabbit pyrogen test, we have been studying the feasibility of substituting this test for the rabbit pyrogen test (Yasuda et al, 1981) and further, to compare the gel-clot method with the synthetic chromogenic substrate (SCS) method of running the test in certain human blood products.

MATERIALS AND METHODS

Rabbit pyrogen test. This test was carried out according to the modified description (Ministry of Health and Welfare, 1981) of the "Minimum Requirements for Biological Products" (Ministry of Health and Welfare of Japan, 1973). The mean temperature rise specified in the "Minimum Requirements for Biological Products" of two, or occassionally more than two, rabbits was measured.

LAL gel-clot method. The reagent used was "Pyrotell" purchased through Wako Junyaku Kogyo Ltd. Tokyo as "Limulus-test Wako". The test was carried out according to the directions supplied in the package insert. However, the method of reading the results was modified slightly in order to rank the firmness of the clot formed during the incubation period. A value of 2 indicated a firm gel that remained intact after inversion of the tube through 180 degrees. A value of 1 indicated a soft gel that broke slowly upon inversion. A positive test was recorded for the diluted sample if the tube was ranked as 1 or greater. Endotoxin content was estimated by comparison with a control standard endotoxin dilution series run in parallel with the samples.

Synthetic chromogenic substrate method. The SCS reagent, "PYRODICK", was supplied by Seikagaku Kogyo Co. Ltd., Tokyo. The chromogenic method described by Iwanaga and his colleagues (Harada et al, 1978, 1979) was improved in cooperation with Seikagaku Kogyo Co. and others. The test was run according to the directions supplied by the manufacturer except the incubation time was shortened from 30 min to 20 min at 37 C. This shorter incubation period gave a linear regression over a wider range of endotoxin concentrations (see results section).

Human blood products. The seven blood products tested are listed in Table 1. A total of 381 samples were analyzed. Manufacturers of these products were arbitrarily assigned the letters A through H. The albumin preparations, human plasma protein fraction (PPF) and human serum albumin (HSA), comprised 66.7% of the samples while other blood products comprised 31.8% and non-blood products 1.5% of the total number of samples tested.

RESULTS

Correlation between the gel-clot and SCS methods. Samples from each blood product class were tested by the gel-clot method and the SCS method. The endotoxin concentration determined from the gel-clot method was compared to the optical density (OD) measured at 405 nm by the SCS method. The results are given in Tables 2-4. The OD values have been grouped into six intervals representing two-fold increases (0-0.025, 0.026-0.05, and so on). It will be explained later why calculated endotoxin content was not used. The corresponding measurements from the gel-clot test were given as concentration of endotoxin in ng/ml. The symbols represent the manufacturers of the product as given in Table 1.

Table 1 List of Human Blood Products tested

Manuf.	Mark	PPF	HSA	N-Ig	Peps-Ig	Sulf-Ig	PEG-Ig	F VIII conc.	Others
A	○	37	42	4			25	20	5
B	▲		10	1	5	22		3	
C	●	21	3	1	5			6	
D	■	60		3				1	
E	×	41	16					5	
F	+	23		1	13				
G	□				5				
H	✱		1	1					1
Total		182	72	11	28	22	25	35	6
%		66.7		31.8					1.5

Grand Total : 381

Abbreviation

PPF	Human Plasma Protein Fraction
HSA	Human Serum Albumin
N-Ig	Human Normal Immunoglobulin
Peps-Ig	Dried Pepsin-treated Human Normal Immunoglobulin
Sulf-Ig	Dried Sulfonated Human Normal Immunoglobulin
PEG-Ig	Dried Polyethylenglycole-treated Human Normal Immunoglobulin
Fvııı conc.	Dried Concentrated Human Anti-Haemophilic Globulin

Human plasma protein fraction (PPF) results are presented in Table 2. A large symbol represents ten smaller ones. Most of the symbols were distributed over a range of 0 to 0.1 ng/ml by the gel-clot method and from 0 to 0.05 OD (405nm) units in the SCS method. Correlation coefficients listed in the same table indicated that significant positive correlations existed between the results of the gel-clot and SCS methods for four of the five manufacturers' products. The samples represented by "+" did not show significant correlation. Although the correlation between the test methods was significant for the solid circles, the results from this manufacturer of PPF were distributed out of the range of the results obtained on samples from the other manufacturers. It is evident that the results from the same type of blood product may vary from manufacturer to manufacturer.

Results from testing human serum albumin (HSA) are

Table 2 The Correlation between The LAL Test and The SCS Test

PPF

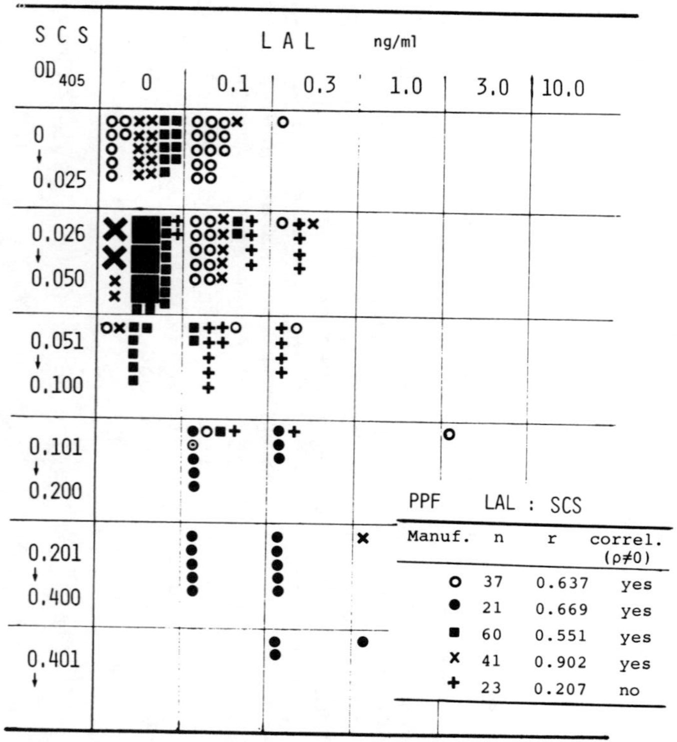

PPF LAL : SCS

Manuf.	n	r	correl. ($\rho \neq 0$)
○	37	0.637	yes
●	21	0.669	yes
■	60	0.551	yes
×	41	0.902	yes
+	23	0.207	no

Table 3 The Correlation between The LAL Test and The SCS Test

HSA

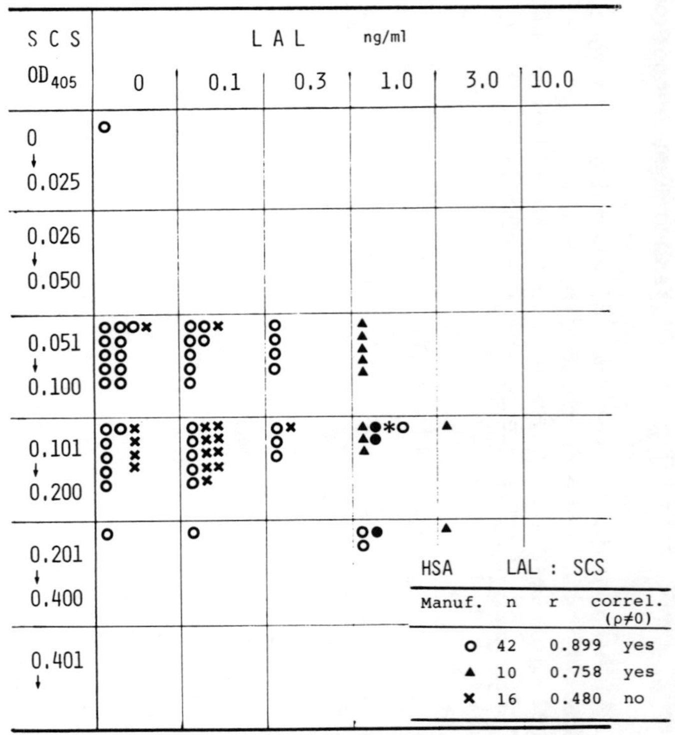

HSA LAL : SCS

Manuf.	n	r	correl. ($\rho \neq 0$)
○	42	0.899	yes
▲	10	0.758	yes
×	16	0.480	no

given in Table 3. These results show a positive correlation between the test methods for HSA from two out of the three manufacturers. Again the values from each method cluster but the range over which they cluster differs by manufacturer.

The number of samples of human normal immunoglobulin (N-Ig) was too small to obtain a correlation between LAL methods according to manufacturer. Neither could a correlation be confirmed when the data were analyzed as a whole; the correlation coefficient (r) was -0.095 on 11 total samples, and therefore, a correlation table was not included.

Table 4 shows the results of testing samples of dried human normal immunoglobulins. The first, pepsin-treated human normal immunoglobulin (Peps-Ig), gave a negative correlation between the gel-clot and SCS methods. There was a tendency to obtain lower values with the SCS than with the gel method. This discrepancy from the results seen for the previous products will be discussed later. Dried polyethyleneglycol-treated immunoglobulin (PEG-Ig) showed no correlation between the test methods. Again, the OD values from the SCS method were low with respect to the gel-clot values. Dried sulfonated human normal immunoglobulin (Sulf-Ig) gave a postive correlation between the two methods, however, the OD values from the SCS method were lower than might be expected from the values obtained from the gel-clot method.

Dried concentrated human anti-hemophilic globulin (FVIII conc) is a concentrate of the blood coagulation factor and might be expected to interfere with the LAL system. No correlation (r=0.092) was demonstrated even when all 35 samples were analyzed together. The correlation table was not included.

Correlation between pyrogen test and the gel-clot or SCS test methods. All the preparations reported on in this paper were submitted to the Department of General Biologics Control and were found to be negative by the rabbit pyrogen test. By this test, material passes if the summed responses of two rabbits does not exceed 0.7 C. As the number of rabbits per test increases up to a total of 9 rabbits, the limiting mean temperature response increases from 0.35 C to 0.56 C. These values represent, within statistical variability, a minimal temperature response, so one would not expect any correlation between this test and one that is significantly more sensitive. Nevertheless, a correlation was attempted by pairing the mean temperature rise with the values obtained from each of the LAL methods for each sample. The correlation coefficients and the number of samples of each product from each manufacturer are listed in Table 5. In general, there were no positive

Table 5

CORRELATION BETWEEN LAL & PY ; SCS & PY

			LAL : PY		SCS : PY	
Product & Manuf.		n	r	Correl. ($\rho \neq 0$)	r	Correl. ($\rho \neq 0$)
PPF	A	37	-0.215	no	0.713	yes
	C	21	-0.036	no	0.088	no
	D	60	0.085	no	0.368	yes
	E	41	0.021	no	-0.310	no
	F	23	0.250	no	0.043	no
HSA	A	42	-0.027	no	0.038	no
	B	10	-0.353	no	-0.314	no
	E	16	-0.031	no	0.009	no
Pep-Ig						
	Total	28	-0.193	no	0.185	no
Sulf-Ig						
	Total	22	0.044	no	-0.040	no
PEG-Ig						
	Total	25	-0.204	no	-0.057	no
FVIII conc.						
	Total	35	0.184	no	0.231	no

LAL: gel-clot method; PY: rabbit pyrogen test;
SCS synthetic chromogenic substrate method;
A-F: symbols of manufacturers shown in Table 1.

correlations between the mean temperature rise and the values obtained by either of the LAL methods. A positive correlation was found between the SCS method and the mean

TABLE 4 CORRELATION BETWEEN THE LAL TEST AND THE SCS TEST

Peps-Ig

SCS OD 405	LAL ng/ml 0	0.1	0.3	1.0	3.0	10.0
0 → 0.025	▲		● + + + ▲ ▲ ▲ ▲	+ + + + + + + + + + □ □ □		
0.026 → 0.050		● ● ●	●	□ □		
0.051 → 0.100						
0.101 → 0.200						
0.201 → 0.400						

Peps-Ig Manuf.	n	r	LAL : SCS correl. ($\rho \neq 0$)
Total	28	-0.387	yes
+	13	0.280	no
▲●□	15	0.112	no

PEG-Ig

SCS OD 405	LAL ng/ml 0	0.1	0.3	1.0	3.0	10.0
0 → 0.025	○ ○ ○ ○ ○ ○ ○	○ ○ ○ ○	○ ○ ○ ○ ○ ○ ○ ○ ○	○ ○ ○		
0.026 → 0.050	○	○				
0.051 → 0.100						
0.101 → 0.200						
0.201 → 0.400						

PEG-Ig	n	r	LAL : SCS correl. ($\rho \neq 0$)
Total	25	-0.204	no

Sulf-Ig

SCS OD 405	LAL ng/ml 0	0.1	0.3	1.0	3.0	10.0
0 → 0.025			▲ ▲ ▲ ▲ ▲ ▲	▲		
0.026 → 0.050			▲ ▲ ▲ ▲ ▲	▲ ▲ ▲ ▲ ▲	▲	
0.051 → 0.100			▲	▲	▲ ▲	
0.101 → 0.200						
0.201 → 0.400						

Sulf-Ig	n	r	LAL : SCS correl. ($\rho \neq 0$)
Total	22	0.647	yes

temperature response for PPF from two out of five manufacturers. This might suggest that the SCS method is more sensitive than the gel-clot method, but the same test would have to be repeated on samples containing pyrogenic levels of endotoxin in order to have a valid comparison.

Dose response of SCS method to E. coli endotoxin preparations. No attempt was made to express the results of the SCS method in concentration of endotoxin since the SCS dose-response varied greatly with the strain of E. coli from which the endotoxin was derived. As shown in Fig. 1, the concentration of endotoxin giving an OD of 0.2 was 0.045 ng with UKT-B(R) endotoxin, 0.125 ng with an O55:B5 (H) preparation, and 0.41 ng with O111:B4 (M) endotoxin.

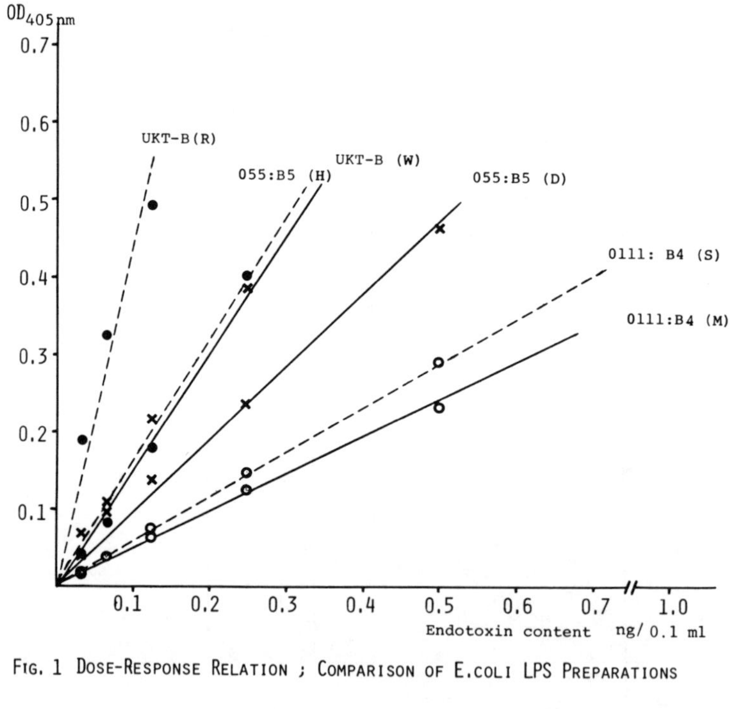

FIG. 1 DOSE-RESPONSE RELATION ; COMPARISON OF E.COLI LPS PREPARATIONS

Furthermore, it was found that different endotoxin preparations from the same strain made by different manufacturers (O55:B5 from manufacturer D and H) gave different dose-response lines.

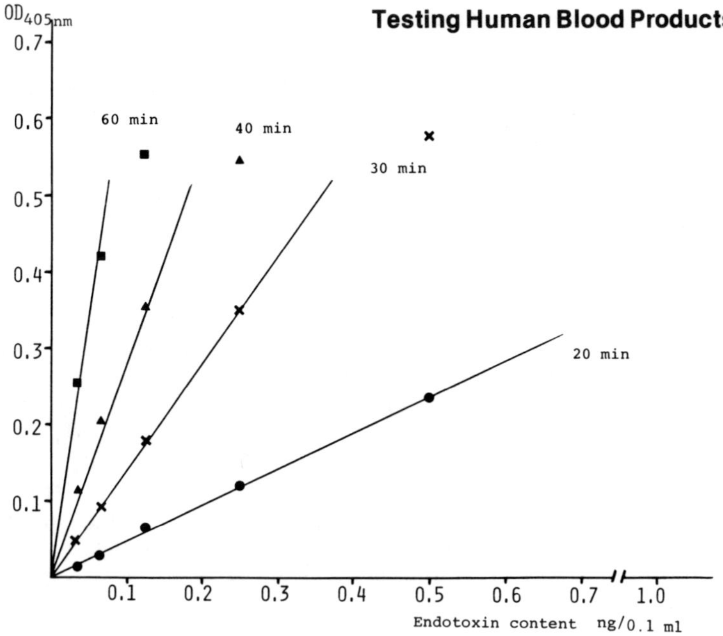

FIG. 2 THE DOSE-RESPONSE RELATION BY VARYING THE INCUBATION TIME

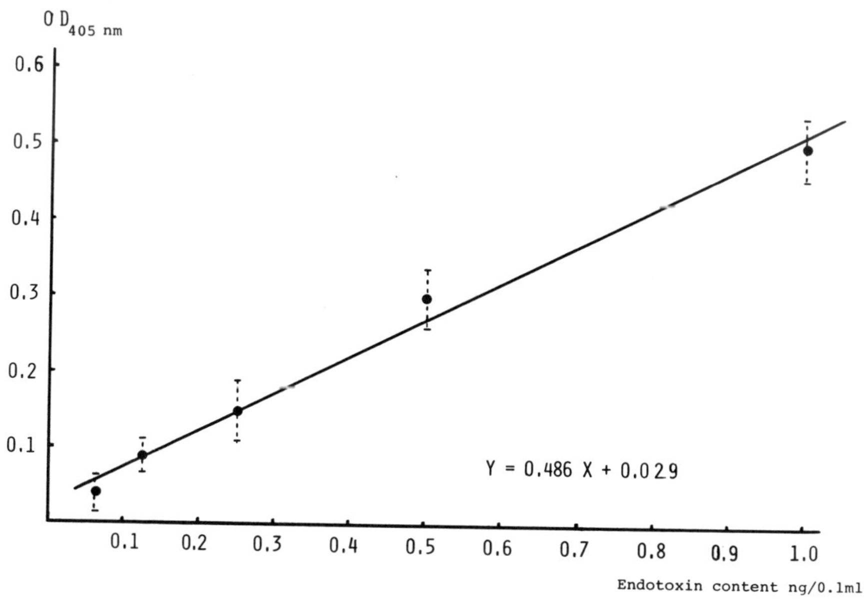

FIG. 3 THE DOSE-RESPONSE RELATION IN THE SCS TEST

Therefore it is important to establish a reference endotoxin in order to compare the concentration of endotoxin in various products. An endotoxin from strain UKT-B(R) prepared by Dr. Kanoh and colleagues (Kanoh et al, 1981) has been proposed as a reference endotoxin and is currently being examined to determine its feasibility for use by our study group.

Dose-response lines derived from the SCS method. Dose-response lines were plotted as OD(405nm) against endotoxin concentration. The effect of varying the incubation period is illustrated in Fig. 2. Known concentrations of the same endotoxin preparation were incubated at 37 C for 20, 30, 40 or 60 minutes. It can be seen that the longer the incubation period, the steeper the slope of the dose-response line. In order to obtain a linear response over a greater range of endotoxin concentrations, and especially over a range of concentrations less than 1.0 ng/ml, we chose to incubate 20 minutes, unless otherwise stated.

Fig. 3 illustrates the linearity of the OD response to endotoxin concentration when the test is incubated for 20 minutes. Endotoxin concentrations were prepared and tested in quadruplicate. The vertical dotted lines represent the 95% confidence intervals at each concentration. The formula of the line obtained from regression analysis is $Y = 0.486X + 0.029$. The linearity of the response makes the SCS method useful for the quantitative determination of endotoxin.

Inhibition of the SCS method by some blood products. As shown in Table 4, the OD values obtained by the SCS method with immunoglobulin preparations (Peps-Ig, Sulf-Ig, and PEG-Ig), appeared to be lower than expected when compared with values obtained from the gel-clot method. In order to demonstrate possible inhibition, five-fold serial dilutions of the immunoglobulins were made and endotoxin, equivalent to the concentration required to give an OD of 0.2, was added to each dilution. The immunoglobulin-endotoxin mixtures were incubated at 37 C for 30 minutes and then tested. The results of these experiments are shown in Fig. 4. Except for Peps-Ig, the responses of the undiluted samples were less than the responses in the samples diluted 5^{-1} and 5^{-2}. This phenomenon suggests that the endotoxin was adsorbed to the immunoglobulins in the undiluted

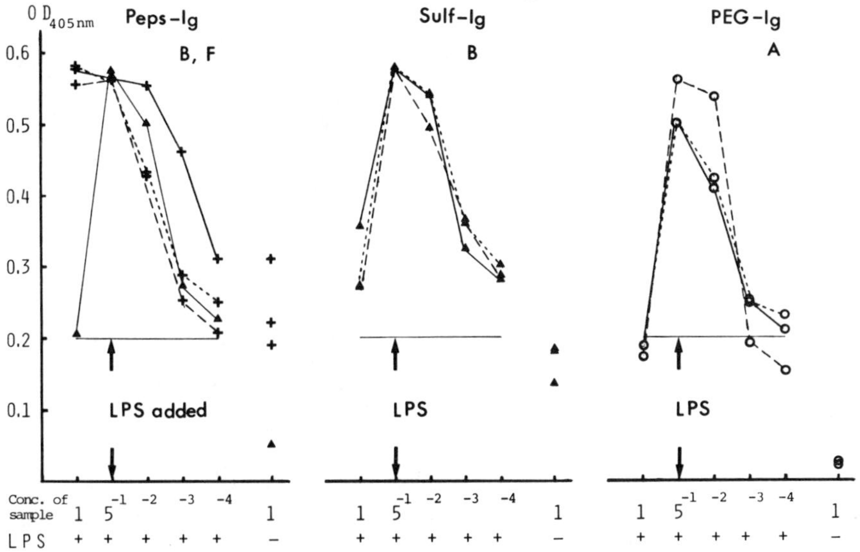

FIG. 4 THE RESPONSE PATTERN OF DILUTED GLOBULIN PRODUCTS AFTER THE ADDITION OF LPS

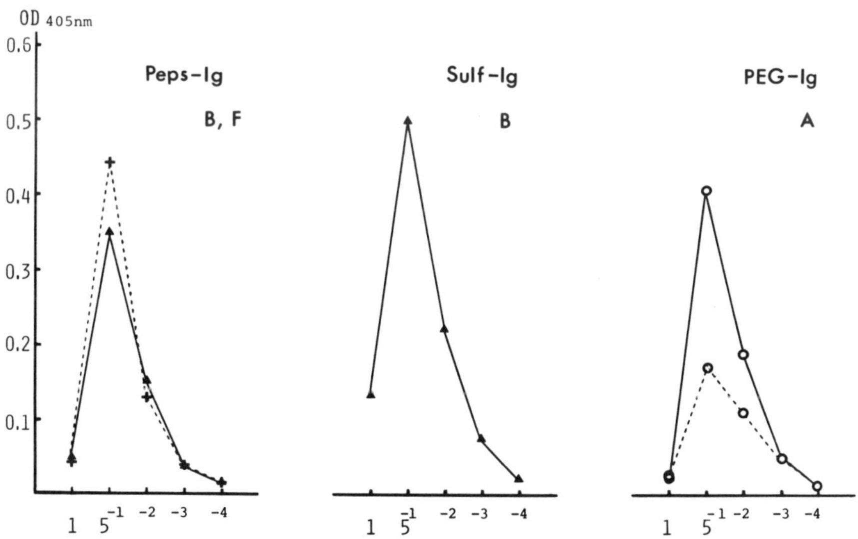

FIG. 5 THE RESPONSE PATTERN OF DILUTED GLOBULIN PRODUCTS

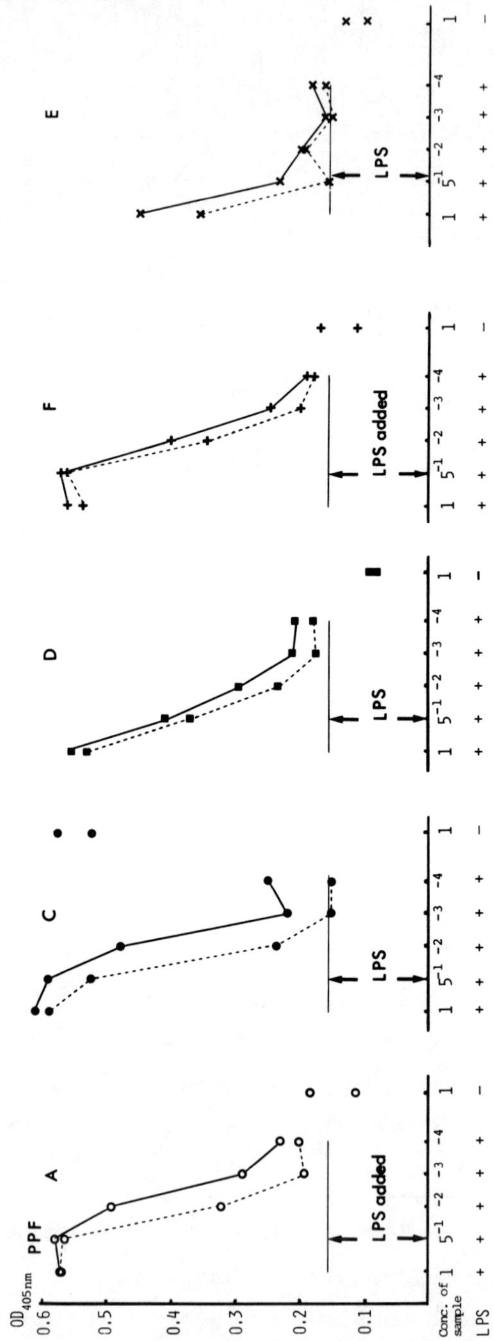

FIG. 6 THE RESPONSE PATTERN OF DILUTED PPF AFTER THE ADDITION OF LPS

state and released when the samples were diluted, or that some inhibiting substances were present and were diluted out. When the same samples were diluted and examined without the addition of endotoxin (Fig. 5), the pattern of responses was similar to that shown in Fig. 4. This indicated that the maximum response to endotoxin already present in the samples was obtained only after dilution, confirming inhibition of the test by the immunoglobulins.

As already shown in the correlation studies, the albumin samples did not give lower OD readings than expected when compared to the gel-clot method. When the albumins were examined for inhibition using the same method reported for the immunoglobulins, the prozone-like phenomenon was not detected in either PPF (Fig. 6) or HSA (Fig. 7).

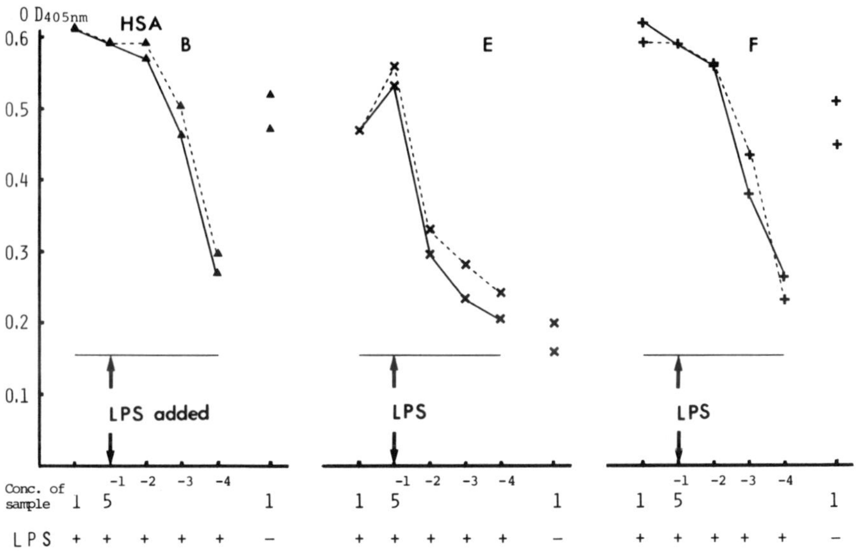

FIG. 7 THE RESPONSE PATTERN OF DILUTED HSA AFTER THE ADDITION OF LPS

<u>SCS response to endotoxin in saline, PPF or HSA</u>. The previous results indicated that either the gel-clot or SCS method of LAL testing could be used as a substitute for the rabbit pyrogen test. In order to compare the in vitro tests directly to the rabbit pyrogen test, the endotoxin had to be diluted in saline and the effects of saline in the SCS method analysed. Dose-response lines were con-

structed using PPF (Fig. 8) and HSA (Fig. 9) as the diluent for the endotoxin and these were compared to the line generated from endotoxin diluted in normal saline. When the dose response patterns of endotoxin diluted in PPF or HSA were compared to those obtained when endotoxin was diluted in saline, it was obvious that the slopes of the lines differed markedly. Additional studies are needed to determine whether or not the regression lines differ from each other within the PPF or HSA samples.

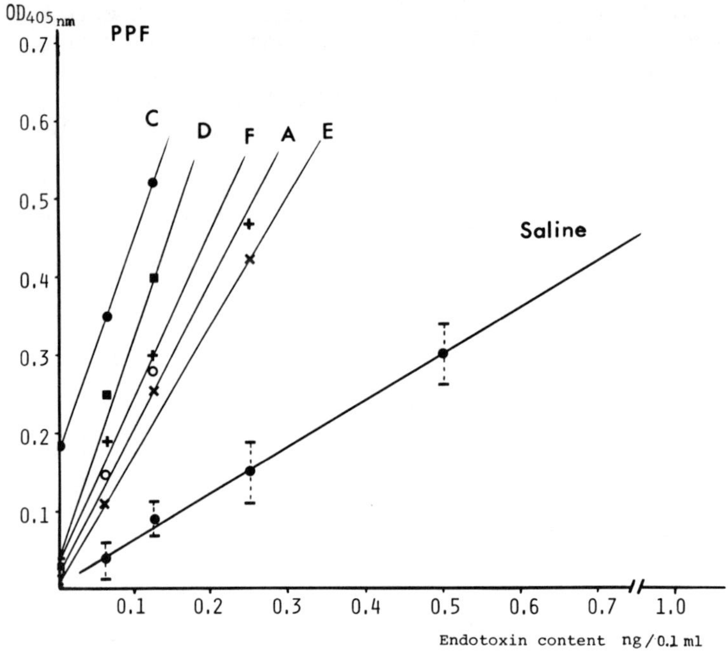

FIG. 8 THE DOSE-RESPONSE LINES OF LPS IN SALINE AND IN PPF

DISCUSSION

Since none of the blood products studied in this investigation failed the rabbit pyrogen test, they were assumed to contain low levels, if any, of endotoxin. A positive correlation was found between the LAL gel-clot and SCS methods for albumins from most of the manufacturers. A positive correlation was found between the SCS method and the rabbit temperature response for only two samples of PPF indicating the greater sensitivity of the

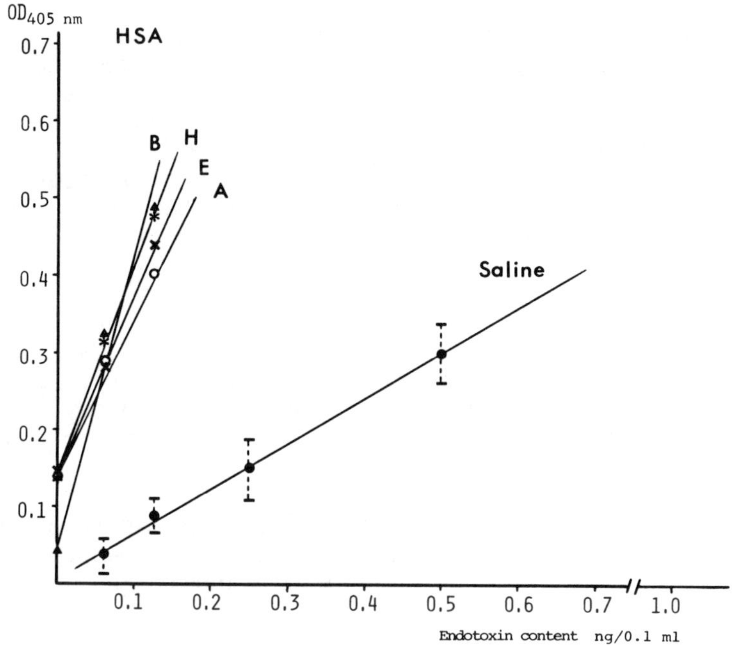

FIG. 9 THE DOSE-RESPONSE LINES OF LPS IN SALINE AND IN HSA

in vitro test over the rabbit pyrogen test for detection of endotoxin.

The prozone-like phenonemon seen in Fig. 4 may have been the result of adsorption of endotoxin to immunoglobulin and/or inhibiting substances present in the undiluted samples. The endotoxin may be released or the inhibitor concentration reduced when the sample is diluted. Additional studies are needed before the SCS method can be used to quantitate endotoxin in immunoglobulins, but it is obvious that the immunoglobulins will have to be diluted prior to testing.

Leukocytic pyrogen, unlike endotoxin, is thermolabile (Atkins, 1960; Kanoh et al, 1967). Since both PPF and HSA were heated to 60 C for 10 hours during the final step of production, false positives stemming from leukocytic pyrogen were ruled out.

The positive correlation between the gel-clot and SCS methods of running the LAL test on albumin products and the unlikely existence of leukocytic pyrogen suggests that these *in vitro* tests could be used to replace the rabbit pyrogen test. However, since in our studies only subpyrogenic concentrations of endotoxins were present, these studies must be repeated using albumin products containing higher concentrations of endotoxins, and the differences between albumin products from various manufacturers must also be studied in more detail.

ACKNOWLEDGEMENT

The authors are grateful to Mr. I. Funasaka, Dr. H. Fujiwara, Dr. H. Goto and members of the Laboratory of the Safety and Pyrogen Tests for their collaboration in the rabbit pyrogen testing.

This study was partially supported by a grant from the Pharmaceutical and Supply Bureau, Ministry of Health and Welfare of Japan which we gratefully acknowledge.

REFERENCES

Atkins E (1960). Pathogenesis of fever. Physiol Rev 40: 580.

Harada T, Morita T, Iwanaga S (1978). A new assay method for bacterial endotoxins using horseshoe crab clotting enzyme. J Med Enzymol 3:43.

Harada T, Morita T, Iwanaga S, Sakakibara S, Niwa M (1979). A new chromogenic substrate method for assay of bacterial endotoxins using Limulus hemocyte lysate. In Cohen E (ed): "Biomedical Application of the Horseshoe Crab (Limulidae)," New York: Alan R Liss, p 209.

Kanoh S, Kawasaki H, Yoshida M, Nishio A (1967). Studies on the pyrogen (IV). Some factors influencing the production of leucocytic pyrogen from polymorphonuclear leucocytes in vitro. Japan J Pharmacol 17:125.

Kanoh S, Ogawa Y, Komuro T, Kawasaki K, Niji K (1981). Studies on the extraction process of *E. coli* LPS and concomitant changes in physical properties, Limulus activity and pyrogenicity. Japan J Med Sci Biol, in press.

Ministry of Health and Welfare, Japanese Government (1973). Minimum requirement for biological products, English edition, p 243.

Ministry of Health and Welfare, Japanese Government

(1981). The Official Gazette, March 20.
Yasuda J, Ueno G, Kuratsuka K, Homma R, Hara S, Kazama M, Ito, M (1981). Japanese minimum requirements for albumin preparations - recent amendments and current problems. Develop Biol Standard 48, p 143.

Endotoxins and Their Detection With the Limulus Amebocyte Lysate Test, pages 319–327

PYROGEN REMOVAL BY ULTRAFILTRATION

Robert S. Tutunjian

Director, Industrial Marketing
Amicon Corporation
Danvers, MA 01923

INTRODUCTION

The need to ensure absence of pyrogens from water or parenteral solutions can be a major problem for pharmaceutical manufacturers. Though defined as any substances which cause a temperature rise when injected, most pyrogens are lipopolysaccharides (LPS) from bacterial cell walls. The sizes of these endotoxins can range from subunits of about 20,000 molecular weight (MW) to aggregates of about 0.1 micron (Hannecart-Pokomi, et al, 1973). The degree of aggregation is reversible and can be affected by the presence of ions and dispersing agents.

The sensitivity of patients to pyrogens is quite high. Concentrations of less than one nanogram per milliliter can cause strong reactions ranging from fever to even death. Pyrogens may be detected by actual injection of sample into a rabbit or by the use of limulus lysate (LAL) tests. The threshold of detectability ranges from 0.01 ng/ml to 0.1 ng/ml with LAL tests.

Pyrogens cannot be eliminated by the standard sterilization techniques of autoclaving or microporous membrane filtration. The LPS molecule is quite stable thermally, requiring exposure to 250°C for over an hour for destruction. The molecule is also relatively insensitive to changes in pH, although high levels of acid or base will cause deactivation.

Processes that have been used for pyrogen removal until now are: distillation, reverse osmosis (RO), and absorption by asbestos or other media. Distillation systems have very high capital and operating costs. RO equipment can reduce expenses, but is difficult to clean and sterilize. Also, neither distillation nor RO can be used to depyrogenate parenteral solutions, since the solute will be removed with the LPS. While asbestos is no longer acceptable, other charged media have shown some degree of effectiveness in removing pyrogens. However, these media are limited in removal efficiency and in their total binding capacity (Geiber, et al, 1980).

ULTRAFILTRATION

Ultrafiltration membranes are skinned, asymmetrical structures with pore sizes larger than those of RO membranes, but smaller than microporous filters. Use of an ultrafilter with a 10,000 MW pore size will provide retention of even LPS subunits. However, unlike distillation and RO, ultrafiltration membranes will allow free passage of parenterals, such as: antibiotics, sugars, radio-opaque markers, and hyperalimentation solutes. Unfortunately, solutes with higher than 10,000 molecular weights, such as plasma proteins, would be excluded along with the LPS.

Hollow fiber ultrafilters have been shown to be particularly effective in removing pyrogens while offering certain advantages over other membrane configurations. The fibers are self-supporting and therefore do not require any support structures which can clog. Cartridges fabricated from hollow fibers are self-contained and are completely sanitary in design. These filters may be cleaned or backflushed with a variety of chemical agents, allowing long life and economical operation.

USE OF ULTRAFILTERS

Ultrafiltration systems may be used in a variety of ways: central or point-of-use purification; batch or continuous; with or without feed pump; etc. This flexibility is actually an advantage, allowing the specific needs of the manufacturer to be met.

Generally, continuous ultrafiltration systems are installed as close to the final use point as possible, following standard prefiltration, carbon, and deionization equipment. This point-of-use filtration avoids possible recontamination by distribution piping. Also, in most cases, the ultrafilters can operate solely on line pressure (25 psi) with a slight bleed (1-2% of feed) to minimize fouling. (Booster pumps and reservoirs may be installed to increase downstream pressure where necessary.) When batch filtration is desired, the ultrafiltration cartridges may be operated like a typical in-line filter, with no bleed stream.

Installation of cartridges must be followed by certain steps to ensure product quality. First, the system should be cleaned with a strong sodium hydroxide solution (0.1 M) to destroy any residual LPS. This base is then flushed out and the filtrate assayed for pyrogens and for pH. The filter cartridges are then integrity tested by use of an air pressure test. These procedures are repeated periodically since the membranes do gradually foul and require cleaning to restore filtration rates.

In actual operation, hollow fiber ultrafilters exhibit dependable removal of pyrogens. Even when challenged with tap water to which a large amount of LPS was added, the filtrate was free of detectable pyrogens (see Table 1). In one instance, the LPS concentration was 10,000 ng/ml prior to ultrafiltration and less than 0.050 ng/ml after. The removal efficiency (inlet concentration/outlet concentration) was over 10^5 in this case.

Ultrafiltration of deionized water has been shown to be effective for removal of bacteria as well as pyrogens. In extended test studies (see Table 2), the water quality exceeded the requirements for U.S.P. Water For Injection essentially at all times. In one instance, low level bacteria counts were observed in the filtrate. A final sterilizing grade microfilter could be placed after the ultrafilter to guarantee sterility. In all cases, the water quality exceeded specifications for U.S.P. Purified Water.

COMPARISON TO OTHER PROCESSES

Ultrafiltration compares very favorably to the other processes used to remove pyrogens (see Table 3). The most

Day	Pyrogens (ng/ml)	
	Before ultrafiltration	After ultrafiltration
1	300	<0.050
2	100	<0.050
3	300	<0.050
4	10,000	<0.050
5	1,000	<0.050
10	1,000	<0.050
15	1,000	<0.050

Table 1

Pyrogen removal from tap water using Amicon H10P10 cartridge. LAL test.

Day	Before ultrafiltration		After ultrafiltration	
	Bacteria count/100 ml	Pyrogens ng/ml	Bacteria count/100 ml	Pyrogens ng/ml
1	12,000	3	0	<0.050
2	25,000	10	3	<0.050
3	9,000	3	0	<0.050
12	7,000	1	0	<0.050
19	4,000	3	0	<0.050
26	6,000	3	0	<0.050
40	8,000	1	0	<0.050
54	4,000	3	0	<0.050
68	7,000	10	0	<0.050
82	5,000	3	0	<0.050

Table 2

Removal of bacteria and pyrogens from deionized water using Amicon H10P10 cartridge

	Hollow Fiber UF	Distillation	Thermo-comp. distillation	Reverse osmosis	Charged filters
Usability					
For water	Yes	Yes	Yes	Yes	Yes
For product	Yes	No	No	No	Yes
Sanitary design	Good	Good	Good	Fair	Good
Removal efficiency	$>10^5$	$>10^5$	$>10^5$	$>10^5$	10^2-10^3
Approximate operating cost, U.S.$ per 1,000 L	0.50-0.80	10-15	1-2	1-2	2.50-10
Approximate capital cost 800 L/hour system	$8,000	$60,000	$140,000	$20,000	$5,000

Table 3

Comparison of ultrafiltration with other methods of pyrogen removal widely used methods--distillation and RO--cannot be used to purify a parenteral product (excluding water). Also, RO systems have been shown to be difficult to sanitize. While charged filters have been successfully used for depyrogenation of products, their removal efficiencies are much lower and hence less reliable.

Finally, the operating and capital costs of ultrafiltration are substantially lower than the other methods. Even Thermocompression Distillation and RO are still double in operating expenses and greater in capital costs.

The advantages and reliability of ultrafiltration have been shown to be significant. As additional manufacturer experience is gained, ultrafiltration should gain widespread acceptance in the parenteral industry.

Geiber C, Hon K, Babineau R, and Fiore J (1980). Pyrogen Control by Depth Filtration. Pharm Tech 4:82.

Hannecart-Pokorni E, Dekegal D, and Dupuydt F (1973). Macromolecular Structure of Lipopolysaccharides from Gram-Negative Bacteria. Eur J Biochem 38:6.

TUTUNJIAN SPEAKER
TITLE: PYROGEN REMOVAL BY ULTRAFILTRATION.

CASE: Have you quantitated endotoxin levels less than 50 pg/ml?
TUTUNJIAN: No, most of our tests are demonstrating to customers that we can remove higher levels of endotoxins.
MILLS: I notice in your slides you show a reduction from 10,000 ng down to about 50 pg/ml. I assume this was endotoxin suspended in water. Have you done similar kinds of work with endotoxin suspended in either a protein solution or in a more complex solution?
TUTUNJIAN: We have not spiked solutions with LPS although we have demonstrated certain cases where radioopaque products were assayed prior to filtration. In some cases endotoxin levels prior to filtration were around 10 ng/ml. Downstream concentrations were below the 50 pg level.
ENZINGER: You imply that the mechanism of removal was size exclusion. How can you explain that with time during ultra filtration, you start to get a bleed through of endotoxin. Secondly do you have a mechanism whereby you test the filter for its integrity if you are going to reuse it?
TUTUNJIAN: Could you specify what you are referring to as far as the bleed through of endotoxin?
ENZINGER: You will see levels of endotoxin coming through the ultrafiltration process.
TUTUNJIAN: The slide I showed, showed a three months record and a bleed through was not observed during this period.
ENZINGER: Was that water for injection?
TUTUNJIAN: Yes.
ENZINGER: With antibiotic solutions you will see a bleed through where several hundred liters are filtered.
TUTUNJIAN: We have done batches in excess of one hundred liters where we have observed pyrogen bleed through on the downstream side. It could have been the case when we overpressurized the cartridge.

ENZINGER: Then how do we test the filter?
TUTUNJIAN: That type of test is basically an air test in which the upstream side of the filter is pressurized and is essentially a pressure (hold) type of test where you look for a drop in pressure. A good cartridge will show no drop in pressure, or rapid bubbling which would also indicate a broken filter and this can be done without disturbing the downstream side. I should point out that in that case you are looking for a broken filter and it doesn't discriminate between a 10,000 and 20,000 weight cutoff. Our tests indicate that over long term use the membrane structure will remain intact and the only thing you have to test is the integrity of the fibers.
SIDOTE: What is the flow rate using tap water?
TUTUNJIAN: A two cartridge system has a capacity of approximately 4 or 5 gals/minute. Using a single cartridge system the capacity is about 2 gal/min.
SABEY: What is the life of the filter?
TUTUNJIAN: It is variable, it can last one to three years in some cases. It depends very strongly on the incoming water quality and how you really treat it. In other words, if you properly regenerate it and clean it periodically the filter will last longer.

Endotoxins and Their Detection With the Limulus Amebocyte Lysate Test, pages 329–342

CYCLIC AND CHRONOBIOLOGICAL CONSIDERATIONS WHEN EMPLOYING THE RABBIT FEVER TEST

Leandro Bellentani, dr Chem., Pharm.

Research Director
Hoechst Italia Sud
Scoppito - L'Aquila - Italy

INTRODUCTION

Greisman and Hornick (1969) found that the sensitivity of man to endotoxin per kg body weight was comparable to that of the rabbit and that an average temperature rise of 0.6 C could be defined as the "threshold pyrogenic dose" in the rabbit. Simon et al. (1976) showed that 3.5 ng of endotoxin from *Shigella dysenteriae*, the lipopolysaccharide (LPS) of the World Health Organization (WHO), per kg body weight caused a mean temperature rise in rabbits that varied with the season of the year; the extremes were 0.286° C in August and 0.574°C in March. Franz Halberg (Hauss et al, 1974) demonstrated that the hour of administration of *Escherichia coli* endotoxin influenced the number of fatalities in the mouse toxicity test. Using the same dose of endotoxin (100 ug/2cc/20g body weight), the mortality varied from 15% at midnight to 85% at 4pm. In August 1980, a rabbit pyrogen test was performed by Dr. Gattone and Dr. Spazzoli (unpublished results) at midnight and, in contrast to the mouse toxicity studies, a higher temperature rise was observed at that time than found during the day with the same dosages.

Because of these studies, the *Limulus* Amebocyte Lysate (LAL) study group of the Associazione Farmaceutici Industria (AFI), Italian Pharmacopoeia, wanted to know how much variability to expect in the temperature response of rabbits injected with a fixed dose of endotoxin, and how much of that variability was a function of the time of day or the month in the year when the dose was administered.

MATERIALS AND METHODS

Endotoxin was prepared from E. coli strain O55:B5 (Difco). Eight concentrations of endotoxin were tested (15.6, 31.3, 62.5, 125, 250, 500, 1000, and 2000 pg/ml) in 1ml/kg doses. Dilutions were made in sterile pyrogen-free water.

Eight laboratories, each with its own rabbit colony, participated in the study. The eight laboratories were divided into two groups; each group administered the same four endotoxin doses at 10am and at 2pm. One of the laboratories also administered all eight doses at midday (12pm) and another of the laboratories administered four of the doses at midnight (12am). The experiments were performed each month, February through April, 1981.

The rabbits used in these studies were New Zealand, Himalaya, Albino, or California races with weights that varied from 1.5 kg to 3 kg. Each test dose was administered to at least one set of three rabbits. The rabbits were standardized with respect to solar time at each laboratory.

RESULTS

In a preliminary experiment to determine the "threshold pyrogenic dose" of E. coli O55:B5 endotoxin, a series of tests were run in February, March, and April of 1978 in which endotoxin was administered at 10am and noon to sets of rabbits. The total rise in temperature was compared to the theoretical rabbit responses given in the European Pharmacopoeia (EP). Figure 1a shows the data points, each representing the sum of the temperature rises plotted for each set of 3 rabbits and for grouped sets of 3 rabbits to give results for 6, 9, and 12 rabbits as a function of endotoxin dose. A smoothed curve was drawn through the data points from the grouped set of 12 rabbits and the values from the smoothed curve were used to calculate theoretical curves (Fig. 1b) for 3, 6, and 9 rabbits. The smoothed curve has been redrawn in figure 1b.

The EP defines "pyrogenic" as the sum of temperature rises for n number of rabbits as greater than some

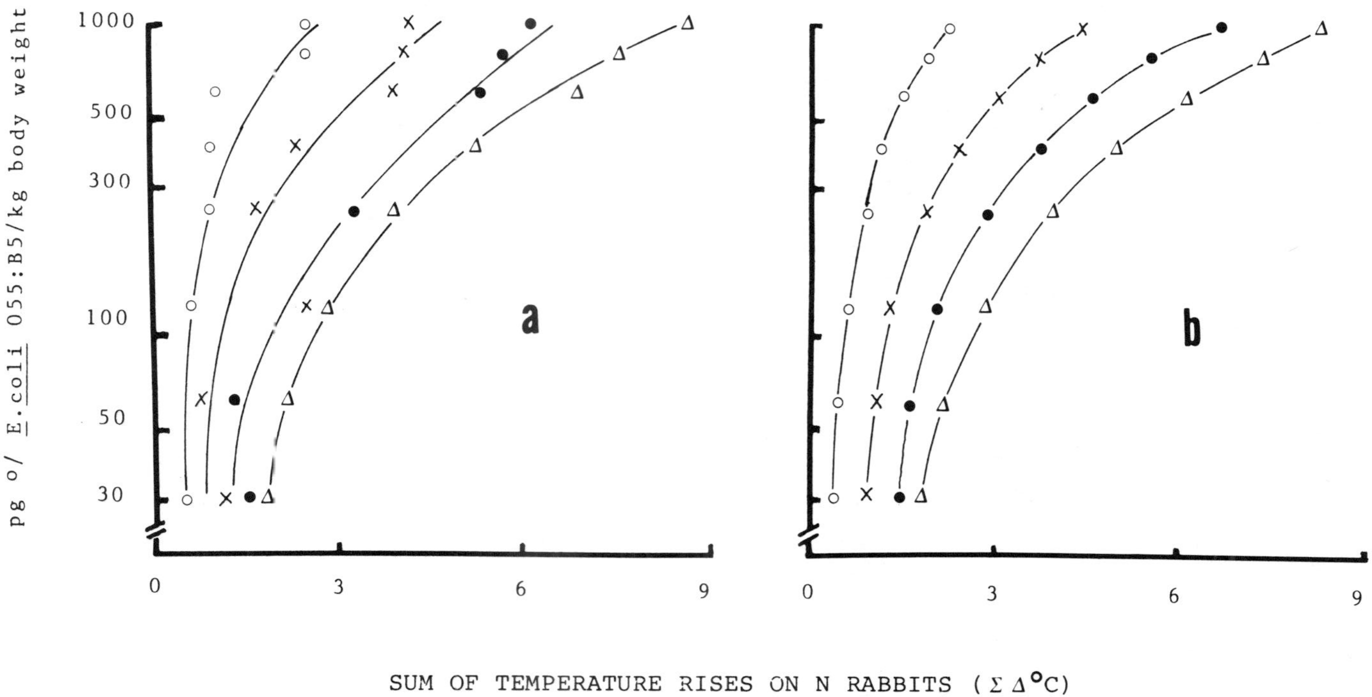

Fig. 1. Experimental (a) and theoretical (b) sums of temperature rises ($\Sigma\Delta^{\circ}C$) for sets of rabbits receiving different concentrations of E.coli 055:B5 endotoxin. Experimental results are from sets of 3(○), 6(X), 9(●), and 12(Δ) rabbits injected between 10am and noon, Feb through Apr 1978. The theoretical curves were calculated from 3, 6, and 9 rabbits using data points taken from the smoothed experimental curve for 12 rabbits.

limiting value, and "non-pyrogenic" as a sum less than some limiting value. As the number of rabbits per test increases, the limiting values approach each other until, with 12 rabbits, the values are the same. Fig. 1a illustrates the rational; the scatter of the data points around the lines decreases as the number of rabbits increases.

Table 1 lists the values in the EP that define pyrogenicity. The corresponding dose of endotoxin was taken from Fig. 1b, and the mean temperature rise was also calculated and listed. These results illustrate variability that may be entirely due to random differences in individual rabbits, the race of rabbits, their sex, and so on; or there may be a superimposed variable, a monthly or hourly difference in sensitivity of the colony to endotoxin, which could be controlled in order to reduce the overall variability of the rabbit pyrogen test.

Table 1. Temperature rises (sum delta°C) for groups of 3, 6, 9, and 12 rabbits that define pyrogenicity (EP) and the corresponding doses (ng/kg) of endotoxin taken from the experimental results in Fig. 1b

	Number of Rabbits per test			
	3	6	9	12
Pyrogenic				
total temp. rise	>2.65	>4.3	>5.95	>6.6
mean temp. rise	>0.88	>0.72	>0.66	>0.55
endotoxin dose	1.0	0.96	0.88	0.68
Non-Pyrogenic				
total temp. rise	<1.15	<2.8	<4.45	<6.6
mean temp. rise	<0.38	<0.47	<0.49	<0.55
endotoxin dose	0.38	0.52	0.55	0.68

Because of the variability demonstrated in the 1978 experiments, the Italian Pharmacopoeia Group requested experiments designed to test the minimum threshold pyrogenic dose when the susceptibility of the rabbit colony was evaluated in terms of time of day and time of year.

Eight laboratories administered the same fixed doses

of E. coli 055:B5 endotoxin to sets of 3 rabbits as described in Materials and Methods. Table 2 lists the sum and mean temperature rises after administration of endotoxin at the hours given for each month of the test. Note that in April, the tests were run at 9 am and 1 pm solar time. The rise in temperature was scalar, increasing with the concentration of endotoxin, but the relationship

Table 2. Temperature rises in response to E. coli 055:B5 endotoxin as a function of month and hour of administration.

Conc. of 055:B5 (pg/kg)	sum Δ°C (3 rabbits) 10am	12pm	2pm	12am	Mean Δ°C 10am	12pm	2pm	12am
FEBRUARY 1981								
62.5	0.73		0.63	0.5	0.24		0.21	0.17
125	0.56		0.83	0.7	0.19		0.28	0.23
250	0.97		1.04	0.9	0.32		0.35	0.3
500	1.24		1.38	1.2	0.41		0.46	0.4
MARCH 1981								
15.6	0.67		0.55		0.22		0.18	
31.3	0.67		0.65		0.22		0.22	
62.5	1.03		1.35*	0.8	0.34		0.45*	0.27
125	1.10		1.37*	0.8	0.37		0.46*	0.27
250	1.18		1.60*	1.0	0.39		0.53*	0.33
500	1.65		1.61*	0.9	0.55		0.54*	0.3
1000	1.83		1.7		0.61		0.57	
2000	2.23		2.22		0.74		0.74	
APRIL 1981								
15.6	0.37	0.35	0.53		0.13	0.12	0.18	
31.3	0.67	1.0	0.49		0.23	0.33	0.16	
62.5	0.62	0.50	0.86	0.40	0.19	0.17	0.23	0.13
125	1.04	0.75	1.03	0.30	0.29	0.25	0.30	0.10
250	0.96	1.05	1.34	0.40	0.30	0.35	0.40	0.13
500	0.97	0.90	1.26	0.60	0.32	0.30	0.37	0.20
1000	1.76	1.85	1.74		0.59	0.62	0.58	
2000	2.42	2.65	2.36		0.81	0.88	0.79	

* - corrected data; extreme values eliminated.
Time change occurred in April. All values reported at 10am were 9am solar time. Four values reported at 2pm, 62.5 - 500 pg/kg, were 1pm solar time.

between temperature rise and endotoxin dose differed when the data were analysed with respect to hour in any month.

Table 3. Linear and logarithmic regressions used to calculate the mean temperature rise (y*) at 1 ng of endotoxin/kg dose. Data are taken from Table 2 (extreme values were omitted except for *times).

	linear regression (y = a + bx)			log regression (y = c + dlnx)		
Feb. 1981	a	b	y* °C	c	d	y* °C
10am	0.1813	0.4686	0.650	0.4530	0.0933	0.453
*10am	0.214	0.2211	0.436	0.4522	0.0735	0.452
2pm	0.1952	0.5460	0.741	0.5285	0.1184	0.528
midnight	0.1559	0.5068	0.662	0.4670	0.1109	0.467
Mar. 1981						
10am	0.3089	0.2468	0.555	0.6167	0.1067	0.616
2pm	0.3553	0.2121	0.567	0.6390	0.1027	0.639
*2pm	0.2866	0.2438	0.530	0.5838	0.1015	0.584
midnight	0.2702	0.0897	0.359	0.3334	0.0243	0.333
Apr. 1981						
10am	0.1949	0.3205	0.515	0.5619	0.1196	0.561
*10am	0.2796	0.3707	0.650	0.7276	0.1520	0.728
noon	0.2028	0.3499	0.552	0.5938	0.1250	0.593
2pm	0.2285	0.2977	0.526	0.5814	0.1180	0.581
*2pm	0.2961	0.2660	0.562	0.6324	0.1176	0.632
midnight	0.0968	0.1906	0.287	0.1999	0.0337	0.199
Aug. 1972 (Simon)						
10am	0.1235	0.2317	0.355	0.2995	0.5265	0.299
Aug. 1980 (Gattone and Spazzoli)						
midnight	0.3970	1.5116	1.908	1.3779	0.3127	1.378

x - dose (ng/kg) of Difco E. coli 055:B5 endotoxin.
y - mean temperature rise.
y* - mean temperature rise calculated for x = 1 ng/kg.
Number of sets of 3 rabbits per dose: Feb. 10am, 8; Feb. 2pm, 7; Feb. noon, 1; Feb. midnight, 1; Mar. 10am, 16; Mar. 2pm, 21; Mar. noon, 1; Mar. midnight, 1; April 10am and 2pm, 26; April noon, 1; April midnight, 1. The values for Feb. noon and Mar. noon have been eliminated because a diluent other than water was used.

In general, the greatest rise in temperature for any given dose of endotoxin occured at 2pm (14 hr) while the least rise occurred at midnight (24 hr). These results were similar to those obtained in February, March, and April of 1978, but opposite to those obtained in August 1980 (Gattone and Spazzoli) when the greatest rise in temperature occured at midnight. Simon (1976) also showed the least rise in temperature to occur in August in the morning. These results suggest that the rabbit response follows a circannual as well as a circadian rhythm.

Linear (y = a + bx) and logarithmic (y = c + dlnx) regression equations were derived from the data in Table 2 and from the Simon data after the endotoxin potency of the WHO endotoxin standard was normalized to the 055:B5 potency (3.5 ng/kg WHO standard = 1.0 ng/kg 055:B5). The parameters of the equations were x = dose expressed as ng 055:B5 endotoxin per kg body weight, y = mean rise in temperature, a and c = y intercept, and b and d = slope. Since the threshold pyrogenic dose is generally accepted to be 1 ng/kg in rabbits, the mean temperature rise corresponding to that dose was calculated from each regression equation and listed in Table 3 along with the calculated a, b, c, and d values that gave the best fit to the data. The mean temperature rises have also been calculated omitting the extreme data points and are indicated by the asterisk. The regression functions were considered to be significant if the correlation or probability coefficient (r) was greater than 90% (0.9) and if the dispersion (syx) of the data points from the line was less than 0.2.

One possible reason for the variability seen in Table 3 may be that the races of rabbits differed in their susceptibility to endotoxin. A mixture of four races (New Zealand, Himalaya, Albino, and California) received the 0.016, 0.031, 1.0, and 2.0 ng/kg doses. Only New Zealand rabbits received the 0.063, 0.13, 0.25, and 0.5 ng/kg doses. The data were grouped according to race of rabbit and regression equations (both linear and logarithmic) were recalculated for the two groups separately. Mean temperature rises were calculated for a dose of 1 ng/kg and given in Table 4. Endotoxin susceptibility does differ between races, but variability with respect to hour of administration is still seen. Therefore, it is likely that a chronobiological influence on susceptibility is greater than any influence of race or individual sensitivity.

Table 4. Mean temperature rise calculated from regression equations at a dose of 1 ng of endotoxin/kg. Data from Table 2 was grouped with respect to race of rabbit and used to calculate linear (y1 = a + bx) and logarithmic (y2 = c + dlnx) equations. Dosages given Group 1 were 0.013, 0.031, 1.0, and 2.0 ng/kg. Dosages given Group 2 were 0.063, 0.125, 0.25, and 0.5 ng/kg.

Group 1.	10am y1	10am y2	2pm y1	2pm y2	midnight y1	midnight y2
March						
Himalaya	0.480	0.618	0.475	0.558		
Albino	0.355	0.394	0.281	0.331		
California	0.759	0.980	0.708	0.936		
New Zealand			0.600	0.771		
April						
Himalaya	0.472	0.605	0.603	0.762		
Albino	0.573	0.715	0.495	0.599		
California	0.708	1.195	0.459	0.589		
New Zealand	0.504	0.656	0.446	0.591		
Group 2.						
February						
New Zealand	0.654	0.454	0.746	0.531	0.663	0.467
March						
New Zealand	0.767	0.575	0.517	0.457	0.360	0.333
April						
New Zealand	0.634	0.540	0.671	0.554	0.287	0.200
noon	0.466	0.392				

These results suggest that a rabbit colony should be tested at the beginning of each month and at the hour when product would normally be tested in order to determine the regression functions and the threshold pyrogenic dose. The FDL (Factor of Dilution) or MVD (Maximum Valid Dilution) may then be calculated as a function of the particular colony at that time.

Many researchers have determined a threshold pyrogenic dose of approximately 1 ng/kg (Hochstein et al, 1973; Tsuji et al, 1979; Wachtel et al, 1977; HIMA 1979) using several different endotoxin preparations, but Simon is the only person to have considered the monthly variability in susceptibility of rabbits to endotoxins. We have calculated threshold pyrogenic doses from our data using four accepted definitions for the limit test with the corresponding confidence limits at greater than 95% (Harrison et al, 1979; Tsuji et al, 1980). "Mean" and "ave." are the terms used although they refer to the same calculation: sum temperature response/number of rabbits:

a) "mean pyrogenic dose" 0.6/1 (USP XX)
b) "ave. pyrogenic dose" 0.46/1; 1.4/3; 3.7/8 (USP XX)
c) "ave pyrogenic dose" 0.55/1; 6.0/12 (EP)
d) "ave. pyrogenic dose" 0.38/1; 1.15/3 (EP)

The regression functions from Table 2 have been plotted as mean temperature rise vs endotoxin dose in Fig. 2. The functions for the different hours and months had different slopes and points of intersection with the y axis. At the delta temperatures specified above, there was wide variability in the corresponding doses with 055:B5 endotoxin. According to the EP (delta°C = 0.55°C), the average pyrogenic dose (APD) was 0.5 ng/kg and the threshold pyrogenic dose (TPD) was 0.4 ng/kg, while according to the USP XX (delta°C = 0.46°C), the APD was 0.35 ng/kg and the TPD was 0.2 ng/kg for the month and at the hour of most sensitivity to endotoxin.

In order to test the cyclic nature of the sensitivity of rabbits to endotoxin, the cosine function,

$$y(t) = M + A\cos(\omega t + \Phi),$$

was used as a chronobiological model (Nelson et al, 1979) where $y(t)$ = the mean rise in temperature calculated for 1 ng/kg dose of endotoxin at the time of day or month of year being tested (Table 2), M = the midline value, A = the amplitude of the wave, Φ = the lag from the defined reference time to the crest time (acrophase), and ω = the angular frequency ie. 360 degrees/24 hr = 15 degrees/hr, and t = time expressed in hours (circadian) or months (circannual) after a reference time point. The rhythm parameters were estimated by least squares fitting to the cosine function. Both the linear and logarithmic regression data were used in the analysis.

Figure 3 illustrates a representative fit of the circadian model to the data from April 1981 (logarithmic values) and Figures 4-5 illustrate a representative fit of the circannual model for the data (logarithmic values)

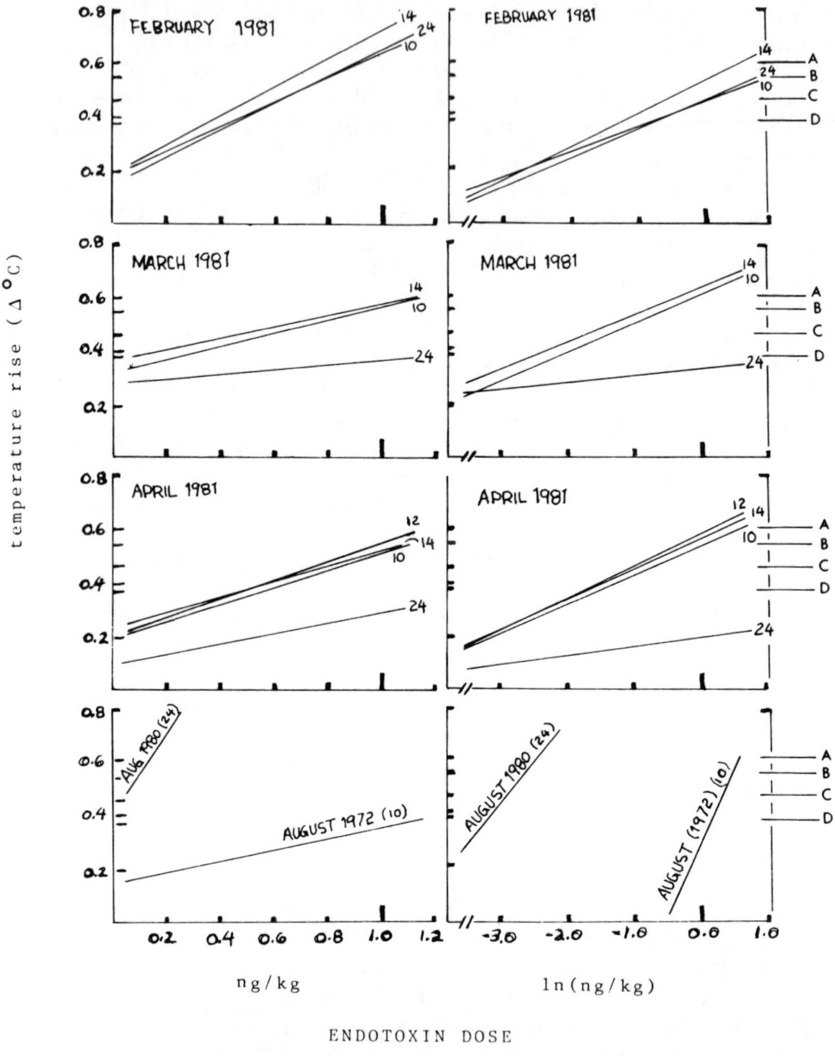

Fig. 2. Mean temperature response of rabbits to different concentrations of endotoxin. Regression equations derived from the data in Table 2 were used to obtain the plotted lines. The hour of administration of endotoxin is indicated in parentheses, ie., 24 refers to midnight. Temperature rises indicated in the righthand margin are the APD's and TPD's given in the EP or USP, (A, 0.6 °C; B, 0.55 °C; C, 0.46 °C; and D, 0.38 °C).

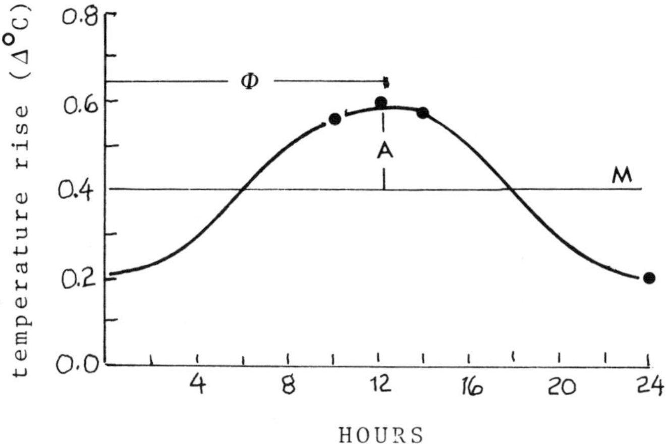

Fig. 3. Theoretical cireadian function(logarithmic) and fit of the data points for April 1981.

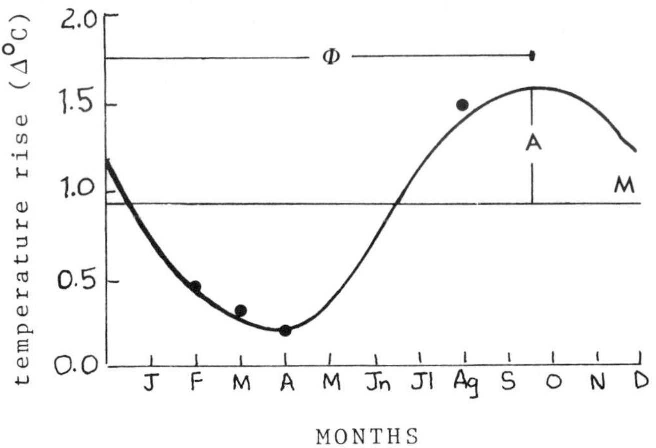

Fig. 4. Theoretical circannual function (logarithmic) and fit of the data points from midnight tests in February-April 1981 and from Gattone's and Spazzoli's August 1980 tests also at midnight.

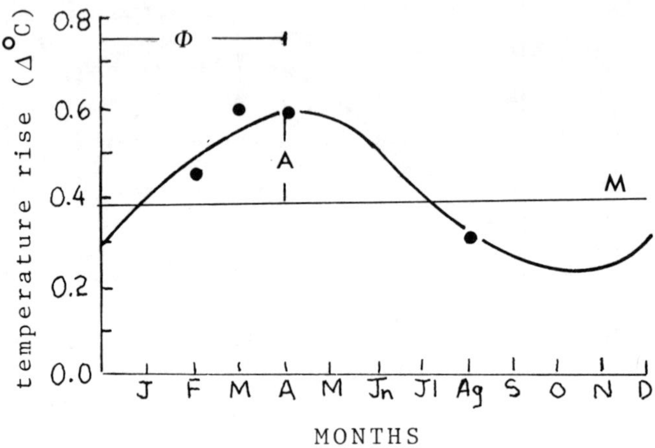

Fig. 5. Theoretical circannual function (logarithmic) and fit of the data points from 10am tests in February - April 1981 and from Simon's August 1972 tests also in the morning.

obtained at midnight and at 10am in February, March, and April 1981 and in August at midnight 1980 (Gattone & Spazzoli) and morning 1972 (Simon et al, 1976). The circadian rhythms in April were chronobiological with a probability of 99.98% (confidence limit within 0.05). The hour of greatest sensitivity was approximately 12:20pm and of least sensitivity at approximately midnight. The rhythms of the year at midnight were also chronobiological with a probability of 99.38% (confidence limit within 0.05) and indicated that the greatest sensitivity was seen at the end of October and the least at the end of April, while at 10am it was the opposite.

DISCUSSION

The sensitivity of rabbits to endotoxin varies with the individual rabbit and with the time of day and year that the dose is administered. Although the data presented here suggest a chronobiological influence, both circadian and circannual, note that the significance of some of the values may be less than that of others. The values used to establish the circadian rhythm rely heavily on the data from the midnight tests, yet the fewest total rabbits were involved per dose at midnight. Likewise, the values used to establish a circannual rhythm rely on data derived from tests done in August, all of which were done by investigators other than the author. The existence of chronobiological rhythms is suggested by the data. By testing a rabbit colony for its threshold sensitivity at the begining of each month and at the hours when products would normally be tested, seasonal variability in sensitivity may be controlled and more data accumulated to substantiate chronobiological rhythms.

ACKNOWLEDGEMENT

The rabbit tests were conducted by the following colleagues of the LAL Group of the Italian Pharmacopoeia: L Becheroni, M Carazzone, MP Luboz, ML Monacelli, C Bianchi, R Genca, F Lattanzi, S Nobilio, P Pantisano, and G Spazzoli. We thank Mrs. ML DiGirolamo for the translation into English.

REFERENCES

Greisman SE, Hornick RB (1969). Comparative reactivity of rabbit and man to bacterial endotoxin. Proc Soc Exp Biol Med 131:154.

Harrison S, Tsuji K, Enzinger RM (1979). Application of LAL for detection of endotoxin in antibiotic preparations. In Cohen E (ed): "Biomedical Application of the Horseshoe Crab (Limulidae)," New York: Alan R Liss, p 358.

Hauss E, Halberg F, Kuehl JFW, Lakatua DJ (1974). Chronopharmacology in animals. Chronobiologia 1 (suppl. 1): 133.

Hochstein HD, Elin RJ, Cooper JF, Seligmann Jr EB, Wolff SM (1973). Further developments of Limulus amebocyte lysate test. Bull Parenter Drug Assoc 27:139.

Nelson W, Tong YL, Lee J-Kuen, Halberg F (1979). Methods for cosinor. Rhythmometry-Chronobiologia 6:308.

Simon S, Toth M, Wallerstein GY, Remeny ZS (1976). Studies with the IPS on the sensitivity and reproducibility of pharmacopoeia pyrogen testing. J Pharm Pharmac 28:111.

Tsuji K, Harrison SJ (1979). Limulus amebocyte lysate - a means to monitor inactivation of lipopolysaccharide. In Cohen E (ed): "Biomedical Applications of the Horseshoe Crab (Limulidae)," New York: Alan R Liss, p 367.

Tsuji K, Steindler KA, Harrison SJ (1980). Limulus amebocyte lysate assay for detection and quantitation of endotoxin in a small-volume parenteral product. Appl Environ Microbiol 40:533.

United States Pharmacopeia (1980). In "United States Pharmacopeia," 20th revision, Mack Publishing Co., Easton, PA.

Wachtel RE, Tsuji K (1977). Comparison of Limulus amebocyte lysates and correlation with the United States Pharmacopeial pyrogen test. Appl Environ Microbiol 33:1265.

Endotoxins and Their Detection With the Limulus Amebocyte Lysate Test, pages 343–356

EXPERIENCES WITH COMPARATIVE EXAMINATIONS FOR PYROGENS BY RABBIT PYROGEN TEST VERSUS THE LAL TEST

Helmut Müller-Calgan, Dr. med. vet.,
Fachtierarzt für Pharmakologie und Toxikologie
Pharma-Qualitätskontrolle Biologie
E. Merck, Darmstadt
D 6100 Darmstadt

We all live in and with a world of microorganisms and are therefore surrounded by bacteria and their constituents, endotoxins (lipopolysaccharides) and other pyrogenic substances (Table 1). In the production of parenterals, these have to be excluded from the end product (SVP, LVP) in accordance with recognized legislation and methods (U.S. Pharmacopeia (USP), European Pharmacopeia (EP), and with more extensive regulations as well as in-house guidelines, for the protection of sick humans or animals. The "state of current scienti-

Table 1. Endotoxins in our spheres of life.

	ng/ml or /g (LAL test, EC 2)	
Pump water, central	0.200	Central Energy Supply,
Town water, central	0.400	E. Merck, 31.8.81, 7.30 a.m.
Drinking water, at home	0.800	31. 8. (mondays), 6.30 a.m.
Drinking water, at the office	1.6	31. 8. and 2.9., 11.30 a.m.
Sugar to sweeten coffee	8	from 1980
Grape sugar (emergency rations)	256	from about 1977
Hands, washed	0.256	rinsed with pyrogen-
Hands after handling animals, washed	16	free aqua dest.
Swimming in natural swimming pool	25600	"Woog", Darmstadt, 30.8.,7 p.m.
Cube sugar, Sweden	10 to 1	
Beet sugar, Sweden	100 to 1	Haskå and Nystrand (1979)
Cane sugar, Import	100	
Beet sugar, Import	10000 (to 100 mg,?)	

fic knowledge" and the special in-house experience and commercial necessities have also to be taken into consideration in the protection of patients. In view of the fact that the limulus test is applicable up to analytically pure substances, it seems necessary to remember that rigid regulations and setting of limiting values by state authorities should as far as possible be avoided. Moreover, the duty of medical diligence and company responsibility cannot be replaced by legislation alone (Göing 1981). The best starting point for such responsible action seems to me to be sovereign handling of pyrogens which has grown from knowledge and experience.

We are among those companies in Europe who have for years been successfully using (Bickel-Meyer 1981; Göing 1981; Krüger 1981; Ronneberger 1974) the official rabbit test on a large scale for the quality control of our products, even beyond the range of end products for which the test is required in the EP and which are mentioned specifically in monographs.

MATERIALS and METHODS.

TEST PROGRAM. In the parent company in Darmstadt alone, we have for years been testing between 2,000 and 4,000 batches or samples of raw materials, and parenterals for very different indications, each year. The spectrum of substances used is very wide, and the volumes of the end products range from 0.5 to 100 ml. The pharmaceutical formulations in question are, in particular, low- to high-concentration aqueous solutions or oily solutions or crystal suspensions in ampoules, lyophilizates, ready-to-use injection syringes, etc. A not inconsiderable number of raw materials are base substances for manufacturers of infusion solutions.

RABBIT TEST-METHOD. Our testing on rabbits is carried out under consideration of recognized rules for housing, care and procedure (Hergott and Klavehn 1971 a,b; Ronneberger 1974, 1977) and of the physiological and the particular pharmacodynamic aspects of our range of parenterals (Müller-Calgan 1980; Müller-Calgan and Henning 1979). Female white New Zealand WN/Ki rabbits weighing from 2.7 to 5.1 kg are predominantly used, and these animals are kept in housing and test rooms under completely climatically controlled conditions (21 ± 1.5°C).

QUANTIFICATION OF ENDOTOXIN CONTENTS IN THE RABBIT TEST AND COMPARISON WITH LAL TEST VALUES.

From the experience gained in 1978 to 1980 relating to use doses of Novo-Pyrexal (from Hermal-Chemie, a company associated with Merck, Darmstadt) in rabbits, it was an obvious assumption that temperature differences measured within a certain dose and volume range (0.31 - 2.5 ng/kg or 0.125 - above 4 ml/kg) are linearly related to the logarithm of the dose administered (Figure 1).*)

WDV/ R. BENDER **Figure 1. Generalized MANOVA with CSE.**

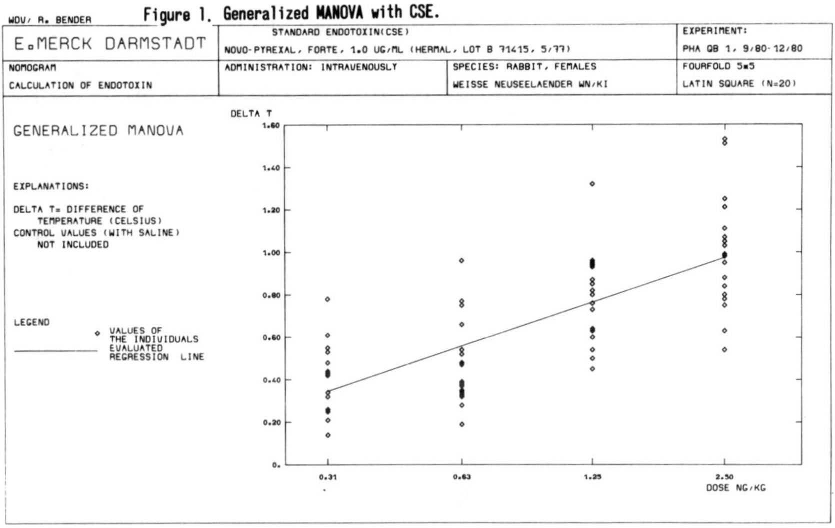

The following model was thus taken as a basis:

$$y = A + B \cdot \log_{10} x$$

y is the increase in the temperature and x is the dose administered. Since in the experimental plan of a latin square several, and therefore dependent, measurements per individual (including those receiving placebo = physiological saline) are made, the individual animals, however, are considered as being independent of

*) Novo-Pyrexal doses and Placebo led to the following increases in temperature in °C ($\bar{x} \pm$ SE (SEM)):

0.31 ng/kg	:	0.367 $\pm$ 0.166 (0.037)
0.63	:	0.483 $\pm$ 0.188 (0.042)
1.25	:	0.773 $\pm$ 0.207 (0.046)
2.5	:	1.011 $\pm$ 0.258 (0.058)
0	:	0.146 $\pm$ 0.080 (0.019)

each other, the parameters A and B were estimated by generalized MANOVA (Multivariate Analysis of Variance;cf. Potthoff and Roy 1964; Grizzle and Allen 1969). In this way A was calculated as being 0.698 and B as being 0.692. For unknown pyrogen contents x , these can thus be calculated from the formula

$$x = 10^{\left(\frac{y - 0.698}{0.692}\right)}$$

in the pyrogen test according to the Pharmacopeias from known temperature differences y (for N = 3, 6, 8, ..) as $\bar{y}$, or they can be roughly estimated rapidly from the nomogram (Figure 1). As we intend to show with a few examples, pyrogen contents can be compared with the endotoxin values obtained in the LAL test. We followed a suggestion of Göing (1981) and selected in particular those examples which were in the threshold area of the previous possibility of evaluation in the rabbit test. For testing of relatively high pyrogen contents, the concentrations or doses were reduced so that the increases in temperature were within the middle range of the regression line. Where possible, the comparison investigations were carried out simultaneously and on the same test quantities.

LAL METHOD. The investigations with end-point estimations were carried out with lysate from Assoc. of Cape Cod Inc., Woods Hole, MA (marketed by Pyrotell Diagnostik, D 6370 Oberursel). The sensitivities were 0.007 ng /ml, 0.008 , 0.013 and in some cases 0.025 ng/ml. The endotoxin standards (CSE) used were 2 ng/ml of E. coli (PP-E- 434 of E. coli O 113)(lot No. 10) of the same origin, and 0.1 µg/ml of Novo-Pyrexal (Hermal-Chemie, D 2057 Reinbek b. Hamburg; CH-B 71415, lot No. 004), and for the comparison investigations with the rabbit test, 1.0 and 0.1 µg/ml of two earlier stocks (5.5. and 3. 5. 1977) of the same batch of Novo-Pyrexal forte and Novo-Pyrexal. Small (13 x 83 mm) control tubes (BIOTEST Serum Institute, D 6000 **Frankfurt**) were used for the solutions, dilutions and pH measurements, and Fisher brand[R] culture tubes, 10 x 75 mm, (kindly supplied by Mr. Kuhl, Pyrotell) for the testing. The glassware, including the tubes, is dried in a hot-air sterilizer and depyrogenized at 240°C for 4 hours. Dilutions are prepared with

Eppendorf pipettes with a discharge device and pipette tips. Doubly distilled water produced in house in a Destamat [R] Bi 18 (Heraeus, D 6450 Hanau) is used for the dilutions.

RESULTS and DISCUSSION.

DISTILLED WATER. As many investigators agree in their reports, the investigation of doubly distilled water for pyrogens presents no difficulties at all in both tests. In the LAL test, the sensitivity corresponds to that of the lysate ($\geqq$ 0.008 ng/ml), and in rabbits the detection limit, derived from figure 1 for water which has been rendered isotonic with sodium chloride is about 0.052 ng/ml, as long as the injection rate does not exceed 0.5 minute/kg at a dose of 10 ml/kg (Table 2). The

Table 2. Detectability of endotoxins (rabbit test) in solutions in dependence of the volume administered / kg body weight (CSE equivalent as Novo-Pyrexal).

	Dosage	Concentration
Water, aqua dest., full deionized water	10 ml/kg	from 0.031 ng/ml (theoretical) from 0.052 ng/ml (more probable)
	2 ml/kg	0.156 ng/ml
	1	0.31
	0.5	0.63
	I	I
	0.125	2.5
	I	I
	0.031	10

sensitivity is thus 6 times higher in the LAL test. For the past 2 - 3 years, we have found no pyrogens either in the rabbit test or in the 247 samples submitted to the LAL test with volumes of 1 to 100 ml ampoules and injection glasses from 6 manufacturers and in 0.5 liter amounts of freshly distilled water from the parenteral plant. Water from our own laboratory distillation apparatus, which is fed with completely deionized water (endotoxin content 28. 8.: 0.128 ng/ml), always corresponded to the highest lysate sensitivity (< 0.008 ng/ml).

COMPLETELY DEIONIZED WATER. As well as the technical center (Table 3, next page), six other plants also regularly investigate their deionized water for pyrogens, although Zimmermann et al (1980) maintain that this is not imperative for Aqua purificata. However, continuous testing in addition to bacteriological examinations has proved useful at Merck in view of the good hygienic sta-

Table 3. Endotoxin contents in a fully deionized water plant (ng/ml endotoxin equivalent [1])

	N	range	median	modal value
town water	29	0.128 - 1.024	0.253	0.256
pump water	28	0.104 - 1.024	0.168	0.128
tap connection II	28	0.104 - 1.024	0.156	0.256
III	27	0.008 - 0.064	0.008 - 0.013	0.008
IV	27	0.104 - 1.024	0.168	0.128

checks once to twice weekly (9. 2. - 15. 6. 81).

[1]) PP-E-434 endotoxin from E. coli 0 113; lysate 0.008 and 0.013 ng/ml.

tus thereby inferred. For some time, testing was conventionally carried out on rabbits with 10 ml/kg, after rendering the samples isotonic. After use of the limulus test, comparative investigations in which the endotoxin content was calculated were also carried out. These investigations showed that the LAL test gives obviously (2-3 times) higher values than the rabbit test. After contamination of fresh, completely deionized water of a particularly low pyrogen content with Novo-Pyrexal, the amount detected at low endotoxin concentrations (0.052 - 0.416 ng/ml) was somewhat less than that added in the LAL test, and significantly less in the rabbit test with volumes above 2.5 ml/kg. The germ content of the water possibly plays a part inasmuch as the RES traps germs and relatively large cell fragments with adhering LPS.

Some results of the LAL test from our central water plant are shown in Table 3 . This table indicates the problems of a large scale industrial plant in which 6 ion exchange units, 4 of which are older units (co-current principle) and the 2 others are newer units (short-cycle unit), produce together. The plant is fed with factory's own pump water and town water. It can be seen that the endotoxin values in the plant are largely dependent on the starting water. In contrast, the difference in the content at points II and IV (short-cycle units) and at III (co-current units) is probably an expression of higher re-contamination of the water in the first units as a result of the greater wear away of the resin as well as rhythmic fluctuations in the flow rates. With the aid of the LAL test, it appears possible to clarify this phenomenon.

CARBOHYDRATES. They are frequently carriers of endotoxins (Haskå and Nystrand 1979; Table 1), but when used for infusion solutions (Bickel and Meyer 1981) they may contain only very small amounts of pyrogens. Bickel and Meyer also state that users of carbohydrates will in the future also approach suppliers with their own idea of a threshold pyrogen content. The draft monographs named Standardzulassung acc. § 36 of the German drug law (AMG) (Suppl. Dtsch. Apoth.Ztg. 121, 3/4; 1981) envisage that the raw material used by infusion solution manufacturers will be subjected to LAL test, in addition to testing of the end product in rabbits, in most of the carbohydrates used in the context of GMP conditions. Manufacturing processes therefore also require careful monitoring from the raw material to the end product within the manufacturer's plant.

SORBITOL. In addition to glucose, mannitol and galactose, we especially subjected sorbitol solution (70%) and sorbitol prepared from it to a comparison in the two tests (Table 5, Figure 2). Table 4 shows the endotoxin

Table 4. Calculated endotoxin content in liquid 70% sorbitol in ng/g equivalent (Novo-Pyrexal) in the rabbit test.

	Specification	N	Range	Mean value	Median value	Pyrogen-free acc. to EP[1]
Manufacturer 1	not specified	20	0.088[2] - 2.77	0.691	0.506	2/20 = 10%
	pyrogen-free	14	0.056[2] - 1.59	0.222	0.101	12/14 = 86%
Manufacturer 2	not specified (if possible, pyrogen-free)	24	0.066[2] - 0.91	0.225	0.161	13/24 = 54%

[1]) Decision after N = 8 animals, further doubtful reactions are considered as not pyrogen-free.

[2]) Values below 0.1 ng/g from regression line (Fig. 1) slightly extrapolated.

contents, calculated by our procedure, of 58 samples for 3 types of raw material which may be supplied for oral and parenteral purposes by 2 other manufacturers. Manufacturer 1 is to a high degree successful in supplying pyrogen-free goods. However the goods supplied must be subsequently checked. In the case of the second manufacturer, operational conditions for pyrogen-free production could be met only with considerable difficulty. To test raw material (and the finished product), a large expenditure on testing has been necessary in the past with rabbits. As a result of the good detectability (about 0.110 ng/g of CSE equivalent; test with 2.8 g/kg and 7 ml/kg of 40% solution), residual pyrogen contents can easily be assayed and, in accordance with the EP frequently necessitate retesting. If the LAL test is incorporated, the expenditure can be reduced considerably. It was necessary to test before-hand to what extend the two methods agree with one another. A correlation and regression calculation from 20 pairs of values was used (Figure 2).

Figure 2. Relations between endotoxins contents in liquid 70 % sorbitol measured in the LAL test (EC 2) and in the rabbit test (Novo-Pyrexal). • Paired values. o values below 0.1 ng/g acc. to Novo-Pyrexal extrapolated.

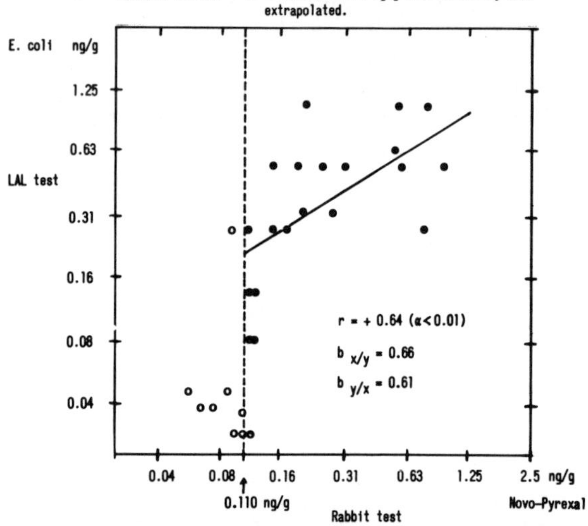

The two processes with different methods and different CSE's correlate. The two regression coefficients and the incorporation of the remaining 9 values (somewhat beyond the detection limit of Novo-Pyrexal equivalents) give a conversion factor of about 1.6, the LAL test values being higher by this factor. The results found in 118 parallel investigations carried out on pyrogen-free sorbitol end product also confirm this (Table 5). In repeat investigations, the combination of the two endotoxin determination methods is the most rational and at the same time most reliable method, regardless of which method has been used first.

Table 5 . Evaluation model for doubtful reactions (example: sorbitol) in the calculative rabbit test and LAL test with endotoxin limits. N = 11 of 118 samples.

	Mean endotoxin contents ng/g (median; N= 118)	Repetitive test 1) USP	EP	LAL 1)	Final decision Fail/Pass after USP	EP 2)	LAL		Endotoxin contents (ng/g equivalent, Novo-Pyrexal) after Fail/Pass evaluation Fail	Pass
Rabbit	< 0.100 (Novo-Pyrexal)	2	11		2/116	0/118		USP	0.3 - 0.65	< 0.1 - 0.35
								EP	—	< 0.1 - 0.65
LAL	0.072 (EC 2)			8			8/110		< 0.1 - 0.65	< 0.1 - 0.16
Combination (preceding rabbit or LAL)			11			4/114 3)			0.3 - 0.65	< 0.1 - 0.16

1) Repetition acc. to the pharmacopeias; in the case of LAL at values over 0.260 ng/g (with lysate 0.008 resp. 0.013 ng/ml).
2) in the case of up to 3 repetitions (N= 12).
3) Limit values set at 0.165 to 0.200 ng/g (equivalent Novo-Pyrexal).

LIVER EXTRACTS. In addition to other criteria, low histamine contents and the absence of pyrogens are required of liver extracts for parenterals. Where the former are favorable, some of the raw materials are still discarded because they contain pyrogens (5/12 = 42 %, Table 6). In spite of repeat investigations, two results

Table 6. Pyrogen content in liver extracts (raw material)1)

	ng/ml pyrogens Rabbit	ng/ml endotoxins LAL
Rabbit + LAL positive	>40	2.1
	20	6.7
	15.2	16.4
	6.3	14.3
	1.0	0.064 (negative)
Rabbit negative	≈ 0.6	0.128
	≈ 0.5	0.832
	≪ 0.6	0.832
	≪ 0.6	1.0 + 0.51 (histamine +)
	≪ 0.6	0.512

1) in rabbits dose 0.5 ml/kg i.v., detection limit 0.63 ng/ml.
≪ 0.6 means: no increase, total values below +0.5°C (N=3).

in the two tests do not agree: one false positive and one false negative in the LAL test. The false positve can be explained. It correlates with a high histamine content, and the sample had to be rejected. The false nagative suggests pyrogenic substances other than endotoxins in the liver. This finding restricts the use of the LAL test for testing liver extracts. In contrast, 10 batches of Merck end products and 4 batches of competitors' end products proved to be free from pyrogens in both methods.

TESTABILITY OF SUBSTANCES - SAFETY FOR HUMANS - CHOICE OF METHOD. The discussion about the increased safety for humans which can be achieved using the LAL technique has been in full swing (Bickel-Meyer 1980; Cooper-Neely 1980; Göing 1980, 1981; Hochstein 1981; Krüger 1981; Ronneberger 1974; Tsuji et al 1980) since the investigations of Greisman and Hornick (1969), who found the same fever threshold sensitivities between rabbits and humans on the basis per kg body weight. At the present high level of lysate development (Hochstein 1981), that is to say, with the possibility of being able to test whether substances are "analytically pure", consideration must also be given to "what is possible and what is necessary" (Naegele, 1981). According to Göing (1980), it would be possible to lay down reliable threshold contents for various products, and always to demand a safety interval if pyrogens cannot be completely avoided in a product. This safety interval has also hitherto been taken into consideration in the rabbit test, for example in the form of doses being laid down in the monographs of the Pharmacopeias and for SVP's in in-house methods (Ronneberger 1974, Müller-Calgan and Henning 1979) cf. Table 7.

Not much has as yet been published relating to the detectability of endotoxins in the LAL test, in particular in connection with the inhibition of gel formation. There is no information regarding the frequency and extent of this phenomenon. Before pyrogen testing, we have therefore tested all the important raw materials in our range of products, and so far about 90% of all parenteral formulations for gel inhibition by the spike method with a double dose of SCE (Cooper and Neely, 1980) and with continous pH measurement in a parallel dilution series. The latter provides information regarding the pH value at which inhibition still or no longer occurs. From these results, we have the strong impression

Table 7. Endotoxin detection and safety margin for humans.

Product	Highest human single dose (volume)	Rabbit test Safety margin to humans factor [1]	Endotoxin detection limit [2] LAL	Rabbits	Decision in favour of LAL or Rabbit test LAL		Rabbit
			ng/human dose				
A	7 ml	10	46	2.2			+
B	L 1 g in 10 ml	3.5	17	6.3			+
C	2 ml	33	0.83	0.63	+	or	+
D	1 ml	17.5	0.208	1.25	+		
E	1.5 ml	6	0.16	3.75	+		(Temp ↓)
F	1 (i.m.)	18 (i.v.) 45 (i.m.)	0.052	0.63	+	or	+ (FE)
G	5 ml	7	0.13	3.15	+		
H	8 (i.m.)	4.5 (i.v.) 11 (i.m.)	0.208	5	– (not as safe)		+
I	1 ml	2	0.026	10	+		
J	5 ml	28	0.065	0.78	+	and	+

1) Relation from ml/kg rabbit and man (70 kg). (i.v.); (i.m.): Since in the rabbit test administration is intravenous the margin after i.m. administration to humans increases by the factor 2.5 (Müller-Calgan 1980/81).

2) Lysate sensitivity 0.013 ng/ml assumed.

L = Lyophilisate. Temp ↓ = decrease in temperature. FE = Follow-up examinations are sometimes necessary.

that a number of products could be tested in even lower dilutions if the pH value can be brought to only slightly below 7.0. Raw materials can frequently be tested in a higher concentration than that in which they are found in the finished formulations, which is all to good of safety in the case of potential pyrogen carriers. This property can also be utilized in the testing of lyophilisates, by testing them in higher concentrations than present in the end product. In the case of water-soluble raw materials, the testability of 2/3 of the substances tested was frequently greater than 1% (1.25 to > 20%), and in 1/3 of the substances tested it was frequently far below 1% (0.02 to 0.53%). From 46 parenteral end products, 7 (15%) were tested in the form of the original solution or in the form of lyophilisates. The remainder of the products required dilutions of 1:2 to at most 1:512, even when buffered. Taking an endotoxin limit of 0.5 ng/ml as a basis, at a lysate sensitivity of 0.025 ng/ml, 28% of our range of products could not be tested by the LAL test, and at a sensitivity of 0.007 ng/ml, 11 % could still not be tested. This would restrict the use of the LAL test in end products for which, from our view, a still higher degree of safety is desirable. In order to illustrate this in more detail, 10 end products were chosen at random from a list of substances with increasing testability (from the top downwards) (Table 7), in order to simulate a decision on the test selection for an assumed lysate sensitivity of 0.013 ng/ml. The rabbit test was based on a sensitivity for 0.31 ng/kg of endotoxin equivalent (Novo-Pyrexal). This list shows that the safety margin from rabbit to humans is sufficiently high (a factor of 2 to 33). Specifically, the reason for deciding on one or the other method (or on both) may vary. It may depend on which end products have already been found to contain pyrogens before or which raw materials are known as potential endotoxin carriers. Of these 10, only product H, in which bovine liver extracts are incorporated may contain pyrogens in commercial samples (see above). Batches of product A have for years no longer contained pyrogens. All the other products showed no endotoxins, even down to the detection limit of the LAL test. Because products B and C (Sodium prednisolone 21-hemisuccinate and L-thyroxin) are used for emergency therapy, and B is sometimes also used intrathecally, these products must be safe as possible. Of examples D to G and I, in which the sensitivity can be increased by 1 to 2 powers of ten, two would be examined by the LAL test: product F because of its initial and therefore non-specific slight temperature-increasing properties requiring a follow-up check in accordance with the EP; the same applies to E (morphine hydrochloride) because of its temperature-decreasing effect.

CONCLUSION and ASSESSMENT.

In the form of the gel formation method, the present 8 months' experience gained on 40 raw materials and 108 end products, the latter mostly in small volumes , in almost 1200 individual investigations, has shown the LAL test to be suitable as an additional, and in the case of some raw materials also as the sole method of investigating pyrogenic endotoxins. From a conservative estimate, 10 - 25 % of animal experiments could at present be saved. It also seems possible that substances with potential risk for animals themselves can be tested more frequently by the LAL test. At the present state of the method, inhibition of gel formation by inhibitors is the greatest obstacle to investigating many of the end products. For some substances, even finer optimization of the method seems to be possible (for example buffering to a narrower pH range of 6.8 to 7.5). For rational testing (utilization of the higher sensitivity, simplified testing with lysate of one sensitivity), it is necessary to have a sufficient amount of highly sensitive lysates (for example, up to 0.005 ng/ml) available at all times. Quantitatively, the LAL test and the rabbit pyrogen test are in good agreement, with few exceptions (about 5 %), even with different standardization by EC 2 and Novo-Pyrexal. False negative and true (not based on contamination) false positive (substantially above the sensitivity threshold of the rabbit) reactions are apparently rare in the LAL test. However, since these must be known, they should be published to a greater extent.

ACKNOWLEDGEMENTS.

All co-workers, especially Mr. K. Prantl and Miss C. Jäckel, are thanked for their skillful and careful technical support. Thanks are also due to Mr. R. Bender, Dipl. Math. of the Biometric Department of Merck for his advice and assistance in biometry.

REFERENCES.

Bickel H, Meyer K-H (1981). Anwendung des LAL-Testes im Rahmen der Inprozess-Kontrolle unter Berücksichtigung quantitativer Aspekte. Pharm Ind 43: 376.

Cooper JF, Neely ME (1980). Validation of the LAL test for endproduct evaluation. Pharmac Technol : 72.

Göing H (1980). Prüfung auf Pyrogene - eine quantitative Betrachtung. forum mikrobiologie 4/80: 213.

Göing H (1981). Zur Frage der Grenzwerte von Pyrogenen. Memo script 2, Nr. 2: 18, Concept Heidelberg.

Greisman SE, Hornick RB (1969). Comparative pyrogenic reactivity of rabbit and man to bacterial endotoxin. Proc Soc Exp Biol Med 131: 1154.

Grizzle JE, Allen DM (1969). Analysis of growth and response curves. Biometrics 25: 357.

Haskå G, Nystrand R (1979). Determination of endotoxins with the limulus test. Appl Environ Microbiol 38: 1078.

Hergott J, Klavehn M (1971 a). Eine modifizierte Pyrogenprüfmethode unter Berücksichtigung bekannter Verfahren. Pharm Ind 33: 385.

Hergott J, Klavehn M (1971 b). Pyrogentest mit neuen Meßgeräten und verbesserten Tierboxen sowie Überblick über Pyrogene. Arzneim Forsch (Drug Res) 21: 553.

Hochstein HD (1981). Pyrogen testing with LAL and endotoxin limits. Arzneim Forsch (Drug Res) 31: 391.

Krüger D (1981). Der Pyrogennachweis mittels des Limulustestes - Methoden, Ergebnisse, Möglichkeiten und Grenzen eines In-vitro-Tests. Pharm Ind 43: 1/81 (im Druck). Dort Literaturübersicht.

Müller-Calgan H, Henning HM (1979). Nicht veröffentlichte Methoden der E.Merck, Darmstadt.

Müller-Calgan H (1980/81). Nicht veröffentlichte Ergebnisse.

Naegele W (1981). Probleme der Standardisierung beim Limulus-Test. Memo script 2, Nr. 2: 30, Concept Heidelberg.

Potthoff RF, Roy SN (1964). A generalized MANOVA model useful especially for growth curves problems. Biometrika 51: 313.

Ronneberger H (1974). Prüfung von Plasmaproteinen auf Freiheit von pyrogenen Substanzen. Pharm Ind 36: 257.

Ronneberger H (1977). Pyrogentest unter standardisierten Bedingungen. Pharm Ind 39: 58.

Tsuji K, Steindler KA, Harrison SJ (1980). LAL assay for detection and quantitation of endotoxin in a small-volume parenteral product. Appl Environ Microbiol 40: 533.

Zimmermann G, Krüger D und Busse M (1980). Die Bakterienflora von destilliertem und vollentsalztem Wasser und ihre Bedeutung bei der Herstellung von Arzneimitteln. Pharm Ind 42: 932.

Endotoxins and Their Detection With the Limulus Amebocyte Lysate Test, pages 357–363

APPARENT INFLUENCE OF ADRENOCORTICOTROPHIC HORMONE (ACTH) IN THE USP RABBIT PYROGEN TEST AND EVALUATION OF LAL AS AN ALTERNATIVE PROCEDURE

R.R. Kleszynski, D. Hoffman, A. Odeneal, and J.P. Aldred
Armour Pharmaceutical Company
P.O. Box 511
Kankakee, Illinois 60901

Several authors have reported that ACTH can produce a decrease in body temperature in both normothermic and febrile rabbits (Douglas and Patan, 1952), and that prior administration of ACTH can reduce the amplitude and duration of fevers produced by pyrogen injection in both man and rabbits (Kass and Finland, 1950). Central administration of ACTH (1-24) was found to produce dose-related decreases in body temperature in the rabbit by Lipton and Glyn (1980). Immunoreactive corticotropin has been found in the hypothalamus of rats, rabbits, dogs, monkeys, and humans (Moldow and Yalow, 1978).

Thus, considerable evidence is available to indicate that ACTH may play a role in central thermoregulatory processes.

The rabbit pyrogen test, as described in USP XX, is routinely used in our laboratory for testing ACTH products for freedom from contamination with fever-inducing substances. For several years, the routine dosage utilized for this test was 0.3 USP u/Kg. body weight, given intravenously. In 1979, in order to provide a greater safety factor between the human dosage and the dosage given in the rabbit pyrogen test, the rabbit dosage was increased to 2.0 USP u/Kg. body weight.

In Table 1 are shown the results of an analysis of pooled data from the rabbit pyrogen test for our ACTH product in 16% gelatin vehicle, by year.

Table 1. Effect of Adrenocorticotrophic Hormone on Rectal Temperature in the Rabbit

Pooled Data For	Dosage (IU/Kg)	N	Rectal Temperature (°C) Mean ± S.E.M. Initial	1 Hr	2 Hr	3 Hr
1976	0.3	100	39.36 ± .02	*39.22 ± .03	39.38 ± .03	39.40 ± .03
1979	2.0	178	39.24 ± .02	*38.93 ± .02	*39.09 ± .03	*39.50 ± .03
1980	2.0	252	39.19 ± .01	*38.92 ± .02	*39.07 ± .03	*39.44 ± .02

*$P < .001$ vs. Initial, paired t-test.

In 1976, when the dosage of ACTH administered was 0.3 USP u/Kg., a total of 100 rabbits was tested with the product, including both three-rabbit and five-rabbit tests. A small but significant ($P < .001$) decrease in rectal temperature of -0.14°C. was observed one hour following drug treatment, when compared to the initial pre-treatment temperature. Rectal temperature was not significantly different from initial at two and three hours post-treatment.

During 1979 and 1980, a total of 178 and 252 rabbits was tested, respectively, at a dosage of ACTH of 2.0 USP u/Kg. The data from each of these years are similar and show a significant ($P < .001$) decrease in rectal temperature one hour post-treatment which averaged -0.29°C. A significant decrease in temperature ($P < .001$) persisted throughout two hours post-treatment. Unexpectedly, at three hours post-treatment, a significant ($P < .001$) increase in rectal temperature was observed in rabbits treated with 2.0 USP u/Kg. ACTH.

Lyophilized ACTH given in normal saline at 2.0 USP u/Kg. was also found to produce the same pattern of response. Also, gelatin vehicle alone did not produce significant alterations in body temperature.

These data indicate that ACTH may produce an apparently dose related hypothermic response in the rabbit, which, at higher doses, is followed by an elevation of body temperature. To our knowledge, this latter effect of ACTH has not been previously reported.

The significant increase observed in mean rectal temperature (0.25°C.) three hours post-treatment in rabbits treated with 2.0 USP u/Kg. ACTH is occasionally the result of an individual rabbit in a three-rabbit test showing an increase of 0.6°C. or more. Such a response requires additional testing of the product in a five-rabbit test.

For example, in 1976, using a dosage of 0.3 USP u/Kg., of thirty lots tested, only two showed individual rabbits with a temperature rise of 0.6°C. or more in the initial three-rabbit test, and required a full eight-rabbit test before both lots met the criteria for release as pyrogen free. In 1979 and 1980, however, using a dosage of 2.0 USP u/Kg., seventeen of thirty-one and eleven of thirty-one lots, respectively, required eight-rabbit tests. In 1979, two of seventeen lots requiring the full eight-rabbit test failed to be confirmed as pyrogen free, and in 1980, five of eleven lots failed. In each case, the failures were due to the increase in rectal temperature which occurred at three hours post-treatment.

We believe that these data indicate that dose related pharmacologic effects of ACTH preparations on body temperature may potentially influence the rabbit pyrogen test, resulting in an occasional false positive temperature response by USP criteria.

In order to gain additional information concerning pyrogen levels of suspect ACTH lots, preliminary studies to validate the use of the LAL assay for ACTH have been performed, and a number of lots have been tested in parallel in both systems.

Table 2. Effect of Endotoxin EC-2 on Rectal Temperature in the Rabbit

EC-2 Dose (ng/Kg)	+/-	Mean Rise (°C) ± S.E.M.
0	0/15	.19 ± .03
0.5	0/6	.30 ± .06
1.0	7/15	.53 ± .07
2.0	11/15	.75 ± .08
4.0	9/9	.83 ± .05

+/- = No. rabbits with at least 0.6°C temperature rise/No. rabbits tested.

In Table 2 are shown the results of studies to determine the sensitivity of our rabbit colony to endotoxin. Endotoxin standard EC-2 was added to pyrogen free normal saline and administered to standardized New Zealand rabbits at varying dosages. The number of rabbits showing a rise in rectal temperature of 0.6°C or more was determined, as well as the mean temperature rise of all observations. The results show that seven of fifteen rabbits (47%) were observed to have a temperature rise of 0.6°C or greater when given a dosage of 1 ng /Kg of endotoxin EC-2. Regression analysis shows a loglinear regression line from 0.5 to 2.0 ng /Kg with a correlation coefficient of 0.98. A 0.6°C rise in temperature corresponded to a dosage of 1.23 ng /Kg . These data compare well with the PD50 value of 1 ng/Kg reported last year at this conference.

In Table 3 are shown the mean results of two recovery studies in which 1 ng /ml of endotoxin EC-2 was added to sterile water, 16% gelatin vehicle, or certain ACTH preparations. Within the one-tube variation inherent in this LAL assay system, the data show complete recovery of added endotoxin.

It should be noted that initial recovery studies with gelatin vehicle and ACTH preparations showed inhibition of lysate gelation at 1:1 and

Table 3. Recovery of EC-2 Endotoxin

1ng/Ml Added To:	N	ng/Ml ± SEM Found
Sterile water	13	1.33 ± .16
Gelatin (16%)	13	1.15 ± .15
ACTH in Gel (40 u/ml)	13	1.16 ± .15
ACTH in Gel (80 u/ml)	12	1.51 ± .18
ACTH lyophilized	12	1.27 ± .16

occasionally 1:2 dilutions. It was determined that this inhibition was a pH related phenomenon. When test preparations with pH's ranging from 4.0 to 6.0 were adjusted to pH 7.0, the inhibition of lysate gelation at high drug concentrations was eliminated.

Following pH adjustment, several ACTH lots determined to be pyrogenic by USP criteria, were tested in the LAL assay at a level ten to twenty times that used in the rabbit assay. No significant endotoxin levels were found in such lots (< 0.06 ng /ml.).

In summary, extensive testing of ACTH preparations in the USP rabbit pyrogen test has resulted in the observation of an early (one hour) decrease in rectal temperature followed by a later (three hour) elevation in rectal temperature. This effect appears to be dose related and is not apparent in vehicle treated animals, but occasionally results in preparations which are determined pyrogenic by USP criteria. Testing of such preparations in the LAL assay has not confirmed the presence of significant levels of endotoxin.

Douglas WW, Paton WDM (1952). The hypothermic and antipyretic effect of preparations of ACTH. Lancet, Feb. 16:342.
Kass EH, Finland M (1950). Effect of ACTH on induced fever. New England J Med 243:693.
Lipton JM, Glyn JR (1980). Central administration of peptides alters thermoregulation in the rabbit. Peptides 1:15.
Moldow R, Yalow RS (1978). Extrahypophysial distribution of corticotropin as a function of brain size. Proc Natl Acad Sci, U.S.A. 75:944.

ALDRED SPEAKER

TITLE: APPARENT INFLUENCE OF ADRENOCORTICOTROPIC HORMONE (ACTH) IN THE USP RABBIT PYROGEN TEST AND EVALUATION OF LAL AS AN ALTERNATIVE PROCEDURE.

ROSS: Have you done any studies in rabbits with endotoxin and ACTH?

ALDRED: We have not done any as yet, however, that is one of the next studies we are going to do. We are planning, based on the results of these data, to try to validate the LAL assay for ACTH.

COOPER: As a raw material, is gelatin a problem with respect to endotoxin contamination?

ALDRED: We have looked at various suppliers and have had difficulties in certain instances, but the gelatin is taken through a series of processing steps which renders it pyrogen free.

ENZINGER: By the rabbit test or the LAL test?

ALDRED: By the rabbit test. We have just started the use of the LAL test for final product testing and have made no attempt to validate a product with it.

ENZINGER: It has been our experience that we find gelatin to contain some levels of endotoxin. I think that does vary a lot with supplier and with the conditions of the process.

Endotoxins and Their Detection With the Limulus Amebocyte Lysate Test, pages 365–379

THE ACTIVITY OF VARIOUS ENDOTOXINS IN THE USP RABBIT TEST AND IN THREE DIFFERENT LAL TESTS

Marlys Weary, Fred C. Pearson III,
Joyce Bohon, and Georgiana Donohue
TRAVENOL LABORATORIES, INC.
6301 Lincoln Avenue
Morton Grove, IL 60053

The activity of a variety of endotoxins was evaluated using two biological pyrogen test systems: the USP rabbit pyrogen test and the _Limulus_ amebocyte lysate (LAL) test. Three manufacturers' lysate products were evaluated simultaneously with each endotoxin. Each endotoxin was prepared in both water and saline. Included were the current US, HIMA, and WHO endotoxin standards. Test results indicated substantial differences in activity among different categories of endotoxin in both test procedures, thus emphasizing the need to establish definitive criteria for the selection of an international endotoxin standard for the LAL test.

INTRODUCTION

When Travenol Laboratories initiated its _Limulus_ amebocyte lysate (LAL) pyrogen test program in the early 1970's, a considerable amount of thought and research effort centered on the selection of an in-house endotoxin standard. A large number of commercially available bacterial endotoxin products were evaluated by using LAL and the rabbit pyrogen test. A sentimental favorite was Piromen®, a _Pseudomonas aeruginosa_ polysaccharide compound, which, at that time, was a licensed biological manufactured by Travenol's Flint Division. However, when Piromen was compared to an _E. coli_ 055:B5 endotoxin (Difco Laboratories, Detroit, MI), differences in the potency of the two endotoxins in the USP rabbit test and LAL gel clot pyrogen test were quite apparent (Table 1). Although the two pyrogen test methods were approximately equal in their ability to detect threshold concentrations of *E. coli* endotoxin, the LAL test was more than 30 times as sensitive in its ability to detect Piromen. Eventually, Travenol selected Difco's *E. coli* 055:B5 endotoxin as its in-house endotoxin standard for five reasons:

1. The biologic activity of *E. coli* endotoxin in man vs rabbit had been well defined (Greisman and Hornick, 1969).
2. The sensitivity of the USP rabbit test and the LAL pyrogen test to this endotoxin was approximately equal.
3. *E. coli* was one of the most potent endotoxins we had investigated.
4. It was commercially available.
5. The FDA had by then announced that its new reference endotoxin would also be an *E. coli*.

Table 1. Difference in Potency of Two Endotoxin Standards Tested with LAL and the USP Rabbit Test*

Endotoxin	USP APD_{50} (ng/kg)	USP APD_{50} (ng/ml, 10 ml/kg)	LAL Gel Clot End Point (ng/ml)	Ratio of Sensitivity USP/LAL
Piromen® (*Pseudomonas* polysaccharide)	530	53	1.6	33
E. coli 055:B5	1.7	0.17	0.15	1

* APD_{50} indicates average pyrogenic dose

However, during the past decade, Travenol's experience indicated that environmental endotoxin is considerably less pyrogenic than the lipopolysaccharide (LPS) preparations currently used to standardize the LAL test (Pearson and Weary, 1981). To further our understanding of the relationship between pyrogenicity and LAL reactivity of various LPS products, we decided to examine a variety of LPS, some that are well defined and currently in use as LAL test standards, and some that have not previously been examined using either test method. Each LPS was evaluated with dose-ranging studies in rabbits and dose-ranging evaluations using three licensed LAL products with gel clot end point methodology.

MATERIALS AND METHODS

USP Pyrogen Test Method

Female, New Zealand white rabbits from the Morton Grove, IL, pyrogen test colony were used to determine the pyrogenicity of each endotoxin by the USP procedure (US Pharmacopeia, 1980). Endotoxins were prepared in sterile, nonpyrogenic 0.9% sodium chloride (Travenol), and were administered intravenously at 10 ml per kilogram of body weight. At least three twofold or four-fold concentrations of endotoxin were administered to establish a dose-ranging curve that bracketed the USP pass/fail test limit of 3.7°C for an eight-rabbit pyrogen test. A minimum of eight rabbits was used at each dose. The following model was used in a nonlinear regression routine to determine the average pyrogenic $dose_{50}$ (APD_{50}), the concentration of endotoxin needed to increase a rabbit temperature beyond 0.46°C (3.7°C ÷ 8 = 0.46°C).

$$Y - 0.46 = B * (X - A)$$

where

Y = Temperature change
X = Concentration
A = Concentration needed to increase temperature 0.46°
B = Slope of line

A typical endotoxin dose-response curve is shown in Figure 1.

Lysates

1. Associates of Cape Cod, Pyrotell, Lot 96-02-264 (clots with 0.05 ng of US Endotoxin Standard EC-2 per ml).

Figure 1

PSEUDOMONAS AERUGINOSA KDC#4 FROM CDC
APD 50 -12.3 ng/ml OR 123 ng/kg

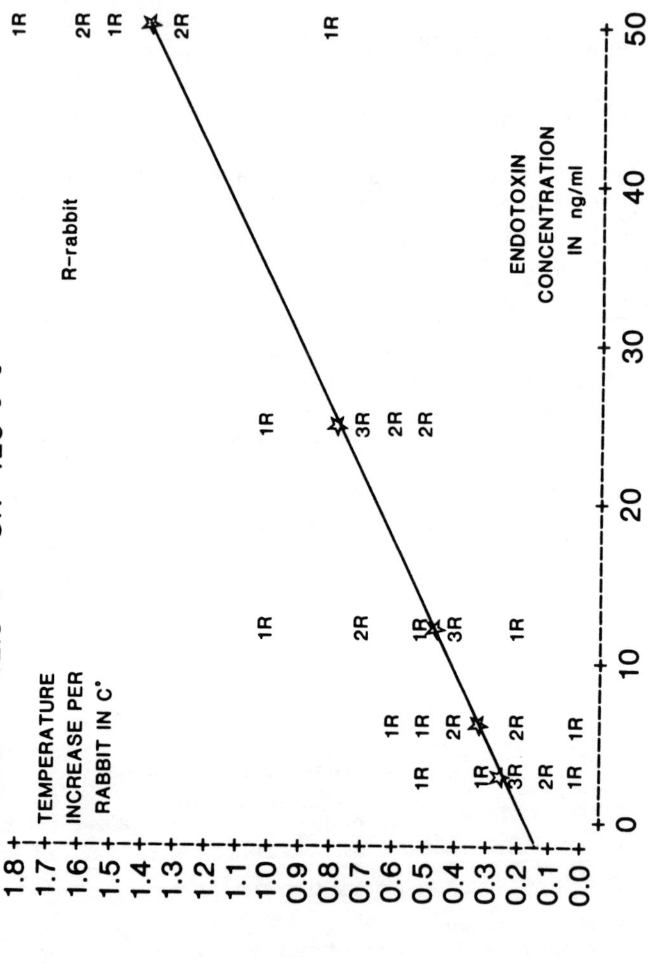

2. Mallinckrodt, Inc., Pyrogent, Lot OKL (clots with 0.18 ng of US Endotoxin Standard EC-4 per ml).

3. Travenol Laboratories, Inc., Lot 3338B001AA (clots with 0.133 ng of US Endotoxin Standard EC-2 per ml). Reconstituted for this study with 3.0 ml of 100 Mm $MgCl_2 \cdot 6H_2O$ and 50 Mm TRIS/TRIS HCl in water.

Endotoxins

1. Escherichia coli 0113:H10:K0, a phenol Westphal-extracted preparation (Rudbach et al., 1976) designated EC-2 and adopted by the Bureau of Biologics (BOB) of the Food and Drug Administration as a national endotoxin reference standard. Obtained from the BOB, Bethesda, MD.

2. Escherichia coli 0113:H10:K0, a phenol Westphal-extracted preparation (Rudbach et al., 1976) herein described as EC-X (an experimental lot), proposed as the next BOB and USP national endotoxin reference standard. Obtained from United States Pharmacopeia (USP), Rockville, MD.

3. Escherichia coli 055:B5, Lot 504089, a trichloroacetic acid Boivan-extracted preparation selected by the Health Industry Manufacturers Association (HIMA) as its reference standard (HIMA, 1979). Obtained from Difco Laboratories, Detroit, MI.

4. Shigella dysenteriae, Lot 79/502 (Davies et al., 1954), a diethyl glycol-extracted preparation designated as the first international reference preparation for pyrogen by the World Health Organization (WHO) Expert Committee on Biological Standardization. Obtained from the National Institute of Biological Standards and Control, London, England.

5. Salmonella abortus equi (Galanos et al., 1979), CH-B. 71415, Article No. 731-3, a phenol-extracted preparation re-extracted by phenol, chloroform, and petroleum ether; electrodialysis; and phenol, chloroform, and petroleum ether. Obtained from Hermal Chemie, Kurt Hermann, Hamburg, W. Germany.

6. Acinetobacter calcoaceticus, var lwoffi, Lot 8/1/79, a phenol Westphal-extracted preparation. Obtained from the Center for Disease Control, Phoenix, AZ.

7. *Pseudomonas aeruginosa*, Lot KDC No. 4, a phenol Westphal-extracted preparation. Obtained from the Center for Disease Control, Phoenix, AZ.

8. *Pseudomonas aeruginosa*, Fisher-Devlin immunotype 1, Lot No. 1, ATCC 27312, a modified phenol Westphal and Jann (1965)-extracted preparation purified by chromatography. Obtained from List Biological Laboratories, Inc., Campbell, CA.

9. *Serratia marcescens*, Lot No. 1, a phenol Westphal and Jann (1965)-extracted preparation purified by chromatography. Obtained from List Biological Laboratories, Inc., Campbell, CA.

10. *Yersinia enterocolitica*, ATCC 27729, Lot No. 1, a modified phenol Westphal and Jann (1965)-extracted preparation further purified by chromatography. Obtained from List Biological Laboratories, Inc., Campbell, CA.

11. *Vitrio cholerae*, Inaba 569B, Lot No. 1, a modified phenol Westphal and Jann (1965)-extracted preparation further purified by chromatography. Obtained from List Biological Laboratories, Inc., Campbell, CA.

LAL Test Procedure

Twofold concentrations of each endotoxin were made in both water and saline. The same preparation of a given endotoxin concentration was tested in quadruplicate using each of the three LAL products. Negative controls of each diluent were run in quadruplicate with each LAL product. The LAL gel-clot procedure involved combining 0.1 ml of endotoxin with 0.1 ml of LAL in depyrogenated test tubes (10 by 75 mm, soda-lime glass), incubating in a 37°C water bath for 60 minutes, and then reading the tests by inverting the tubes. A test was considered positive if a solid gel clot was formed that remained intact after the test tube was inverted 180°. A test was considered negative if the clot broke apart, or if the contents of the tube remained liquid. Geometric mean end points were determined from the smallest concentration of endotoxin that produced positive tests for the four samples.

RESULTS AND DISCUSSION

The APD_{50} for each endotoxin is found in Table 2. The comparative geometric mean end points for each endotoxin with the three LAL preparations are found in Table 3. In 10 of the 11 comparison LAL tests, the Associates of Cape Cod and Mallinckrodt LAL test end points in water were within one twofold dilutional concentration of one another. The Travenol LAL was typically one half to one quarter less sensitive. Travenol lysate is normally used in a quantitative colorimetric optical density test procedure that measures protein precipitation and does not require a more sensitive gel clot end point. Thus, for purposes of comparison with the rabbit test, the mean of the Associates of Cape Cod and Mallinckrodt LAL test end points in water was computed and divided into the APD_{50} endotoxin concentration to determine the ratio of sensitivity of the two test systems. Results are shown in Table 4.

Table 2. Average Pyrogenic Dose (APD_{50}) for Various Endotoxin Preparations

Endotoxin	APD_{50} (ng/kg)	Standard Error of Estimate
E. coli 0113 EC-2	0.94	(3.3)
E. coli 0113 EC-X	0.82	(16.9)
E. coli 055:B5 HIMA	1.21	(5.2)
S. dysenteriae WHO	3.29	(156.2)
S. abortus equi Novo Pyrexal	0.57	(4.8)
A. calcoaceticus CDC	2.53	(40.2)
P. aerugenosa CDC	123.0	(1.5)
P. aeruginosa LIST	5.99	(90.9)
S. marcescens LIST	2.87	(59.4)
Y. enterocolitica LIST	0.61	(13.4)
V. cholorae LIST	17.29	(265.7)

Table 3. Endotoxin LAL Gel-Clot Geometric Mean End Points (in pg/ml)

Endotoxin	Assoc. of Cape Cod		Mallinckrodt		Travenol	
	H_2O	0.9% NaCl	H_2O	0.9% NaCl	H_2O	0.9% NaCl
E. coli 0113 EC-2	50	336	42	336	100	336
E. coli 0113 EC-X	12	25	18	35	50	59
E. coli 055:B5 HIMA	25	84	21	50	100	119
S. dysenteriae WHO	149	707	125	500	1000	1682
S. abortus equi Novo Pyrexal	11	25	6	25	21	50
A. calcoaceticus CDC	93	78	78	78	312	625
P. aeruginosa CDC	625	186	313	625	743	313
P. aeruginosa LIST	8	31	8	8	26	31
S. marcescens LIST	8	12	6	12	50	50
Y. enterocolitica LIST	31	37	22	26	74	88
V. cholorae LIST	44	63	13	16	250	250

Table 4. Sensitivity Ratio: USP Rabbit Test and LAL Clot End Point

Endotoxin	LAL* End Point in H_2O (pg/ml)	APD_{50} (pg/ml, 10 ml/kg)	Sensitivity Ratio USP/LAL
E. coli 0113 EC-2	46	94	2.0
E. coli 0113 EC-X	15	82	5.5
E coli 055:B5 HIMA	23	121	5.3
S. dysenteriae WHO	137	329	2.4
S. abortus equi (Novo Pyrexal)	9	57	6.3
A. calcoaceticus CDC	86	252	2.9
P. aeruginosa CDC	469	12,300	26.2
P. aeruginosa LIST	8	599	74.8
S. marcescens LIST	7	287	41.0
Y. enterocolitica LIST	27	61	2.3
V. cholerae LIST	29	1,729	59.6

* Average of geometric mean end points from tests of Associates of Cape Cod and Mallinckrodt

The 11 LPS endotoxin preparations appear to fall into two distinct classes based on their relative potency with the rabbit and LAL pyrogen tests. The first class of products are those that are two to six times more potent in the LAL test than in the USP rabbit pyrogen test. This group includes all three *E. coli* preparations, *S. dysenteria*, *S. abortus equi*, *A. calcoaceticus*, and *Y. enterocolitica*. The second class of products are those that are from 26 to 75 times more potent in the LAL test than in the USP rabbit pyrogen test. This group includes both *P. aeruginosa* preparations, *S. marcescens*, and *V. cholerae*. It is of interest to note that this relationship is relatively similar in both *P. aeruginosa* preparations, even though the individual rabbit and LAL potency data indicate that the List Biological Laboratories product is 20 to 60 times more potent than the CDC preparation. The Piromen *Pseudomonas* polysaccharide data in Table 1 also fit into this class of products.

How is the LAL test user to resolve the dilemma of selecting an endotoxin standard from these two classes of endotoxins? If it is true that enviromental endotoxin is consistently less pyrogenic than the LPS endotoxin standards currently used for LAL testing, perhaps it is because environmental endotoxin contaminants are more likely to be the Pseudomonas variety commonly found in industrial water systems. If this is the case, perhaps this class of endotoxin would be a more realistic LAL test standard for the pharmaceutical industry.

Obviously there is a great deal still to be learned concerning the relationship of environmental endotoxin pyrogenicity and LAL test sensitivity limits as defined by current LPS endotoxin standards. It was for this reason that the HIMA Guideline for the Use of the LAL Test for Pyrogen Testing of Medical Devices (HIMA, 1978) recommended that the eight-rabbit USP test be retained as the official reference test in the event of LAL failure. We concur with HIMA for the following reasons:

1. The LAL test has proven to be an extremely sensitive method for the detection of bacterial endotoxin. For this reason, it can be best used by industry as a screening test for the absence of endotoxin. When properly controlled, a negative LAL test will ensure the absence of a given level of bacterial endotoxin from the product. However, the occurrence of a positive LAL test does not necessarily indicate the presence of pyrogenic endotoxin levels. The rabbit pyrogen test should continue to be used as a reference test to elucidate the significance of a positive LAL test until a larger data base on environmental endotoxin is generated by industry and regulatory agencies.

2. It is recommended that a reference rabbit test be done with eight rabbits rather than with three. It has been shown that the probability of a false negative response in a three-rabbit test can be uncomfortably high at threshold levels of pyrogenicity (Mascoli and Weary, 1979). Furthermore, retests of this sort will be necessary infrequently and deserve the more thorough examination that the expanded test will provide.

3. The pharmaceutical industry has had 40 years of experience with the USP rabbit pyrogen test. Our remarkable record of product safety with regard to pyrogenicity has given us confidence that products passing this test can be administered to man with the highest degree of safety. This outstanding track record highly recommends the USP rabbit pyrogen test as the official reference test to be used when differences appear, or in the event of dispute.

4. The purified and refined LPS preparations currently in use as LAL test standards have been shown to be highly pyrogenic in rabbit and man. Manufacturers who have been using the LAL test routinely for in-process screening of drug products indicate that naturally occurring bacterial endotoxins detected by LAL are frequently less pyrogenic than are the LAL test LPS endotoxin standards (Pearson and Weary, 1981). It is important that industry and regulatory agencies understand more clearly the relationship of the rabbit and LAL test systems with regard to "environmental" endotoxin. Retaining the USP eight-rabbit test as a reference test will enable both groups to compare the two pyrogen test systems, side by side, with identical samples of product. In this way, we will be able to build a data base with which to establish sensible LAL test endotoxin reference standards that equate to the rabbit pyrogen test.

REFERENCES

Davies DAL, Morgan WTJ, Mosemann W (1954). Studies in immunochemistry 13. Preparation and properties of the 'O' somatic antigen of _Shigella dysenteriae_ (SHIGA). Biochem J 56:572.

Galanos C, Luderitz O, Westphal O (1979). Preparation and properties of a standardized lipopolysaccharide from _Salmonella abortus equi_ (Novo-Pyrexal). Zentralbl Bakteriol Parasitenkd Infektionskr Hyg Abt 1 Orig Reihe A 243:226.

Greisman SE, Hornick RB (1979). Comparative pyrogenic reactivity of rabbit and man to bacterial endotoxin. Proc Soc Exp Biol Med 131:1154.

Health Industry Manufacturers Association (1979). HIMA collaborative study for the pyrogenicity evaluation of a reference endotoxin by the USP rabbit test. Washington, DC:

Health Industry Manufacturers Association Document Series, vol 1, no. 7.

Health Industry Manufacturers Association (1978). Guideline for the use of Limulus lysate test (LAL) for pyrogen testing of medical devices. Washington, DC: Health Industry Manufacturers Association Report No. 78-8.

Mascoli CC, Weary ME (1979). Limulus amebocyte lysate (LAL) test for detecting pyrogens in parenteral injectable products and medical devices. Advantages to manufacturers and regulatory officials. J Paren Drug Assoc 33:81.

Pearson FC, Weary ME (1981). The relative pyrogenicity of environmental endotoxin as measured by the Limulus amebocyte lysate assay and the USP rabbit pyrogen test. Intl Conf on Endotoxin Standards and LAL Use with Parenteral Drugs, Woods Hole Oceanographic Inst.

Rudbach JA, Akiya FI, Elin RJ, Hochstein HD, Luoma MK, Milner ECB, Milner KC, Thomas KR (1976). Preparation and properties of a national reference endotoxin. J Clin Microbiol 3:21.

Rabbit Pyrogen Test (1980). United States Pharmacopeia XX, Rockville, MD. United States Pharmacopeial Convention, Inc., Rockville, MD. p 902.

Westphal O, Jann K (1965). Bacterial lipopolysaccharides extraction with phenol-water and further applications of the procedure. In Whistler RL (ed): "Methods in Carbohydrate Chemistry," vol V. New York: Academic Press. p 83.

WEARY SPEAKER
TITLE: THE ACTIVITY OF VARIOUS ENDOTOXINS IN THE USP RABBIT TEST AND THREE LIMULUS AMEBOCYTE LYSATES.

REICHLER: If those endotoxins were related back to endotoxin units, how do you think the data would compare? You are using weight and we have heard throughout these past couple of days that this is a poor way to describe endotoxins. What would your opinion be on how that comparison chart would turn out if you went back and normalized those endotoxins?
WEARY: Endotoxin units are assigned to endotoxins based on their LAL activity. Since the LAL endpoint is shown for each endotoxin on the comparison chart (TABLE 4), all of these endotoxins would share the identical endotoxin unit value. However, the USP/LAL sensitivity ratio would remain the same for each endotoxin. Thus, each endotoxin's LAL endpoint in EU/ml multiplied by its sensitivity ratio would give the APD_{50} in EU/ml.
REICHLER: Except that some of those endotoxins by weight are certainly going to be greater than the present pass-fail criteria endotoxin units if you do weight/weight. You have some very hot endotoxins.
WEARY: That's true, but these are purified LPS compounds. You are not going to find them in the product.
REICHLER: But you are also not going to pass-fail on those weights, you are going to pass-fail on reactivity against a standard
WEARY: That is why selection of the standard is so important.
MORRISON: The enormous difference between the CDC Pseudomonas endotoxin and the List Biological endotoxin which varied between 20-and 60-fold in terms of activity is very interesting. You attribute this at least in part to the fact that one preparation is highly purified and one is not, but it is highly unlikely that the unpurified contains only between 2 and 5% of the endotoxin as opposed to the highly purified List preparation. Does that, therefore, suggest the presence of factors in less purified preparations which may inhibit either the limulus response and/or the pyrogen response? It seems to me that it would be easy to test by taking a mixture of the two and looking for inhibition.

WEARY: That is something to think about.
KURATSUKA: How many rabbits do you use per one sample?
WEARY: There were from 8 to 18 rabbits per dose. There were at least three doses to generate the curve and in many cases there were 5.
MUNSON: From the FDA's viewpoint, I have to question your proposal that LAL failures be tested in rabbits and released if they pass. We have seen about three incidents where the clinical evidence supported an endotoxin reaction in the patient. In all cases the product was shown, by LAL, to have endotoxin in excess of the limits proposed in the guidelines, but all passed a valid rabbit test. For this reason I must disagree that your proposal is in the best interest of the consumer.
WEARY: These are cases that would have to be examined on a test-by-test basis.
MUNSON: Yes, but you are making a blanket statement.
WEARY: Well I think these are the cases that are most interesting obviously from a regulatory point of view. Had the product been tested in more than one rabbit colony?
MUNSON: There was no need to, they had already passed the manufacturers release rabbit test.
WEARY: When the product was retested after the product complaint, was it still negative?
MUNSON: It was still negative.
WEARY: Were they confirmed cases of pyrogenicity in the patients?
MUNSON: As far as we could determine from the clinical records, they showed the signs of pyrogenicity and that was the diagnosis made by the physician in every case.
WEARY: For 40 years we have been using the rabbit pyrogen test and what is the incidence of these circumstances during that time period because prior to the lysate test we had nothing else to go on but the rabbit test.
WATSON: Why does the PDR list pyrogenicity as a possible side effect for so many parenteral drugs if clinicians had never observed a pyrogenic response?

MUNSON: This also came up when we were discussing the guidelines. When a patient gets a fever in the hospital, the average clinician doesn't associate this fever with endotoxins. Unless an epidemiologist or people working in the hospital are really on the ball, they will probably not pick up these cases most of the time because fever associated with an illness or a patient in surgery usually is overlooked. This is the information that came to our attention when we tried to find out just how good was the rabbit test. It appears that it has served a useful purpose and I think done its job of protecting the public to a great degree. But, as has happened time and time again when we start looking for the problem, we find it.
WEARY: We just don't have enough information. What I would like to see is a repeat rabbit test, preferably with 8 rabbits. Possibly we would get a better fix on the problem if such studies were conducted.
MUNSON: As far as release criteria goes, we would like to err on the side of safety and stick with the LAL test rather than a rabbit retest.

Endotoxins and Their Detection With the Limulus Amebocyte Lysate Test, pages 381–384

THE SIGNIFICANCE OF THE PENETRATION OF ENDOTOXIN FROM THE INTESTINAL TRACT

B. Urbaschek, B. Ditter, and R. Urbaschek

Abt. f. Immunologie u. Serologie, Institut für Hygiene u. Med. Mikrobiologie, Klinikum Mannheim der Univ. Heidelberg,D 6,5,68 Mannheim,W.-Germany

When sufficient amounts of endotoxins enter the circulation they induce pathophysiologic alterations which may lead to fragments of shock or shock. Very minute amounts of endotoxins in the circulation induce nonspecific resistance and maintain the status of nonspecific host defense.

Besides the drop in white blood cell and platelet counts, the first manifestations of circulating endotoxins occur in the microvasculature including increased permeability (Urbaschek 1967, Urbaschek 1971). Biogenic amines, such as histamine and serotonin, are increased up to 100-fold in their activity in the presence of small amounts of endotoxins (Urbaschek, Versteyl 1965). In the capillary bed combined threshold doses of endotoxins and histamine or serotonin lead to severe changes which are comparable to those of an LD_{75} of endotoxin including the permeation of white blood cells and petechial hemorrhages. Endotoxins initially generate C3a with histamine-like activity as well as the induction of degranulation of periadventitial mast cells.

Biogenic amines may cause permeation of endotoxins from the intestinal tract into the circulation (Urbaschek 1966). Before the development at the LAL-test to measure endotoxins we detected endotoxin in the peripheral blood of 90 % of guinea pigs after gavaged endotoxin and histamine administration using the Kolmer complement fixation test (Urbaschek 1967). Twenty percent of the guinea pigs had detectable endotoxemia after gavaged endotoxin without concomitant histamine injection.

The LAL-test was employed to determine circulating endotoxins in the mouse. An optimal model in the mouse was established after preliminary studies with various concentrations of serotonin and endotoxin as well as different time intervals between treatment and collection of blood (Ditter, Urbaschek, Urbaschek, 1981). One mg of gavaged endotoxin concomitant with an intravenous injection of 0.5 mg of serotonin induced endotoxemia in 80 % of the mice after four hours. Either substance alone did not result in endotoxemia (Table 1). Dilution and heating of the samples (Watson, S.W. personal communication) improved the determination of pg amounts of endotoxins in blood (Table 1). In these experiments blood was collected by heart puncture. The LAL-test was performed in tubes with LAL from Associates of Cape Cod* (sensitivity 9pg/ml endotoxin) in whole blood.

Table 1: % positive LAL-tests of blood samples obtained 4 hours after treatment

Blood (samples diluted 1:10)	Serotonin i.v. (0.5mg/mouse)	Oral Endotoxin (1mg/mouse)	Serotonin + Endotoxin
	% of mice with endotoxemia		
Non heated	0 %	0 %	10 %
Heated	0 %	10 %	80 %

A comparison between the content of endotoxin in portal and cardiac blood clearly revealed clearance of endotoxin by the liver in this model, when blood was obtained simultaneously at different times from the portal vein and by heart puncture (Figure 1).

---

* LAL was kindly provided by Dr. S.W.Watson

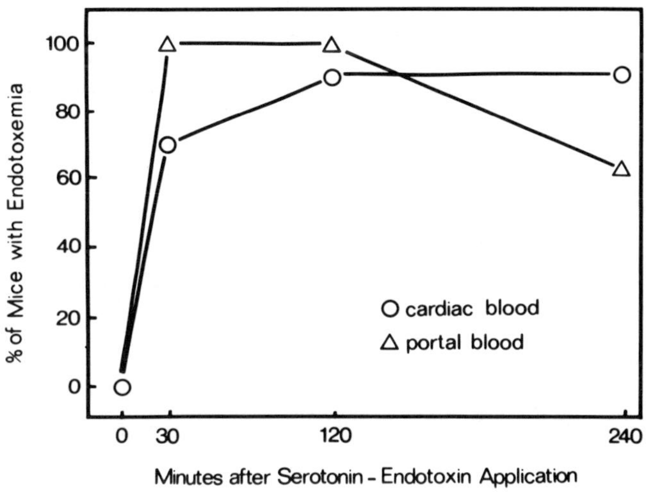

Fig. 1. Endotoxin detected in mice in cardiac and portal blood at different times after gavaged endotoxin and i.v. injected serotonin.

This detection of circulating endotoxin resulting from penetration through the intestinal tract also allows study of the prevention of this endotoxemia. The following substances were tested for their capacity to bind endotoxin in the gut: Kaolin, Pectin, Kaolin/Pectin-mixture, Bentonite, charcoal particles, Kaopectate^{R} and Lactulose (Bifiteral^{R}). Thirty minutes after simultaneous administration of serotonin and endotoxin, different concentrations of these substances were gavaged in 0.4 ml water.

Complete prevention of the induction of endotoxemia was obtained by 75 mg of Bentonite, 150 mg of Kaopectate and 300 mg of Kaolin. The other substances - used in the highest possible amount - were not able to prevent endotoxemia completely. The concentrations of the tested substances which prevented endotoxemia in 50 % of the animals (ED_{50}) is summarized in Table 2.

Table 2: Comparison of in vivo endotoxemia ratio- ED_{50} - of various gavaged reagents after administration of serotonin and endotoxin

Reagent	Estimated ED_{50} (µg/mouse)
Bentonite	31 µg
Kaopectate[R]	54 µg
Carbo medicinalis	62 µg
Kaolin/pectin (49:1;w/w)	100 µg
Kaolin	160 µg
Lactulose	295 µg

In vitro experiments using ^{14}C-labeled endotoxin the binding capacity for endotoxin was measured. The sequence of the capacity to bind endotoxin in vitro was identical to that obtained from in vivo studies.

This work was supported by the Deutsche Forschungsgemeinschaft within the SFB 90.

Ditter B, Urbaschek, R, Urbaschek B, 1981. The ability of various adsorbants used in the treatment of diarrhoea to bind endotoxin in mice. In preperation.

Urbaschek B, Versteyl R, 1965. Increase of the effect of histamine by E. coli endotoxin on the smooth muscle. Nature 207: 763-764.

Urbaschek,B, 1966. The histamine-sensitizing effect of endotoxins. Ann NY Acad Sci 133: 709-711.

Urbaschek B, 1967. Zur Frage des Wirkungsmechanismus bakterieller Endotoxine uns seiner Beeinflussung. Habilitationsschrift, Med. Fakultät der Universität Heidelberg.

Urbaschek B, 1971. The effects of endotoxins in the microcirculation. In Kadis S, Weinbaum G Ajl Sj (eds): "Microbial Toxins", Vol V, New York: Academic Press, p. 261-275.

Endotoxins and Their Detection With the Limulus Amebocyte Lysate Test, pages 385–392

DETECTION OF ENDOTOXIN IN BLOOD AND OTHER SPECIMENS BY EVALUATION OF PHOTOMETRICALLY REGISTERED LAL-REACTION-KINETICS IN MICROTITER PLATES

B. Ditter, K.-P. Becker, R. Urbaschek, and B. Urbaschek
Abt. f. Immunologie u. Serologie, Institut für Hygiene u. Med. Mikrobiologie, Klinikum Mannheim der Univ. Heidelberg,D6,5,68 Mannheim,W.-Germany

The difficulties involved in the Limulus Amebocyte Lysate (LAL) endotoxin reaction in blood were studied using an automated method. This provided subtle recognition of the course of the reaction of LAL with endotoxin concentrations even in samples that contained factors which mask the reaction.

To accomplish this, a quantitative microtiter test was developed based on the measurement of the course of the LAL-endotoxin reaction. The kinetics of this reaction were registered photometrically with a modified 8 channel photometer (Titertek Multiskan, Flow Laboratories) at 380 nm over a period of about 80 minutes. The maximal increases in optical density within 5 minutes (ΔOD max 5') of sample dilutions were compared with those of an endotoxin standard. The samples as well as the endotoxin standard were titrated in twofold dilutions in a volume of 50 µl in U-bottom microtiter plates (Linbro^R, Flow Laboratories).To all wells 50 µl of LAL was added, which was diluted 1:2 with 0.025 mol/l $MgSO_4$. The sensitivity of LAL (Pyroquant^R, Associates of Cape Cod*) was 50 pg/ml endotoxin. A LAL in distilled water was used as negative control; distilled water alone served as a blank. The microtiter plates were kept in the measuring chamber of the photometer at 37°C which was reached by a thermoregulated air flow. Figure 1 represents the equipment used for registration of and processing the measured data.

---

*LAL kindly was provided by S.W.Watson.

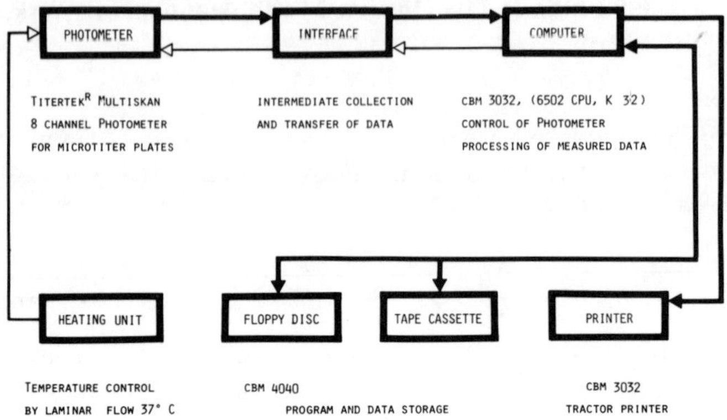

Fig. 1. Diagram of equipment used for automated detection of endotoxin

Using this procedure with a logarithmic scale, a linear standard endotoxin curve is obtained covering a wide concentration range (2000 pg to 3.9 pg/ml endotoxin). This would not have been possible using endpoint determinations at 60 minutes. A comparison of a standard curve of Novo Pyrexal^R (Hermal Chemie) with EC-2, a reference standard endotoxin from FDA, Bethesda, revealed almost identical LAL-reactivity. Comparing Novo Pyrexal^R with TCA endotoxin, extracted in our laboratories from E. coli 0111 (ETU), a considerable higher activity of ETU was observed with parallel linearity (Fig. 2).

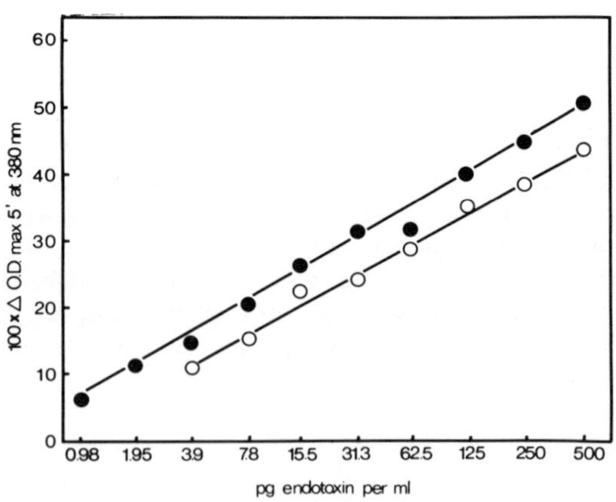

Fig. 2. Comparison of △ OD max 5' of LAL-reaction with two different endotoxins. ●—● TCA-endotoxin (ETU); ○—○ Novo Pyrexal^R

These results clearly indicate the need for selecting one endotoxin preparation as a standard in order to eliminate noncomparable reference values.

A further advantage of this quantitative LAL microtiter test is the fact that without endotoxin spiking, inhibitors are detectable in the samples at several dilution steps. With increasing dilution of the samples the Δ OD max 5' increases until a point is reached where further dilution results in measurements which decrease according to the dilution ratio of the endotoxin.

This became apparent in experiments that were performed to study endotoxin clearance during the state of nonspecific resistance to endotoxic effects. Resistance was induced by 24 hours pretreatment with a detoxified endotoxin. This rapidly inducible endotoxin tolerance was described (Urbaschek 1967; Urbaschek, Nowotny 1968) using detoxified endotoxin obtained after treatment with potassium methylate according to the method of Nowotny (1963). Endotoxin clearance markedly increased in the status of nonspecific tolerance compared to that of nonpretreated controls when determined by measuring the endotoxin content of heated platelet-rich plasma (PRP). In this experiment inhibitors of the LAL reaction were detectable in PRP of all animals two hours after endotoxin injection (Fig. 3). Before this time and in the group

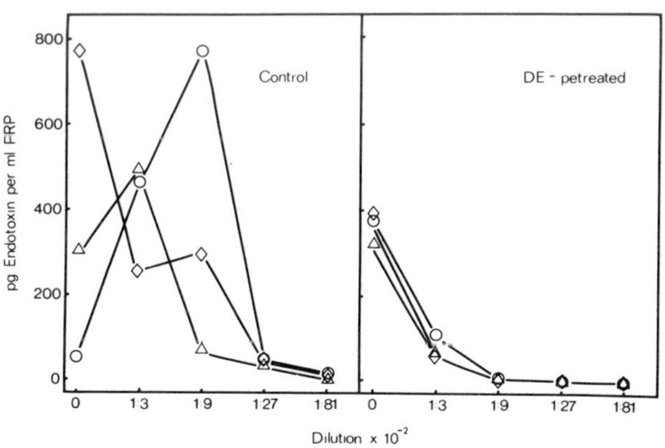

Fig. 3. Individual endotoxin levels in PRP of guinea pigs two hours after endotoxin (5µg/100 g i.v.) without pretreatment (control) and 24 hours after pretreatment with detoxified endotoxin(DE).

of guinea pigs that were pretreated with detoxified endotoxin no inhibition occured.

In general it was observed that storage of blood samples at -80° C results in lowered endotoxin content. However, the inhibitors of the LAL-endotoxin reaction also are decreased.

A study of the ability of lactulose (Bifiteral[R], Thomae) to bind endotoxin (Liehr 1980) revealed that lactulose alone caused inhibition of the LAL reaction which was pH-independent. Lactulose concentrations of 2.5 mg/ml and less repeatedly yielded the expected Δ OD max 5' for all specimens containing known endotoxin concentrations both for 50 pg/ml or 500 pg/ml of standard endotoxin. Higher concentrations of lactulose showed continuously increasing inhibition of the reaction (Fig. 4). With varying endotoxin and constant lactulose concentrations, inhibition of the LAL-reaction was shown by a distinct decrease of the inhibited standard curve in comparison to the standard endotoxin curve (Ditter, Urbaschek, Urbaschek 1981).

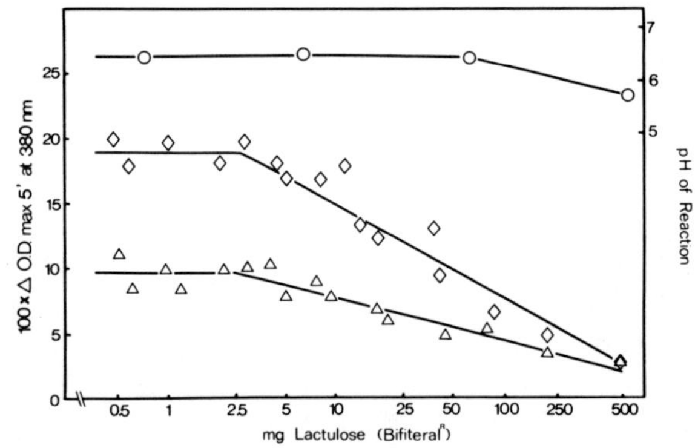

Fig. 4. Influence of lactulose (Bifiteral[R]) on LAL-endotoxin reaction. ◇–◇ 500 pg/ml endotoxin or △–△ 50 pg/ml endotoxin is included in each dilution step of lactulose (500 mg to 0.5 mg). ○–○ pH in the wells after addition of LAL.

The endotoxin concentrations in human whole blood, serum, plasma, and PRP were compared. The measurements in nonheated samples did not represent the actual endotoxin content as determined in heated samples (10 minutes at 100° C). Therefore, all further assays were performed with heated samples.

The detection of the endotoxin content of whole blood was possible in the LAL tube test, after heating and diluting 1:10. However, using this photometric method, determinations could not be performed at dilutions lower than 1:50 due to the light absorption of whole blood. While serum endotoxin levels were lower than those detected in plasma or PRP, the latter were comparable to each other.

The complexity of the detection of endotoxin in plasma became apparent when an endotoxin curve in normal plasma was used as standard. In comparison to an endotoxin standard curve in distilled water, a distinct amplification of the reaction was observed in plasma with an increasing loss of linearity in the dilutions containing higher plasma concentrations. In this study 200 pg of endotoxin was added to 1:10 diluted plasma before or after heating or to distilled water and then tested in twofold dilutions (Fig. 5). Differences in amplification for each sample in plasma occurred to such an extent that a single comparison to a common plasma standard curve was impossible. To compensate for the amplification of individual samples, an internal endotoxin standard 200 pg/ml was diluted in two different constant plasma concentrations of three samples (Fig. 6). The individual differences are most pronounced in the 1:10 plasma dilutions. From these internal standardizations the endotoxin content was determined including the value of the plasma without addition of endotoxin by fitting a nonlinear equation

$$\Delta \text{OD max } 5' = a + b \times \log(u + s)$$

u = unknown endotoxin content to be determined (pg/ml)
s = endotoxin spike (pg/ml)

to the measured points under the prerequisites that the dependence of Δ OD max 5' versus concentration of endotoxin is logarithmic.

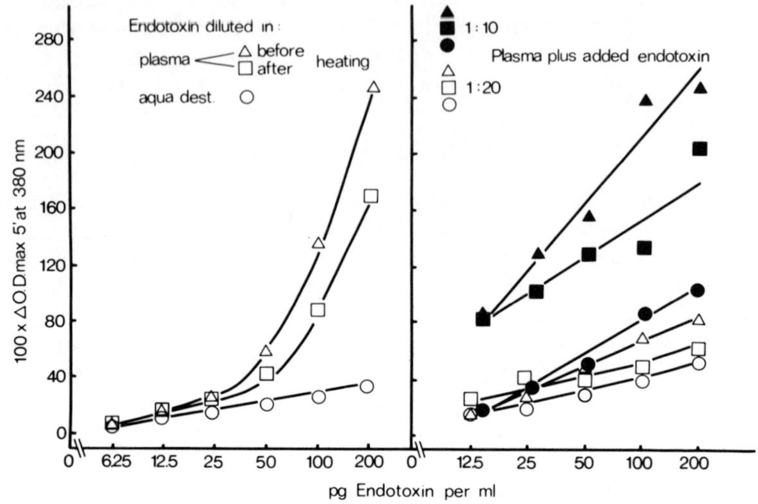

Fig. 5

Fig. 6

Fig. 5. Comparison of LAL reaction with endotoxin diluted in plasma or distilled water. Standard endotoxin curve in distilled water ○—○. Twofold dilutions of 1:10 plasma with addition of 200 pg/ml endotoxin either before △—△ or after □—□ heating.

Fig. 6. Comparison of LAL reaction with endotoxin in plasma. Various concentrations of endotoxin (200 pg/ml to 12.5 pg/ml) were added to constant 1:10 or 1:20 plasma of three individuals.

The method used illustrates the complexity of the LAL-endotoxin reaction in blood and demonstrates the advantages of the use of the automated quantitative LAL-microtiter test for the recognition of inhibitors concomitant with the detection of the actual endotoxin concentration in the sample. A report is in preparation on the usefulness of this method for the detection of endotoxemia in patients.

Acknowledgment: We thank U. Ell and R. Breunig for excellent technical assistance.
This work was supported by the Deutsche Forschungsgemeinschaft within the SFB 90.

Ditter B, Urbaschek R, Urbaschek B, 1981. The ability of various adsorbants used in the treatment of diarrhoea to bind endotoxin in mice. In preparation.

Liehr H, Englisch G, Rasenack U, 1980. Lactulose - A drug with antiendotoxin effect. Hepato-Gastroenterology 27: 356-360.

Nowotny A, 1963. Endotoxoid preparations. Nature 197: 721-722.

Urbaschek B, 1967. Zur Frage des Wirkungsmechanismus bakterieller Endotoxine und seiner Beeinflussung. Habilitationsschrift, Med. Fakultät der Universität Heidelberg.

Urbaschek B, Nowotny A, 1968. Endotoxin tolerance induced by detoxified endotoxin (endotoxoid). Proc Soc Exp Biol Med 127: 650-652.

DITTER SPEAKER
TITLE: DETECTION OF ENDOTOXIN IN BLOOD AND OTHER SPECIMENS BY EVALUATION OF PHOTOMETRICALLY REGISTERED LAL - REACTION KINETICS IN MICROTITER PLATES.

LEVIN: Your last figure shows that diluted plasma increases the ability to detect endotoxin in water, in contrast to what is generally believed.
DITTER: Yes, this is the case when the plasma is diluted 1:10 and then heated.
LEVIN: How do you explain this finding?
DITTER: The amplification factor decreased with increasing the plasma dilution.
KURATSUKA: Dr. Urbascnek, in your experimental system I want to know the effect of amines and steroids. Are the effects of these substances excluded or not?
URBASCHEK: Biogenic amines are among the first mediators effective in the initial phase of endotoxemia. In the model described serotonin together with gavaged endotoxin induced endotoxemia. The effect of steroids was not studied in this model. In vivo microscopic studies we found that glucocorticoids prevent the severe endotoxin-caused disturbances of the microvasculature when given no later than 20 minutes after endotoxin.
DITTER: We first used 405 nm and then found that 380 nm gives higher delta O.D. maximum values so that we now measure at this wavelength.

Endotoxins and Their Detection With the
Limulus Amebocyte Lysate Test, pages 393-397

CHARACTERIZATION OF PEG PRECIPITATED ANTIBODIES IN PATIENTS WITH LIVER CIRRHOSIS AND SEPTICEMIA

Peter C.Fink, Klaus D.Schultze

Medizinische Hochschule Hannover, Institute for Clinical Chemistry,Karl-Wiechert-Allee 9, D-3000 Hannover, Germany

INTRODUCTION

Circulating immune complex-like material (ICLM) consisting of IgG, IgM, C1q and C3 has been demonstrated in patients with liver cirrhosis (Thomas et al. 1978; Abrass et al. 1980) and gram-negative septicemia (Koster et al.1978). For both disease entities the question often arises whether ICLM has pathogenic significance or solely acts as a humoral epiphenomenon. In this context the coincidence of ICLM and endotoxemia in these patient groups (Sodomann et al. 1978; Liehr 1979; Fink et al. 1981 a; Koster et al.1978) may indicate a possible role of endotoxin with respect to the antigen specificity of the immunoglobulins involved in ICLM.

We used the polyethylene glycol 6000 (PEG) precipitation technique (Gauci et al. 1978) which allows, in addition to detection, further characterization of ICLM in order to confirm the previous observations and to investigate class and antigen specificity of PEG precipitated antibodies. To demonstrate a relationship between antibodies found in ICLM and endotoxemia, lipid A, the least variable moiety of bacterial endotoxins, was used as the antigenic target in a hemolysis test system.

MATERIALS AND METHODS

Sera from patients with bioptically confirmed liver cirrhosis and proven gram-negative septicemia, which yielded LAL positive reactions, were tested with a standardized PEG precipitation technique (Fink et al. 1981b). IgG, IgM and C3 in sera and corresponding PEG precipitates were determined by rate nephelometry using the Beckman Immunochemistry System.

Anti-lipid A activity in sera and corresponding PEG precipitated antibodies was detected by passive hemolysis test using lipid A coated sheep red blood cells as described elsewhere (Fink et al. 1981c).

RESULTS AND DISCUSSION

In comparison to the amounts of PEG precipitable IgG (mean value: 175.7 ± 44.6 ug/ml) and IgM (mean value: 356.7 ± 127.5 ug/ml) found in sera of 19 healthy blood donors, significantly more IgG and IgM could be PEG precipitated in sera from 29 patients with liver cirrhosis (PEG precipitable IgG: mean value: 425.3 ± 275.6 ug/ml; PEG precipitable IgM: mean value: 618.0 ± 368.6 ug/ml; for both parameters $P<0.005$) and from 30 septicemic patients (PEG precipitable IgG: mean value: 421.0 ± 258.0 ug/ml; PEG precipitable IgM: mean value: 568.2 ± 231.1 ug/ml; for both parameters $P<0.0005$) (Fig. 1).

Determining C3 in the PEG precipitates, significantly higher ($P<0.0025$) quantities of precipitable C3 were found in sera from 21 patients with liver cirrhosis (mean value: 131.9 ± 44.9 ug/ml) and from 19 patients with gram-negative septicemia (mean value: 141.3 ± 54.8 ug/ml) than in 18 sera from the control group (mean value: 93.8 ± 17.1 ug/ml) (Fig. 2).

In preliminary experiments the specificity of the immunoglobulins involved was investigated by passive hemolysis test which measured the anti-lipid A activity in endotoxemic and control sera as well as in PEG precipitates. Whereas no titers were found in sera and PEG precipitates of the blood donor group, in about 50 % of the liver cirrhotic and septicemic sera titers up to 1 : 32 did occur. It was of interest to find that in PEG precipitates from patients with liver cirrhosis 50 % revealed significant titers up to 1 : 40. In the PEG precipitates

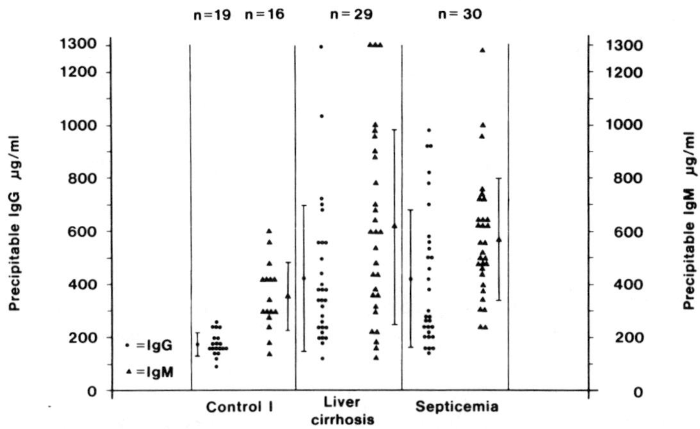

FIGURE 1: Determination of precipitable immunoglobulins in sera from healthy blood donors and patients with liver cirrhosis and gram-negative septicemia using the PEG precipitation technique. —•— and —▲— indicate mean value ± SD.

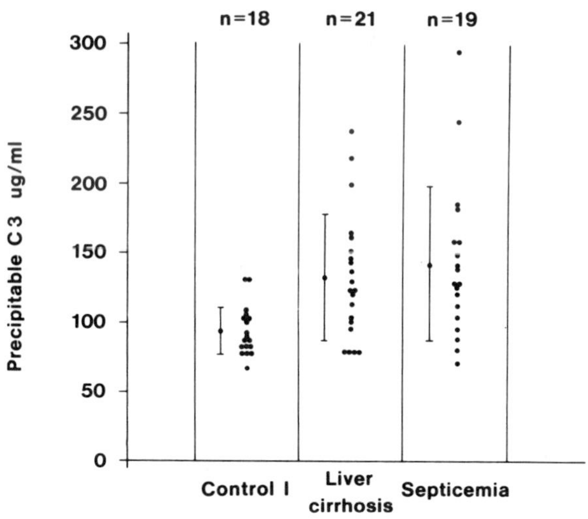

FIGURE 2: Determination of precipitable C3 in sera from healthy blood donors and patients with liver cirrhosis and gram-negative septicemia using the PEG precipitation technique. —•— indicates mean value ± SD.

from patients with gram-negative septicemia anti-lipid A activity was detectable only in one case.

In conclusion these results show that ICLM composed of immunoglobulins of different classes and complement components can be PEG precipitated. The detection of antibodies with lipid A specificity in this material may reflect a humoral reactivity of the host against gut-derived or exogenous endotoxins.

REFERENCES

Abrass CK,Border WA,Hepner G (1980): Non-specificity of circulating immune complexes in patients with acute and chronic liver disease. Clin Exp Immunol 40: 292.

Fink PC,Lehr L,Urbaschek R,Kozak J (1981a): Limulus amebocyte lysate test for endotoxemia: Investigations with a femtogram sensitive spectrophotometric assay. Klin Wochenschr 59: 213.

Fink PC,Schultze K-D (1981b): A standardized polyethylene glycol precipitation technique for determining immune complex-like material in serum. J Clin Chem Clin Biochem 19: 666.

Fink PC,Galanos C (1981c): Determination of anti-lipid A and lipid A by enzyme immunoassay. Immunobiology 158: 380.

Gauci L,Ursule F,Serrou B (1978): A semi-automated system permitting analysis of immune complex components. In Peeters H (ed): Proceedings of the twenty-sixth colloquium. Protides of the biological fluids. Oxford: Pergamon Press, p.29.

Koster F,Levin J,Walker L,Tung KSK,Gilman RH,Rahaman MM,Majid MA,Islam S,Williams RC Jr (1978): Hemolytic-uremic syndrome after shigellosis. Relation to endotoxemia and circulating immune complexes. N Engl J Med 298: 927.

Liehr H (1979): The Limulus assay for endotoxemia as applied in gastroenterology. In Cohen E (ed): Biomedical applications of the horseshoe crab (Limulidae). New York: Alan R Liss, p.309.

Sodomann CP,Prokein K,Schmidt H,Martini GA (1978): Zirkulierende Immunkomplexe bei Lebererkrankungen. Z Gastroenterol 16: 501.

Thomas HC, De Villiers D, Potter B, Hodgson H, Jain S, Jewell DP, Sherlock S (1978): Immune complexes in acute and chronic liver disease. Clin Exp Immunol 31: 150.

Acknowledgment

This work was supported in part by grant Fi281/3-3 from the Deutsche Forschungsgemeinschaft.

Endotoxins and Their Detection With the Limulus Amebocyte Lysate Test, pages 399–409

AIRBORNE ENDOTOXINS IN VARIOUS OCCUPATIONAL ENVIRONMENTS

Ragnar Rylander, M.D. and Janet Vesterlund, M.A.

Department of Environmental Hygiene
University of Gothenburg
400 33 Gothenburg, Sweden

1. Introduction

Airborne endotoxins have been found to be present in a variety of industrial environments. In this presentation, the techniques used to assess airborne endotoxin in dust are reported and typical values in some environments are given. The effects after inhalation are reviewed, together with symptoms reported among exposed workers.

2. Industrial environments with endotoxin

Vegetable fibers such as grass, cotton and flax contain an abundance of gram-negative bacteria. While cotton in the unopened boll is sterile, it rapidly becomes contaminated once it opens in the field. Quantities ranging from 10^2 to 10^9 gram-negative bacteria/g material have been found in bale cotton from different countries (Rylander and Lundholm 1978). Bagasse, hemp and mouldy hay also contain large amounts of gram-negative bacteria (Rylander et al., 1975). The amount of bacteria on the plants mentioned depends largely upon climatic conditions--rain and frost have been shown to increase the bacteria contamination up to 1000-fold overnight (Clark et al., 1947).

When cotton and flax are processed in industry, dust and bacteria become airborne. Airborne gram-negative bacteria in cotton mills have been measured by Cinkotai et al. (1977), Cinkotai and Whitaker (1978), Fischer (1979), Bergström et al. (1980) and Haglind et al. (1981). The number in cardrooms has been found to range from 10^2 to 10^5

colony forming units (c.f.u.) per cubic meter.

The main species found in cotton and flax dust is *Enterobacter agglomerans*, which constitutes 40 to 100% of all gram-negative organisms found in cotton (Haglind *et al.*, 1981). Of other fermenting species, *Klebsiella* is often present. Among the nonfermenting gram-negative bacteria, strains of *Pseudomonas* are found, as well as *Acinetobacter* and *Agrobacterium*. Dutkiewicz (1978) demonstrated that high numbers of *Erwinia herbicola* (*Ent. agglomerans*) were present in the air of grain mills.

If a correct estimate of the amount of airborne endotoxin is to be made, the measurement of airborne viable bacteria alone is not sufficient. The bacterial population on dusts consists of dead as well as viable bacteria (Jorgensen 1979), and a specific test for endotoxin activity must be used.

For this sampling, Millipore filters (0.8 um) are used as for dust sampling. They are thereafter shaken in 10 ml of pyrogen-free water, and serial dilutions are prepared. All glassware used in the analysis is rendered endotoxin-free by heating it at 180° C for 3 hours. Equal aliquots of LAL (Assoc.of Cape Cod) and dilutions of the filter water extracts are mixed in glass tubes and incubated in a water bath at 37° C for 60 minutes. The tubes are inversed, and the last dilution with a stable clot is related to a similar clot formation in a dilution series of known amounts of commercial *Escherichia coli* LPS (Difco). Taking into consideration the dilution and amount of air drawn through the filter, the results can be expressed as ug of endotoxin/m^3.

Cinkotai *et al*. (1977) were the first to employ this method to determine the concentration of airborne endotoxin in cotton mills in England. Values ranging from 0.2 to 1.6 ug/m^3 were found in cardroom air. Similar concentrations were later reported in the United States by Fischer (1979). Bergström *et al*. (1980) reported values ranging from 0.3 to 2.5 ug/m^3 in an experimental cardroom where ordinary bale cotton was carded under conditions resembling those in cotton mills. Symington (1980) found levels of up to 2 ug/m^3 in the air of flax mills in Scotland.

Table 1 reports the amounts of airborne endotoxin at different working sites in two different cotton mills. It is seen that the amount of endotoxin decreased with the continued processing of cotton in the two plants. The levels of endotoxin found at various sites in the cotton mills were highest in carding areas and considerably lower in areas for spinning or weaving.

Location	Mill A	Mill B
carding	0.3	0.02
roving	0.6	0.02
spinning	0.06	0.05
weaving	0.01	0.001

Table 1: Airborne endotoxin (ug/m^3) at different working sites in two cotton mills.

Considerable differences were present between equivalent working sites in the two plants. This could be due to differences in the bacterial contamination of cotton bales processed, but processing characteristics and ventilation are also important.

The level often found in carding--0.5 ug/m^3--can be translated into a human dose. Considering that a 70 kg man will breathe about 15 m^3 during an 8-hour working shift, and assuming a 50% deposition of particles in the size range of cotton dust, the endotoxin dose will be about 0.05 ug/kg body weight.

Table 2 reports levels of airborne endotoxin found in some non-textile industries. It is seen that levels equivalent to those present in cotton and flax mills are found in swine confinement buildings, poultry processing plants and in office buildings with humidifiers in which the water reservoir is contaminated with gram-negative bacteria.

Location	Airborne endotoxin
cotton gin	0.2 - 2.0
cotton seed oil mill	1.7
swine confinement buildings	0.3 - 1.2
chicken slaughterhouse	0.1 - 0.8
offices with contaminated humidifiers	0.1 - 1.2$^+$

Table 2: Airborne endotoxin (ug/m^3) in various non-textile industries.

$^+$estimated airborne value

In view of the presence of biologically significant amounts of airborne endotoxin in a variety of industrial settings, the inhalation toxicity of endotoxin will now be reviewed and related to symptoms among workers in environments where airborne endotoxins are present.

3. Inhalation effects

3.1 Background

It has already been shown that one of the environments where airborne endotoxin is particularly common is the cotton and flax industry. In 1713, Ramazzini reported a disease affecting the health of workers in textile mills, and described a relation between dust from flax and hemp and workers' cough and "asthmatic troubles". In 1877, Proust introduced the term byssinosis, derived from the Egyptian word "bessus" for fine linen or flax, as the name for the respiratory disease peculiar to textile workers.

In the 1950's, Schilling (1956) demonstrated that byssinosis was a significant problem in Lancashire cotton mills. It was established soon afterwards that byssinosis also occurred in other cotton processing countries, such as Egypt, the Netherlands, Sweden, Uganda, India and Israel.

Schilling (1963) developed the classical grading of byssinosis based on two pulmonary symptoms traditionally seen in affected cotton workers: a subjective feeling of chest tightness and breathlessness. The symptoms initially occur on Mondays or the first day of a return to work but, over a period of years, may appear also on following work days during the week. The presence of this reaction--byssinosis--can be assessed using questionnaires to reveal chest tightness and by respiratory function measurements made before and after the shift.

3.2 Pulmonary function

A decrease in the air flow may be found after inhalation exposure to endotoxin. Measurements of the forced expiratory volume in the first second ($FEV_{1.0}$) is the most commonly used measure of this decrease, which occurs a few hours after exposure.

Cavagna _et al_. (1969) exposed normal human subjects and subjects with mild chronic bronchitis to an aerosol containing 40 and 80 ug of purified _E.coli_ endotoxin in

2 ml of saline. In 2 out of 8 normal subjects, a significant reduction of $FEV_{1.0}$ was seen after inhalation of 80 ug of endotoxin.

Exposure to considerably lower doses of endotoxin was performed by Muittari _et al_. (1980a, 1980b) in a study of shower taker's disease. Using tap water containing about 1 ug of endotoxin/ml, a 1 ml inhalation challenge was found to cause a decreased lung diffusion capacity 4 to 6 hours after the exposure. No studies of $FEV_{1.0}$ were performed in this experiment.

With reference to conditions in the cotton industry, extensive pulmonary function measurements have been reported by several researchers, e.g. Merchant _et al_. (1973), Haglind _et al_. (to be published). An airway flow decrease over the working shift appears both among previously unexposed persons and those who have been exposed at work for several years. The $FEV_{1.0}$ reduction over the shift is less pronounced further down the production line, even when the dust levels are high.

Rylander _et al_. (1979) measured the average decrease in $FEV_{1.0}$ over the Monday shift, respirable dust levels and the number of Gram-negative bacteria in bale cotton being processed in 23 cotton mills in the United States. The average decrease in $FEV_{1.0}$ over the shift correlated better to the amount of bacteria on cotton than to airborne dust levels.

3.3 Inflammation

Exposure to airborne endotoxin causes an invasion of polymorphonuclear leukocytes (PMN) into the airways of animals and man. This effect, indicative of early inflammation, has been demonstrated in several experiments (Rylander and Nordstrand 1974, Walker _et al_., 1975, Rylander and Snella 1976, Hudson _et al_., 1977, Snella and Rylander 1981). The reaction is dose-related and differs in magnitude among different species of animals. The number of PMN in lung lavage preparations from guinea pigs exposed to airborne endotoxin increases rapidly up to 12 hours after exposure and remains elevated up to 24 hours afterwards. The value is then progressively lower at 48 and 72 hours (Snella and Rylander 1981).

The degree of PMN migration is dose-dependent (Helander _et al_., 1980) and differs between endotoxins from different strains of gram-negative bacteria found in cotton. Some endotoxins (_Enterobacter_, _Klebsiella_, _Pseudomonas_) have a large capacity for mobilizing PMN, whereas other bacteria have a very low capacity (_Agrobacterium_). This LPS also has a lower potency to clot in the LAL test. We have recently shown that the PMN reaction is caused by the Lipid A part of the endotoxin molecule (Helander, to be published).

A long term exposure to endotoxin will cause an increase in the mucus producing goblet cells and cell infiltration in the respiratory epithelium (Snell 1966) resembling chronic bronchitis.

As regards industrial situations, a PMN invasion into the airway epithelium over the work shift has been demonstrated in humans among workers in a cotton mill (Merchant _et al_., 1975) and cotton mill workers and naive subjects exposed in an experimental cardroom (Bergström _et al_., 1980). After several years' exposure to high dust levels, the incidence of chronic bronchitis will increase, particularly among smoking cotton mill workers.

3.4 Fever

Exposure to airborne endotoxin will cause fever in man and animals. After repeated exposure, a tolerance develops. Although information on threshold doses in humans is scarce, fever reactions have been reported at i.v. doses of 0.002 ug/kg b.w., which is more than 10 times less than the dose received in cotton cardrooms.

When employees enter a cotton mill for the first time, or when naive subjects are experimentally exposed to cotton dust, some develop an acute fever. This occurs between 4 and 6 hours after exposure and normally disappears during the following 6 to 12 hours. The fever symptoms become progressively milder at repeated exposure and eventually disappear. They may return when the worker has been absent from the mill environment for a prolonged period of time.

4. Comments

This review of endotoxin inhalation effects demonstrates that several of the symptoms found among persons with byssinosis closely resemble those caused by the inhalation of purified endotoxin. So far, the exact pathogenesis of the subjectively judged chest tightness has not been explained. Several laboratories have however shown that intravenous endotoxin may cause aggregation of neutrophils and platelets in pulmonary capillaries. We have recently demonstrated that bacteria from cotton can cause production of platelet activating factor (PAF) from human PMN (Holt et al., 1981). It is thus our current hypothesis that chest tightness and probably also the $FEV_{1.0}$ change is caused by a neutrophil-platelet interaction induced by endotoxin from the Gram-negative bacteria.

Symptoms similar to the mill fever reaction in cotton workers--fever, malaise, chills and leukocytosis--are general features of another group of diseases commonly referred to as allergic extrinsic alveolitis, or hypersensitivity pneumonitis (HP).

A close examination of case histories published on HP reveals that two different groups of symptoms may be distinguished. One group of symptoms is the classical HP picture, with a disease lasting for several days and lung X-ray changes demonstrating fibrosis.

The other group of symptoms is characterized by rapid onset of fever, chills and influenza-like symptoms. The lung X-ray is negative, and there is generally no history of progressive deterioration in work capacity over the years. The patient recovers relatively rapidly, often the day following the acute episode.

In the environment of several such patients whom we have examined, significant amount of airborne endotoxin can be found. Most of the cases relate to the aerosolization of water containing endotoxin, for instance from humidifiers.

The results from experiments in our laboratories as well as experiments reviewed here demonstrate that biologically significant amounts of endotoxin are present in several industrial environments. The LAL method is easy to apply on the water extracts from dust filters and inhibition is not present. When the values of airborne endotoxins are compared to those of other workers, methodological differences in the performance of the LAL test must be taken into consideration. While values similar to the ones reported here for cotton cardrooms were also reported by Fischer (1979), considerably higher values were reported for swine confinement buildings by Thedell et al (1980). However, Fischer used the clot evaluation method as used here, whereas Thedell et al used a turbidimetric method.

In conclusion, the similarities in the reactions seen after exposure to endotoxin and byssinosis, as well as certain types of hypersensitivity pneumonitis, suggest that endotoxins play an important part in the pathogenesis of these diseases. Further research is needed in order to evaluate the exact cellular reactions responsible for the disease, as well as to obtain information on dose-response relationships. Determinations of airborne endotoxin are an important part of this work.

Acknowledgement

The work reported in this presentation has been supported by Cotton Inc., and the Swedish Work Environment Fund.

References

Bergström R, Haglind P, Rylander R (1980). Experience from human exposures in an experimental cardroom. In Wakelyn PJ (ed): Proceedings of the Fourth Special Session on Cotton Dust Research, Beltwide Cotton Production Research Conferences, pp. 22-25.

Cavagna G, Foa V, Vigliani EC (1969). Effects in man and rabbits of inhalation of cotton dusts or extracts and purified endotoxin. Br J Ind Med 26:314.

Cinkotai FF, Lockwood MG, Rylander R (1977). Airborne microorganisms and prevalence of byssinotic symptoms in cotton mills. Am Ind Hyg Ass J 38:554.

Cinkotai FF, Whitaker CJ (1978). Airborne bacteria and the prevalence of byssinotic symptoms in 21 cotton spinning mills in Lancashire. Ann Occup Hyg 21:239.

Clark FE, Hervey RJ, Blank LM (1947). Occurrence of cotton fiber contaminated by Aerobacter cloacae. US Department of Agriculture Technical Bulletin 935:1-28.

Dutkiewicz J (1978). Exposure to dust-born bacteria in agriculture. I. Environmental studies. Arch Environ Health 33:250.

Haglind P, Lundholm M, Rylander R (1981). Byssinosis prevalence in Swedish cotton mills. Br J Ind Med 38:235.

Haglind P, Bake B, Belin L (1981). Is mild byssinosis associated with small airways disease? to be published.

Helander I, Salkinoja-Salonen M, Rylander R (1980). Chemical structure and inhalation toxicity of lipopolysaccharides from bacteria on cotton. Infect Immun 29:859.

Helander I (1981). Unpublished data.

Holt PG, Holt BJ, Rylander R (1981). Stimulation of human polymorphonuclear leukocytes by cotton dust bacteria. I. Production of platelet activating factor by PMN from normal volunteers. To be published.

Fischer JJ (1979). The microbial composition of cotton dusts, raw cotton lint samples and the air of carding areas in mills. In Wakelyn PJ (ed): Proceedings of the Third Special Session on Cotton Dust Research, Beltwide Cotton Production Research Conferences, pp. 8.

Hudson A, Kilburn K, Halprin G, McKensie W (1977). Granulocyte recruitment to airways exposed to endotoxin aerosol. Am Rev Respir Dis 115:89.

Jorgensen JH, Lee JC, Alexander GA, Wolf HW (1979). Comparison of Limulus assay, standard plate count and total coliform count for microbiological assessment of renovated wastewater. Appl Environ Microbiol 37:928.

Merchant JA, Halprin GG, Hudson AF, Kilburn KH, McKensie WN, Hunt DJ, Bermazohn P (1975). Responses to cotton dust. Arch Environ Health 30:222.

Merchant JA, Lumsden J, Kilburn K, Germino M, Hamilton J, Lynn W, Byrd H, Baucom P (1973). Preprocessing cotton to prevent byssinosis. Br J Ind Med 30:237.

Muittari A, Rylander R, Salkinoja-Salonen M (1980a). Endotoxin and bath water fever. The Lancet ii:89.

Muittari A, Kuusisto P, Virtanen P, Sovijärvi A, Grönroos P, Harmoinen A, Antila P, Kellomäki L (1980b). An epidemic of extrinsic allergic alveolitis caused by tap water. Clin Allergy 10:77.

Proust A (1877). Traité d'Hygiène Publique et Privée, Paris, G Massan, p. 171.

Ramazzini B (1713). Diseases of the Workers (De Morbis Artificium).

Rylander R, Nordstrand A (1974). Pulmonary cell reactions after exposure to cotton dust extract. Br J Ind Med 31:220.

Rylander R, Nordstrand A, Snella MC (1975). Bacteria contamination of organic dusts: effects on pulmonary cell reactions. Arch Environ Health 30:137.

Rylander R, Snella MC (1976). Acute inhalation toxicity of cotton plant dusts. Br J Ind Med 33:175.

Rylander R, Lundholm M (1978). Bacterial contamination of cotton and cotton dust and effects on the lung. Br J Ind Med 35:204.

RYLANDER SPEAKER
TITLE: AIRBORNE ENDOTOXIN IN VARIOUS OCCUPATIONAL ENVIRONMENTS.

MULLER-CALGAN: Do you find endotoxin in tobacco?
RYLANDER: In tobacco there are no endotoxins.
WATSON: Is there any evidence that airborne endotoxin enters the circulatory system?
RYLANDER: Yes, certain preliminary data from inhalation studies suggest that endotoxin does enter the circulatory system. An interesting feature here is the symptom of excessive fatigue reported by many persons exposed to dust containing endotoxins. If endotoxin enters through the lung, it shortcuts the liver; it is not immediately detoxified, and it circulates throughout the body . This may be the reason for fatigue that is so common in these cases. I would like to follow up on the report by Pearson on the discrepancy between the rabbit test and the LAL test found when material for dialysis was analyzed. I suggested last night that the endotoxin in this raw material was denaturated resulting in free Lipid A which is not soluble in water; thus, this product would not cause fever but was able to react with the LAL. To test this hypothesis I called our lab in Sweden and asked them to do an LAL test on pure Lipid A from bacteria isolated from cotton. These indeed show that free Lipid A which is insoluble can react with the LAL test but at a concentration which is 100 times below that of Lipid A which is attached to BSA to make it soluble. My suggestion is that what happened to Pearson's products was that endotoxins were denaturated; the threshold for the fever production was not reached but the threshold for the limulus activity was still present in spite of the decrease in pyrogenic activity.

Rylander R, Imbus HR, Suh MW (1979). Bacterial contamination of cotton as an indicator of respiratory effects among cardroom workers. Br J Ind Med 36:299.

Schilling RSF (1956). Byssinosis in cotton and other textile workers. The Lancet ii:261 and 319.

Schilling RSF, Vigliani EC Lammers B, Valic F, Gilson JC (1963). A report on a conference on byssinosis. In: Proceedings of the XVI International Congress on Occupational Health, Amsterdam: Excerpta Medica Foundation, vol. 2, p. 137.

Snell JC (1966). Effects of inhaled endotoxin. J Lab Clin Med 67:624.

Snella MC, Rylander R (1981). Airway cell reactions after acute and subacute inhalation of bacterial lipopolysaccharide. To be published.

Symington I (1980). Unpublished data.

Thedell TD, Mull JC, Olenchock SA (1980). A brief report of Gram-negative bacterial endotoxin levels in airborne and settled dusts in animal confinement buildings. Am J Ind Med 1:3.

Walker RF, Eidson G, Hatcher JD (1975). Influence of cotton dust inhalation on free lung cells in rats and guinea pigs. Lab Invest 33:28.

Index

PROGRESS IN CLINICAL AND BIOLOGICAL RESEARCH

Vol 1: **Erythrocyte Structure and Function,** George J. Brewer, *Editor*

Vol 2: **Preventability of Perinatal Injury,** Karlis Adamsons and Howard A. Fox, *Editors*

Vol 3: **Infections of the Fetus and the Newborn Infant,** Saul Krugman and Anne A. Gershon, *Editors*

Vol 4: **Conflicts in Childhood Cancer: An Evaluation of Current Management,** Lucius F. Sinks and John O. Godden, *Editors*

Vol 5: **Trace Components of Plasma: Isolation and Clinical Significance,** G.A. Jamieson and T.J. Greenwalt, *Editors*

Vol 6: **Prostatic Disease,** H. Marberger, H. Haschek, H.K.A. Schirmer, J.A.C. Colston, and E. Witkin, *Editors*

Vol 7: **Blood Pressure, Edema and Proteinuria in Pregnancy,** Emanuel A. Friedman, *Editor*

Vol 8: **Cell Surface Receptors,** Garth L. Nicolson, Michael A. Raftery, Martin Rodbell, and C. Fred Fox, *Editors*

Vol 9: **Membranes and Neoplasia: New Approaches and Strategies,** Vincent T. Marchesi, *Editor*

Vol 10: **Diabetes and Other Endocrine Disorders During Pregnancy and in the Newborn,** Maria I. New and Robert H. Fiser, *Editors*

Vol 11: **Clinical Uses of Frozen-Thawed Red Blood Cells,** John A. Griep, *Editor*

Vol 12: **Breast Cancer,** Albert C.W. Montague, Geary L. Stonesifer, Jr., and Edward F. Lewison, *Editors*

Vol 13: **The Granulocyte: Function and Clinical Utilization,** Tibor J. Greenwalt and G.A. Jamieson, *Editors*

Vol 14: **Zinc Metabolism: Current Aspects in Health and Disease,** George J. Brewer and Ananda S. Prasad, *Editors*

Vol 15: **Cellular Neurobiology,** Zach Hall, Regis Kelly, and C. Fred Fox, *Editors*

Vol 16: **HLA and Malignancy,** Gerald P. Murphy, *Editor*

Vol 17: **Cell Shape and Surface Architecture,** Jean Paul Revel, Ulf Henning, and C. Fred Fox, *Editors*

Vol 18: **Tay-Sachs Disease: Screening and Prevention,** Michael M. Kaback, *Editor*

Vol 19: **Blood Substitutes and Plasma Expanders,** G.A. Jamieson and T.J. Greenwalt, *Editors*

Vol 20: **Erythrocyte Membranes: Recent Clinical and Experimental Advances,** Walter C. Kruckeberg, John W. Eaton, and George J. Brewer, *Editors*

Vol 21: **The Red Cell,** George J. Brewer, *Editor*

Vol 22: **Molecular Aspects of Membrane Transport,** Dale Oxender and C. Fred Fox, *Editors*

Vol 23: **Cell Surface Carbohydrates and Biological Recognition,** Vincent T. Marchesi, Victor Ginsburg, Phillips W. Robbins, and C. Fred Fox, *Editors*

Vol 24: **Twin Research, Proceedings of the Second International Congress on Twin Studies,** Walter E. Nance, *Editor*
Published in 3 Volumes:
Part A: **Psychology and Methodology**
Part B: **Biology and Epidemiology**
Part C: **Clinical Studies**

Vol 25: **Recent Advances in Clinical Oncology,** Tapan A. Hazra and Michael C. Beachley, *Editors*

Vol 26: **Origin and Natural History of Cell Lines,** Claudio Barigozzi, *Editor*

Vol 27: **Membrane Mechanisms of Drugs of Abuse,** Charles W. Sharp and Leo G. Abood, *Editors*

Vol 28: **The Blood Platelet in Transfusion Therapy,** G.A. Jamieson and Tibor J. Greenwalt, *Editors*

Vol 29: **Biomedical Applications of the Horseshoe Crab (Limulidae),** Elias Cohen, *Editor-in-Chief*

Vol 30: **Normal and Abnormal Red Cell Membranes,** Samuel E. Lux, Vincent T. Marchesi, and C. Fred Fox, *Editors*

Vol 31: **Transmembrane Signaling,** Mark Bitensky, R. John Collier, Donald F. Steiner, and C. Fred Fox, *Editors*

Vol 32: **Genetic Analysis of Common Diseases: Applications to Predictive Factors in Coronary Disease,** Charles F. Sing and Mark Skolnick, *Editors*

Vol 33: **Prostate Cancer and Hormone Receptors,** Gerald P. Murphy and Avery A. Sandberg, *Editors*

Vol 34: **The Management of Genetic Disorders,** Constantine J. Papadatos and Christos S. Bartsocas, *Editors*

Vol 35: **Antibiotics and Hospitals,** Carlo Grassi and Giuseppe Ostino, *Editors*

Vol 36: **Drug and Chemical Risks to the Fetus and Newborn,** Richard H. Schwarz and Sumner J. Yaffe, *Editors*

Vol 37: **Models for Prostate Cancer,** Gerald P. Murphy, *Editor*

Vol 38: **Ethics, Humanism, and Medicine,** Marc D. Basson, *Editor*

Vol 39: **Neurochemistry and Clinical Neurology,** Leontino Battistin, George Hashim, and Abel Lajtha, *Editors*

Vol 40: **Biological Recognition and Assembly,** David S. Eisenberg, James A. Lake, and C. Fred Fox, *Editors*

Vol 41: **Tumor Cell Surfaces and Malignancy,** Richard O. Hynes and C. Fred Fox, *Editors*

Vol 42: **Membranes, Receptors, and the Immune Response: 80 Years After Ehrlich's Side Chain Theory,** Edward P. Cohen and Heinz Köhler, *Editors*

Vol 43: **Immunobiology of the Erythrocyte,** S. Gerald Sandler, Jacob Nusbacher, and Moses S. Schanfield, *Editors*

Vol 44: **Perinatal Medicine Today,** Bruce K. Young, *Editor*

Vol 45: **Mammalian Genetics and Cancer: The Jackson Laboratory Fiftieth Anniversary Symposium,** Elizabeth S. Russell, *Editor*

Vol 46: **Etiology of Cleft Lip and Cleft Palate,** Michael Melnick, David Bixler, and Edward D. Shields, *Editors*

Vol 47: **New Developments With Human and Veterinary Vaccines,** A. Mizrahi, I. Hertman, M.A. Klingberg, and A. Kohn, *Editors*

Vol 48: **Cloning of Human Tumor Stem Cells,** Sydney E. Salmon, *Editor*

Vol 49: **Myelin: Chemistry and Biology,** George A. Hashim, *Editor*